TELEPEN
82 8720162 6

AF598496

Evolutionary Genetics of Invertebrate Behavior

Progress and Prospects

Evolutionary Genetics of Invertebrate Behavior

Progress and Prospects

Edited by

Milton Davis Huettel

Insect Attractants, Behavior, and Basic Biology Research Laboratory
Agricultural Research Service
U.S. Department of Agriculture
Gainesville, Florida

Plenum Press • New York and London

Library of Congress Cataloging in Publication Data

Evolutionary genetics of invertebrate behavior.

"Based on a colloquium on evolutionary genetics of invertebrate behavior, held March 21-24, 1983, in Gainesville, Fla."
Bibliography: p.
Includes index.
1. Insects—Genetics—Congresses. 2. Insects—Behavior—Congresses. 3. Insects—Evolution—Congresses. 4. Invertebrates—Genetics—Congresses. 5. Invertebrates—Behavior—Congresses. 6. Invertebrates—Evolution—Congresses. I. Huettel, Milton Davis.
QL493.E96 1986 592′.015 87-6950
ISBN 0-306-42488-6

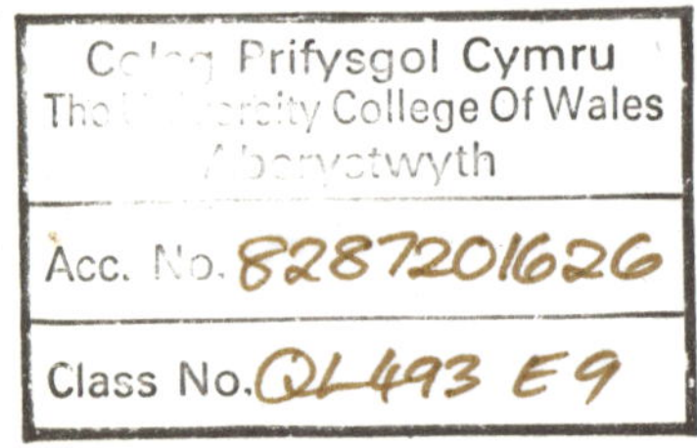

Based on a colloquium on Evolutionary Genetics of Invertebrate Behavior, held March 21-24, 1983, in Gainesville, Florida

Mention of commercial or proprietary products in this publication does not constitute an endorsement of these products by the U.S. Department of Agriculture or any other governmental agency.

A Division of Plenum Publishing Corporation
233 Spring Street, New York, N.Y. 10013

Printed in the United States of America

Preface

In the preface to Sir Vincent B. Wigglesworth's classic 1939 book on insect physiology he asserted that insects provide an ideal medium in which to study all the problems of physiology. A strong case can be made as well for the use of insects as significant systems for the study of behavior and genetics. Contributions to genetics through decades of research on *Drosophila* species have made this small fly the most important metazoan in genetics research. At the same time, population and behavioral research on insects and other invertebrates have provided new perspectives that can be combined with the genetics approach. Through such integrated research we are able to identify evolutionary genetics of behavior as a highly significant emerging area of interest. These perspectives are ably described by Dr. Guy Bush in the introductory chapter of this book.

During March 21-24, 1983, many of the world's leading scientists in invertebrate behavioral genetics were drawn together in Gainesville, Florida, for a colloquium entitled "Evolutionary Genetics of Invertebrate Behavior." This conference was sponsored jointly by the Department of Entomology and Nematology, University of Florida, chaired by Dr. Daniel Shankland, and the Insect Attractants, Behavior and Basic Biology Research Laboratory, U.S. Department of Agriculture, directed then by Dr. Derrell Chambers.

The program was divided into three major areas, "Behavioral Variation in Natural Populations," "Molecular and Biochemical Genetics of Behavior," and "General and Theoretical Aspects of Behavior in Evolution and Speciation." This book is organized in the same manner as the colloquium with chapters written by nearly all the speakers at the conference.

The interest and enthusiasm that attended the colloquium is transmitted through the chapters written by the participants. Dr. Milton Huettel edited the book with scientific reviews provided by a number of colleagues, and with editorial assistance by Dr. Herbert Oberlander, Laboratory Directory, Insect Attractants, Behavior, and Basic Biology Research Laboratory, Gainesville, Florida.

Local arrangements for the conference were coordinated by Dr. James Tumlinson with the assistance of Ms. Marianne Donato, who also prepared typescripts of many of the manuscripts. Mrs. Elaine Turner was responsible for preparation of the final manuscripts and worked closely with Mrs. Barbara Schmit, Department of Printing and Graphics, University of Florida.

Special thanks go to IFAS, Univesity of Florida, and the Insect Attractants, Behavior and Basic Biology Research Laboratory, USDA for financial and organizational sponsorship of many aspects of the colloquium and preparation of the Proceedings. Most of all we thank all the participants who contributed to the success of the conference and to the completion of this book.

Milton D. Huettel
Beltsville, Maryland
August, 1986

Contents

I. INTRODUCTORY LECTURE

II. BEHAVIORAL VARIATION IN NATURAL POPULATIONS

A. Non-reproductive Behavior

B. Courtship and Mating

C. Strategies for Success of Progeny

III. MOLECULAR AND BIOCHEMICAL GENETICS OF BEHAVIOR

IV. GENERAL AND THEORETICAL ASPECTS OF BEHAVIOR IN EVOLUTION AND SPECIATION

V. CLOSING ADDRESS

Evolutionary Behavior Genetics

Guy L. Bush

Department of Zoology
Michigan State University
East Lansing, MI 48824

It has often been said that behavior is one of the most labile traits in animal evolution. Whether this is so remains to be demonstrated, but it is clear that rather minor genetic changes can sometimes have profound effects on behavior without affecting morphology (Ehrman and Parsons 1981). It is also common for closely related species to be reproductively isolated by premating isolating mechanisms alone, indicating that genes affecting behavior often may be the first and most important components of the genome to undergo divergence in speciation. Although I recognize that in some taxa sterility barriers and other postcopulatory factors are important in some modes of speciation, I would like to focus my attention today on the contribution that behavior genetics can make to our understanding of the evolution of reproductive isolation and speciation, and explore where this approach is likely to take us.

Invertebrates are exceptionally well suited for evolutionary, behavior genetic studies. Their behavioral repertoire includes species specific, hierarchically organized, fixed action patterns. The modification of behavior within an individual's lifetime, as a result of habituation or associative and exploratory learning, appears to be strongly restricted by sensory and neurological constraints that limit response only to a narrow range of stimuli (Matthews and Matthews 1978). Certain tasks can be mastered while others cannot. Invertebrate behavior is highly stereotyped and predictable, thus facilitating the study of the genetic basis of specific behavioral traits.

Much of the research on the behavioral genetics of invertebrates has dealt with intraspecific variation found in various general maintenance activities such as taxis, feeding and learning, and the response of this variation to selection. Although the results of these investigations have demonstrated that many aspects of behavior are under genetic control and can respond rapidly to selection, they provide little insight into how behavioral differences among species evolve, or what role the evolution of these differences plays in speciation. Comparative genetic studies directed toward unraveling the genetic bases for behavioral differences between and within species or biologically distinct races are more useful in this regard. Especially important are studies of habitat selection and various types of communicatory activity usually associated with mating success. This comparative evolutionary approach to the study of the genetics of precopulatory reproductive isolating mechanisms is now beginning to reveal the role such traits play in the speciation process, and as a result we should eventually be able to develop more precise and realistic models of speciation.

Quantitative genetic studies of mating behavior have been carried out on only a few organisms such as *Drosophila* and crickets. By studying the pattern of inheritance of several components of the calling song in F^1 hybrids produced by crossing related crickets, Hoy and his colleagues (Hoy 1974) were able to demonstrate that pulse rates, intervals, and progressions are controlled by a complex, polygenic, multichromosomal system. Polygenic control of mating was also inferred in a quantitative genetic study of hybrids and backcrosses made between closely related "semispecies" in the *Drosophila paulistorum* complex (Ehrman 1965). The details of these polygenic control systems have yet to be worked out.

Thus, although differences in mating behavior are recognized as a major cause of reproductive isolation, we know little about their genetic bases, control, or even which genes are important in the evolution of sexual behavior in most organisms. Furthermore, sexual behavior may not be the only behavioral factor responsible for reproductive isolation, and in some cases may play no direct role. There is no *a priori* reason to assume that differences that may now exist in sexual activity between two closely related species played a key role in the development of reproductive isolation during speciation. Such traits could have evolved long after the speciation event that was initiated by, and dependent upon, some other cause of reproductive isolation. The response to ecological cues, for instance, may be much more important in initiating and maintaining reproductive isolation than mating behavior. Monophagous parasitic species living on different hosts may be more strongly isolated from one another by host selection behavior than by sexual behavior (Diehl and Bush 1984); sometimes, the genes controlling such behavior may be few in number (Huettel and Bush 1972, Gould 1983).

The lack of appreciation by population geneticists of the role and great diversity of precopulatory reproductive isolating mechanisms in speciation has resulted in the development of biologically untenable and untestable genetic models of speciation. In the absence of hard data on the genetic basis for these traits, model builders and theorists have been free to pick and choose the genetic criteria for speciation that best suited their needs and computational limitations without reflecting on how their models relate to specific natural speciation events. The value of these theoretical scenarios as evidence for or against one mode of speciation or another is, therefore, questionable (Bush and Diehl 1982).

A more direct approach to the problem of speciation is to establish the genetic basis of specific behaviors identified as key traits in reproductive isolation. Since many closely related species pairs are isolated strictly by differences in behavior, particularly sympatric sibling species in the early stages of divergence (Diehl and Bush 1984), an understanding of the genetic basis for precopulatory reproductive isolation can provide unique insight into the speciation process. I do not advocate this strictly reductionist view as an answer to all our problems, but only as a necessary and rewarding tactic designed to advance our understanding of the genetics of speciation and provide a basis for the development of more biologically meaningful speciation models.

It is, therefore, encouraging that some of the contributions to this volume deal with the genetic basis of reproductive isolation. But how do we identify key behavioral traits important in reproductive isolation and speciation? Although most questions concerning the origin and evolution of reproductive isolation arise as a result of a systematic or phylogenetic investigation, such studies provide little more than a few indirect clues, inferred from morphological characters, as to which behavioral traits might be important. The only way to pinpoint key behavioral traits is to conduct a thorough analysis of the ecology and mating behavior of the species in question. Anyone who undertakes such a task is immediately struck by the apparent complexity of the behavioral repertoire and by how intricately these behaviors are tied to ecological factors. The immediate reaction is to conclude that differences that exist between even very closely related species must represent the accumulation of a great many gene changes during the course of speciation. On closer examination, this necessarily may not be the case.

For example, in the fruit fly genus *Rhagoletis*, the group with which I am most familiar, mating follows a sequence of steps, each encompassing a set of fixed action patterns, that are undoubtedly under the control of many genes. As mating occurs on the host fruit, both sexes must first locate and identify the right host plant before courtship can be initiated. This orientation phase requires the recognition of several long- and short-range visual and olfactory cues emanating from the plant and fruit. Once on the fruit, the adults, which are brightly marked with distinct wing patterns, use highly stereotyped wing and body displays as well as short-range contact pheromones to complete the mating sequences. This is the general pattern of mating behavior displayed by all *Rhagoletis* and many other Tephritidae as well (Boller and Prokopy 1976, Zwolfer 1983).

Each step leading to a successful mating in these flies is essential and requires a complicated set of genetically controlled behavioral traits. However, only a few genetic changes, such as in loci affecting host selection, actually play a critical role in reproductive isolation (Huettel and Bush 1972, Bush and Diehl 1982). Although courtship includes a series of stereotyped wing waves, posturing, and contact pheromones, we have found no evidence that any of the four *pomonella* group sibling species can discriminate between one another either in the laboratory or once they are on a fruit. Apparently, the courtship behavior of the four is so similar that they are unable to tell each other apart. Because their host plant serves as a rendezvous for courtship and mating, host selection is the key behavior separating species. In these insects, and many other parasites as well, a change in genes controlling chemosensory or other responses involved in host selection can have a profound and evolutionarily significant effect on mate choice.

As for host selection, we have found that although visual, tactile, and chemical cues are important components of searching behavior in these flies, it is the chemical cues emanating from the fruit that are used primarily to discriminate between potential host plants (Prokopy et al. 1973). Recently, Fein et al. (1982) and Reissig et al. (1983) have reported that the apple maggot, *Rhagoletis pomonella* (Walsh), is attracted to five volatile compounds found in apple fruits. A different, but related, set of attractants has recently been identified in blueberry fruit which attracts *R. mendax* (Curran), a closely related sibling species to *R. pomonella* (Silk, personal communication).

We know from past experience that these two species can be hybridized easily and backcrossed in the laboratory and are interfertile in all combinations. Thus, it is now possible to carry out a detailed genetic analysis of interspecific host preference differences and establish the number and type of genes directly involved in the expression and control of this important behavioral trait. A genetic model of speciation that focuses on the relationship between host and mate selection should be far more meaningful, as we have pointed out elsewhere (Bush and Diehl 1983), than one based exclusively on mechanisms of assortative mating without host or habitat selection, such as the model of sympatric speciation recently proposed by Felsenstein (1981). Previous sympatric speciation models have generally failed to incorporate specific biological attributes and unique genetic systems characteristic of a particular group of organisms and, thus, usually bear little resemblance to the natural process of speciation.

Because genetic studies often indicate polygenic control of mating and other behavioral patterns, it is usually assumed that the evolution of species differences in such traits proceeds by mutations that have small effects (Mayr 1963). Furthermore, these genes are thought to be part of highly coadapted gene complexes resistant to selection and change. This view has led evolutionists to reject relatively simple speciation models based on two or three polymorphic genes. They accept instead the concept that speciation results from differences that accumulate in many genes during long periods of geographic separation, or by a genetic revolution resulting from inbreeding and drift in small founder populations. An examination of the evidence that adaptation always occurs by microevolution indicates that much of it is circumstantial at best (Bush 1982), and the widely held theory of broad coadaptation of the genome has been seriously challenged (Hedrich et al. 1978).

We actually know very little about the genetic control of behavior or what kind of mutations are necessry for altering key adaptive traits such as those involved in reproductive isolation. To assume that changes at many loci are necessary for such alterations seems premature. In fact, a significant difference in mating behavior or habitat choice could result from a change in one or two genes active at different periods during development that alter the hard-wired neuronal pattern in the central nervous system by simply shifting the timing of a specific developmental cue. Such a mutation, possibly involving a regulatory locus, could, in this way, simultaneously affect the expression of many genes associated with a particular pattern of behavior.

If the control of genes is hierarchial, as now seems likely (Hunkapiller et al. 1982), and certain controlling elements have multigenic effects on the expression of a trait, the role and importance of single major gene mutations for speciation and the evolution of behavior can no longer be ignored. The recent discoveries in molecular genetics are revolutionizing our view of how the eukaryotic genome is organized and functions. The genetic complement of an organism is far more mobile and flexible than we previously suspected and there appears to be a continuous repatterning of gene arrangements within the genome. This reorganization is likely to have far reaching evolutionary implications.

Many of the topics to be covered in this symposium focus directly on the questions raised here concerning the genetics of mate and habitat selection, and all are pertinent to our understanding of the genetics of speciation. We have come a long way since the early selection experiments on the genetics of photo- and geotaxis in *Drosophila*. Obviously, behavior genetics has come of age. It should be interesting to see what sort of fruit this maturity bears.

Literature Cited

Boller, E. F., and R. J. Prokopy. 1976. Bionomics and management of *Rhagoletis*. Annu. Rev. Entomol. 21: 223.

Bush, G. L. 1982. What do we really know about speciation? *In* R. Milkman, ed. Perspectives in Evolution. Sinauer Assoc., Inc., Sunderland, Massachusetts.

Bush, G. L., and S. R. Diehl. 1982. Host shifts, genetic models of sympatric speciation and the origin of parasitic insect species. *In* J. H. Visser and A. K. Minks, ed. Proc. 5th Int. Symp. Insect-Plant Relationships. Pudoc, Wageningen, The Netherlands.

Diehl, S. R., and G. L. Bush. 1984. An evolutionary and applied perspective of insect biotypes. Annu. Rev. Entomol. 29: 471.

Ehrman, L. and P. A. Parsons. 1981. Behavior Genetics and Evolution. McGraw-Hill Book Co., New York.

Fein, B. L., W. H. Reissig, W. L. Roelofs. 1982. Identification of apple volatiles attractive to the apple maggot, *Rhagoletis pomonella*. J. Chem. Ecol. 8: 1473.

Felsenstein, J. 1981. Skepticism towards Santa Rosalia, or why are there so few kinds of animals? Evolution 35: 124.

Gould, F. 1983. Genetics of plant-herbivore systems: interactions between applied and basic study. *In* R. F. Denno and M. S. McClure. Impact of Host Quality on Herbivorous Insects. Academic Press, New York.

Hedrick, P., S. Jain, and L. Holden. 1978. Multilocus systems in evolution. *In* M. K. Heckt, W. C. Steere, and B. Wallace, eds. Evolutionary Biology. Plenum Press, New York.

Hoy, R. R. 1974. Genetic control of acoustic behavior in crickets. Amer. Zool. 14: 1067.

Huettel, M. D., and G. L. Bush. 1972. The genetics of host selection and its bearing on sympatric speciation in *Procecidochares* (Diptera: Tephritidae). Entomol. Exp. Appl. 15: 465.

Hunkapiller, T., H. Unang, L. Hood, J. H. Campbell. 1982. The impact of modern genetics on evolutionary theory. *In* R. Milkman, ed. Perspectives on Evolution. Sinauer Assoc., Inc., Sunderland, Massachusetts.

Matthews, R. W., and J. R. Matthews. 1978. Insect Behavior. John Wiley & Sons, New York.

Mayr, E. 1963. Animal Species and Evolution. Harvard Univ. Press. Cambridge, Massachusetts.

Prokopy, R. J., V. Moerike, and G. L. Bush. 1973. Attraction of apple maggot flies to odor of apples. Environ. Entomol. 2: 743.

Reissig, W. H., B. L. Fein, and W. L. Roelofs. 1982. Field tests of synthetic apple volatiles as apple maggot (Diptera: Tephritidae) attractants. Environ. Entomol. 11: 1294.

Zwolfer, H. 1983. Life systems and strategies of resource exploitation in tephritids. *In* Proc. Int. Symp. Fruit Flies. CEC Press, Athens (in press).

Behavior Genetics of Flexible Life Histories in Milkweed Bugs *(Oncopeltus fasciatus)*

Hugh Dingle*

James F. Leslie

James O. Palmer

Program in Evolutionary Ecology and Behavior
Department of Zoology
University of Iowa
Iowa City, Iowa 52242

Introduction

Insects often face uncertain or unpredictable environments and must make appropriate adjustments in their life histories. Adaptations to predictable environments are well-known, and there is an extensive literature on the use of photoperiod, the most reliable seasonal cue, to time diapause, migration, reproduction and related responses. The difficulty comes when photoperiod (or any other cue) is unreliable. Under such conditions, the best strategy for an organism is to remain flexible, and the most obvious way to do so is via behavior, in particular those behaviors such as migration and diapause that allow choices of where and when to breed. Behavior can thus be an important element of a life history "strategy" (Dingle 1982, 1984).

We have been concerned with life history evolution in the large milkweed bug *Oncopeltus fasciatus* (Dallas)(Hemiptera: Lygaeidae). Several field studies provide the broad outlines of its natural history in North America and the Caribbean (Dingle 1981, Evans 1982, Miller and Dingle 1982). In the North the bug is a seasonal migrant making its first appearance in late spring or summer. There is a late summer adult reproductive diapause (Dingle 1974) leading to migratory exodus during the period of suppressed reproduction, as confirmed by marking experiments in the field (Dingle 1981). The bugs cannot overwinter in the Northeast, but are able to do so in California (Evans 1982). During the summer from one to three generations are produced depending on latitude and weather. Populations are present throughout the winter in southern Florida, and some breeding may occur (Miller and Dingle 1982). Uncertainty enters in with respect to both the number of generations in the North and the success of winter reproduction in Florida. On the island of Puerto Rico, successful breeding takes place throughout the year with local movement among milkweed (*Asclepias curassavica L.*) patches.

*Present address: Hugh Dingle, Department of Entomology, University of California-Davis, Davis, California 95616.

Populations of *O. fasciatus* over the extent of their ranges thus face different environments having varying degrees of uncertainty. As might be expected, there are differences among populations with respect to an assortment of life history traits (e.g. Dingle 1981). Some of these traits are the expression of genetic systems having considerable variability (Dingle et al. 1977; see also Tallamy and Dingle, this volume). As an example, bugs from Iowa are large, enter diapause on short days, and display considerable long distance flight. Bugs from Puerto Rico, on the other hand, are relatively small, do not diapause, and fly only for short periods. Several life table statistics also differ, both absolutely and in response to temperature and photoperiod (Dingle et al. 1980a and 1980b, Dingle 1981). In this paper we shall examine the genetics of life histories in *O. fasciatus*, first for the Iowa population, and then for the differences among Iowa, Puerto Rico and another northern population from Maryland. We shall address the question of how the genetic structure of life histories incorporates behavioral flexibility into interactions among elements to make the life history "work" for the insect in environments differing in predictability.

Methods

To analyze life history genetics in *O. fasciatus*, we have used three methods: sibling analysis, selection for wing length, and crosses between geographically separated populations. The sibling analysis provides a way to examine the genetic variance and covariance structure within a population and a baseline for further studies of complex life history adaptations. The basic design is outlined fully by Falconer (1981) and involves the generation of both full-sib and half-sib families. In our experiments, we reared Iowa bugs in a LD 16:8 and 23 °C regimen in which they did not diapause. The anaysis was run in three blocks; in each block 10 males were mated to each of 4 females, i.e., each male mated with 4 females while each female mated with only 1 male, for a total of 40 pairings per block and 120 pairings overall. Ten female offspring were collected from each full-sib family to yield 1200 offspring. Data on development time, age at first reproduction, clutch size, fertility, and body and wing length were recorded for each parent and offspring. This design allows estimates of variance among fathers (unrelated progeny), among mothers within fathers (half-sibs), and within mothers (full-sibs). One can then obtain estimates of heritabilities and genetic correlations for the characters from the components of variance and covariance.

The selection experiment involves Iowa bugs reared at LD 14:10 and 27 °C and consists of three replicated lines: a long-winged, a short-winged, and randomly selected control with each line consisting of 20 families per generation. Wing-length was chosen as the character for selection because heritability estimates indicated sufficient additive genetic variance for a significant response. Each line is replicated to avoid, as far as possible, complications arising from drift and other extraneous sources of variance (Falconer 1977, 1981, Rose and Charlesworth 1981a,b). A "within-family" design, was used in which the largest and smallest members of each family of 30-40 offspring are the selected parents. This design minimizes the effects of environmentally induced variation between families resulting from heterogeneities in growth chamber performance and the effects of inbreeding. Differences among lines in both wing length and correlated life table characters were assessed following five generations of selection.

Finally, we made crosses between populations of bugs originally collected in Iowa, Maryland, and Puerto Rico. Ten lines, each founded by three pairs of wild-captured bugs, were established from each of the three geographic sources and reared at LD 14:10 and 27 °C for two generations. Seven strains from each of the three areas were chosen randomly and two males and two females from each were paired randomly to generate new outcrossed lines for each source area. Eight to eleven fertile parental pairs from each of these crosses were also chosen randomly, and their offspring used in the population crosses. Pairs were maintained individually and measurements made of a series of life history traits on four female offspring per pair mated to male offspring of the same cross. A similar selection process within lines produced the F_2 "purebred" and "hybrid" pairs on which measurements were likewise made.

In all experiments, individual pairs of bugs were kept separately in petri dishes. Each pair was provided with a water bottle with a cotton wick, a supply of milkweed seeds (*Asclepias syriaca* L.), and a wad of cotton which served as an oviposition site. The bugs were examined daily until the death of the female; if the male died first, he was replaced by a male of similar age and genetic stock. Cotton in which eggs had been deposited was removed and replaced with a fresh wad.

Results

Sibling Analysis

A brief summary of the results of the sibling analysis is given in Table 1 which is excerpted from the complete genetic correlation matrix of 12 life history characters (see Hegmann and Dingle, 1982, for the complete table). Some particular points worth noting are the following. First, the heritability of age at first reproduction (α, measured as the interval between adult eclosion and oviposition of the first clutch of eggs) is 0.25, a value which is considerably lower than the value of .0.70 estimated from response to selection for early reproduction in a diapause inducing photoperiod (Dingle et al. 1977). This perhaps suggests that there is a separation of response to critical photoperiod and age at first reproduction per se, thus introducing flexibility into the life cycle. Secondly, both wing length (WL, which also indicates body size) and development time (DT, egg to adult eclosion) display high heritabilities indicating considerable additive genetic variance available for natural selection. Thirdly, there are significant genetic correlations among wing length, clutch size (CS), and development time. The negative correlations between wing length and clutch size on the one hand and development time on the other indicate that large bugs and bugs producing large clutches develop faster, and the positive correlation between wing length and clutch size indicates big bugs have big clutches. To summarize: long winged (large) bugs develop rapidly and have large clutches as a result of the three traits sharing genes in common (pleiotropy), or sharing linked parts of the genome (linkage).

A major exception to this array of genetic correlations is age at first reproduction (α) which appears to be uncorrelated with any of the other traits. Since age at first reproduction is also a function of diapause and migration (Dingle 1978), it is a character directly involved with life history flexibility in the face of environmental uncertainty. The lack of correlation of α with other life history traits means that α is free to vary under selection without corresponding variation in those other characters. In other words, there are no genetic correlations to "tie-up" the genetic variance contributing to life history flexibility by forcing variation in, say, fecundity characters to follow variation in α. Rather, the latter can vary independently. We interpret this to mean that for this population of *O. fasciatus*, natural selection has produced a life history strategy with an array of genetically correlated traits that do not include α. Age at first reproduction is thus free of a "cost of correlation" and can vary to provide considerable flexibility of reproductive timing without incurring consequences for other fitness characters. This "freedom" of α from costs of genetic correlation may also occur in other insects with close association between migration and diapause (Derr 1980, Dingle 1983). For a fuller discussion of this topic see Hegmann and Dingle (1982) and Dingle (1984). The association between body size, clutch size and development time is what one might expect of an "r-selected" migrant colonizer (Safriel and Ritte 1980, Dingle 1984).

Table 1. Heritabilities (on leading diagonal) and genetic correlations (± s.e.) for some life history traits of *Oncopeltus fasciatus* calculated from sibling-analysis (Hegmann and Dingle 1982).

Character	WL	α	CS	DT
Wing length (WL)	.55 ± .22			
Age at first reproduction α	−.29 ± .28	.25 ± .12		
Size of first clutch (CS)	.73 ± .13	−.22 ± .29	.25 ± .10	
Development time (DT, egg to adult)	−.55 ± .19	−.13 ± .28	−.57 ± .18	.89 ± .32

Selection for Wing Length

To see if we could repeat the above results if genetic correlations were assessed in a different way, we initiated a selection experiment with wing length as the targeted trait. Our experiments are thus similar to those of Rose and Charlesworth (1981a,b) who analyzed life histories in *Drosophila melanogaster* Meigen using both sibling-analysis and selection also. In our case, however, the sibling-analysis and selection were done under different environmental regimens. The prediction from the data in Table 1 was that age at first reproduction would not respond to selection on wing length (because there was no apparent genetic correlation between the two traits), while traits such as body size and fecundity would.

The data in Figure 1 show the response to selection for wing length in the replicated long wing, short wing, and control (unselected) lines. Clearly over the five generations included here, wing length has responded to selection in both directions. Differences between both long and short lines and controls are highly significant as indicated by the standard errors (see legend, Fig. 1). Further response in selected lines continued in the sixth and seventh generations; a complete analysis of these data through the seventh generation, including realized heritabilities, will appear elsewhere (J. O. Palmer, unpublished).

As indicated in Figures 2 and 3, we obtained correlated responses to selection on wing length in both body length and fecundity. In Figure 2, randomly chosen females from each line in generation 5 are compared to females of the base population. (This is an initial and conservative test for correlated response to selection.) Selection for wing length has brought about correlated responses in body length with long-winged females having longer bodies and vice versa. Changes in fecundity are indicated in Figure 3. Here the selected parents in generation 5 are compared to selected parents from the generation 0 base population (data from randomly chosen seventh generation females will be analyzed at the completion of the experiment). Again there is strong evidence for correlated responses. Fecundity of parent females remained unchanged between generations 0 and 5 in the control lines, increased in the long wing lines, and decreased in the short wing lines, consistent with a positive genetic correlation between wing length and fecundity.

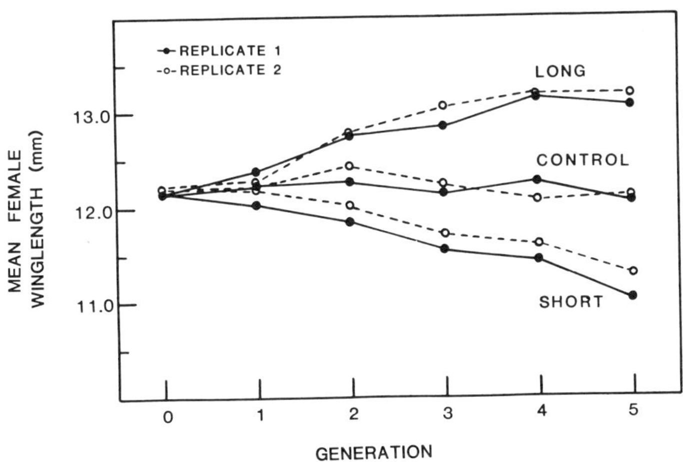

Fig. 1. Response to selection for wing length in an Iowa population of *O. fasciatus*. Points indicate mean wing lengths for all females within each line for each of the five generations. Sample sizes for each mean range from 244-313 females. Standard errors range between ± .04 (less than the width of a point).

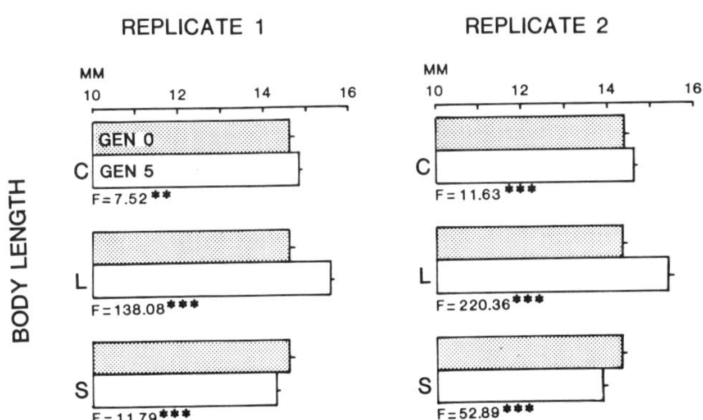

Fig. 2. Correlated response in body length to selection on wing length. Mean body length of 163-233 randomly chosen females from each line is compared to generations 0 (shaded bars) and 5 (open bars). One standard error is attached to each mean. One, two, or three asterisks indicating significance at $p < .05$, $p < .01$, and $p < .001$, respectively, by ANOVA.

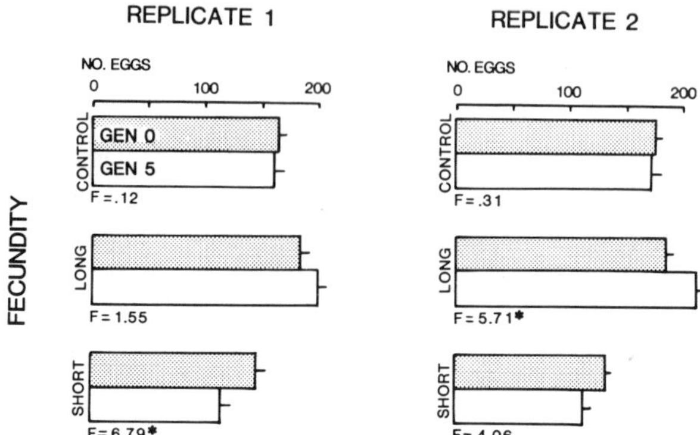

Fig. 3. Correlated response in fecundity among selected female parents. Fecundity here is the total number of eggs produced during the first five days of reproduction. Sample sizes range from 13-19 females. Labeling as in Fig. 2.

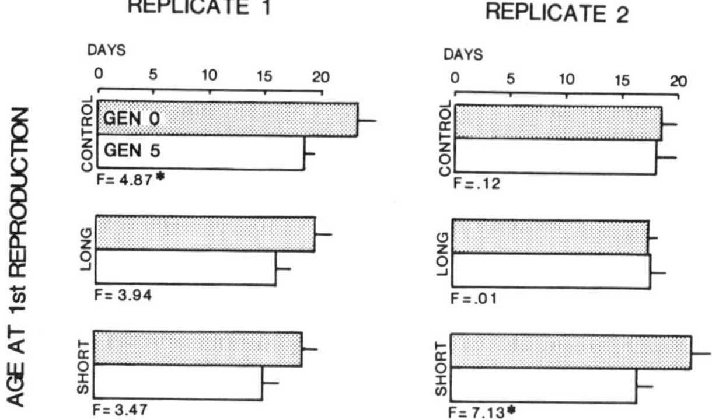

Fig. 4. Changes in age at first reproduction between generations 0 and 5 of selection for wing length. Sample sizes range from 17-20 females. Labeling as in Fig. 2. No correlated response to selection on wing length is apparent.

The data for age at first reproduction (α) are presented in Figure 4. It should first be noted that there seems to be a slight but consistent decrease in α between generations 0 and 5 in all lines. This could be due to drift, but is more likely the result of weak uncontrolled selection across all lines due to a culturing procedure which unavoidably favors earlier reproducing females (see Dingle et al. 1977). Because of this, the significantly reduced α in the short line of replicate 2 does not indicate response to selection. Note that there is also significantly reduced α in the control line of replicate 1. Thus no consistent trends in α were noted in response to selection on wing length, and we interpret this to mean that, as predicted from the results of the sibling analysis, α is not genetically correlated with wing length and so does not respond to selection on the latter.

Population Crosses

In a previous study of population crosses between *O. fasciatus* from Iowa and Puerto Rico at 24 °C under long and short day conditions, three general results were obtained (Dingle et al. 1982). For age at first reproduction and wing length, hybrid offspring were intermediate to the parents in the expression of the traits, suggesting between population additive genetic variance. Secondly, for an array of traits associated with fecundity, the hybrids expressed overdominance, suggesting coadaptation in the two original population genomes (Endler 1977). This was even true in some instances where parental phenotypes were not significantly different, suggesting that they were nevertheless the result of different genotypes. Finally, for clutch size there was dominance deviation toward the Puerto Rico parent in one environment but not the other (short day *vs.* long day), and for development time overdominant and intermediate hybrids occurred in the two environments. The type of expression was thus environment specific in both traits. Overall the results indicated that the phenotypic life history differences between Iowa and Puerto Rico were the result of differentiated gene pools. The present set of population crosses was designed to see if the Iowa-Puerto Rico results were the same in a different environment and to examine gene differences between Iowa and another northern population from Maryland and between Maryland and Puerto Rico.

We consider here age at first reproduction (α), eggs per day produced during the reproductive period (a measure of fecundity) and clutch size because these three traits showed different responses in the original Iowa X Puerto Rico cross. In the case of α (Fig. 5) first note that in the F_1 generation Iowa X Puerto Rico cross, hybrids were intermediate, similar to the previous results at 24 °C, but with a suggestion of some dominance deviation toward Puerto Rico. In the F_2, hybrids were intermediate. In the Maryland X Puerto Rico cross, the hybrid values more strongly suggest deviation toward the Puerto Rico parent; again intermediate hybrids occur in the F_2. These results indicate that gene differences occur between either of the northern populations and Puerto Rico. There is some tendency for the Puerto Rico genome to be "dominant," but the intermediate F_2 hybrids indicate that polygenic differences separate the populations. The Iowa and Maryland populations do not differ phenotypically. The apparent influence of direction of cross (MI *vs.* IM) may indicate some differences in the two source genomes, but it does not seem to be as great as those producing phenotypic differences between the northern and Puerto Rican bugs.

In the case of our fecundity measure of eggs per day, results for the F_1 generation Iowa or Maryland X Puerto Rico crosses reveal overdominance in the hybrids (Fig. 6), duplicating previous results. The presence of overdominance indicates that parental gene pools differ even when, as in the case of Maryland and Puerto Rico, there is no statistically significant difference between parental phenotypes. No significant trends or differences occurred between Iowa, Maryland, or their hybrids, suggesting the populations were similar enough genetically with respect to fecundity that our crosses did not detect between-source genetic variation.

Finally, clutch size in both Iowa and Maryland X Puerto Rico crosses showed a dominance deviation toward Puerto Rico in the F_1 hybrids, but were intermediate, indicating polygenes, in the F_2 (Fig. 7). The F_1 results are similar to the F_1 results from Iowa X Puerto Rico in short day at 24 °C, while the F_2 results are like those in long day (Dingle et al. 1982). Taken together all these results indicate a Puerto Rico dominance component in a polygenically determined trait. It is interesting that differences in oviposition site preference between Puerto Rico and other populations seem to be determined similarly (Leslie and Dingle 1983) so that there may be a "reproductive syndrome" involving both clutch size and egg laying behavior. As with fecundity, we detected no genetic or phenotypic differences between Iowa and Maryland bugs with respect to clutch size.

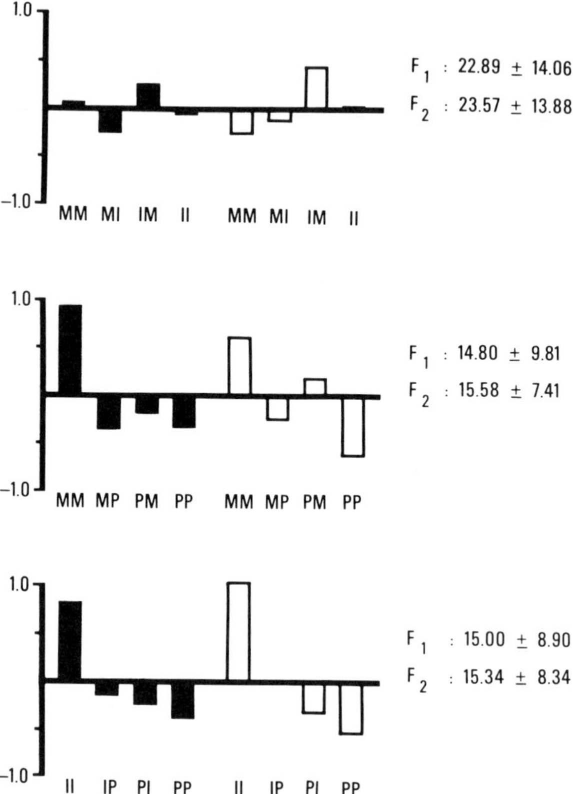

Fig. 5. Expression of age at first reproduction (alpha) in crosses between bugs of Iowa (I), Maryland (M), and Puerto Rico (P) origin. Female listed first in each cross, e.g., MP is Maryland X Puerto Rico. Solid bars are F_1's, and open bars F_2's. Mean values in days for each cross are plotted in standard deviation units (ordinate) as deviations from the overall generation means given at the right (± s.d.), e.g., the MM mean alpha for the F_1 in the M X P series is approximately 1 standard deviation (about 9 days) longer than the overall F_1 mean of 14.80 days. Sample sizes range from 28-42 females here and in Fig. 6 and 7.

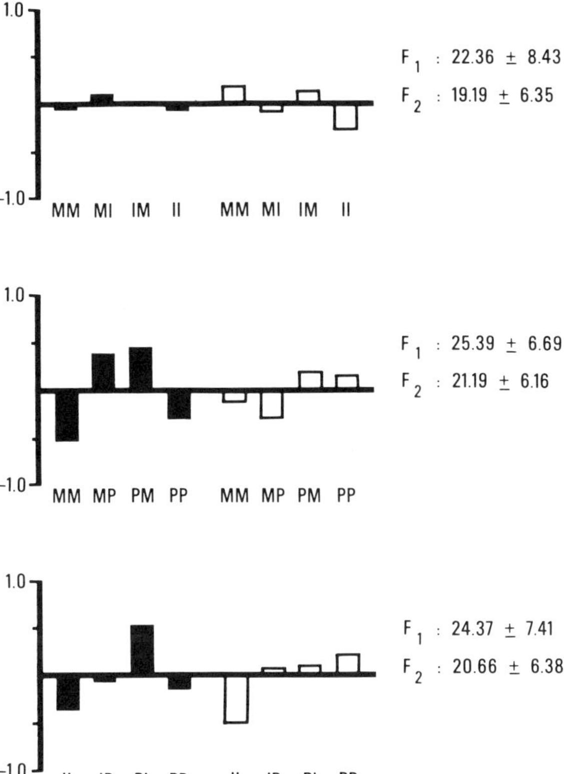

Fig. 6. Expression of the fecundity measure eggs per day over the reproductive period for the various crosses. Plotted as in Fig. 5.

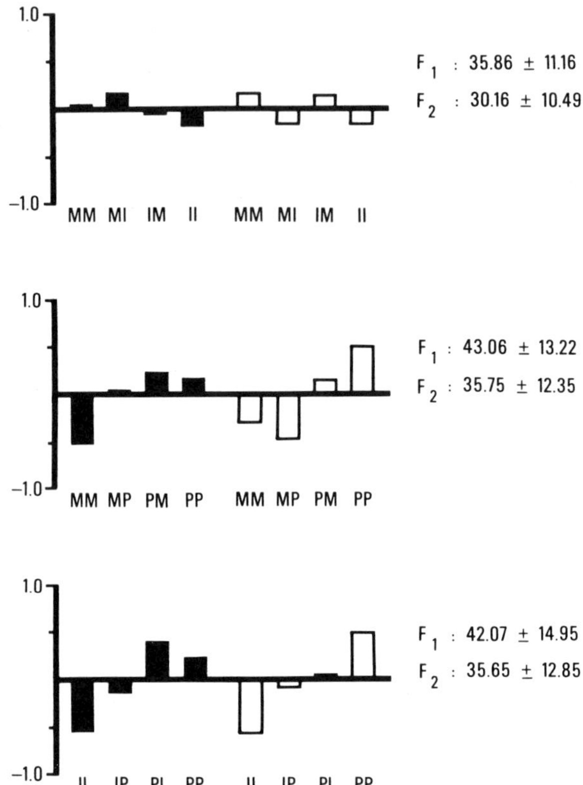

Fig. 7. Expression of clutch size in the different crosses. Plotted as in Fig. 5.

Discussion

Our original sibling analysis suggested that in an Iowa population of *O. fasciatus* there was a life history ''strategy'' consisting of interrelationships among life history traits involving genetic correlations. Of particular interest was the fact that age at first reproduction seemed to be free of such correlations and could therefore vary under natural selection independently of other fitness characters. The importance of this to the bug would be that it could delay reproduction to migrate or enter diapause without adverse consequences resulting from genetic correlations with other traits. This behavioral and genetic flexibility allows an array of options in the face of environmental exigencies.

We have begun a further examination of this life history strategy by testing with selection experiments the predictions arising from the sib-analysis results. In general the predictions seem to hold up. Age at first reproduction does not seem to be correlated genetically with wing length, since we observed no correlated response to α when the former was the target of selection. Conversely, and as expected, there were positively correlated responses in body length and fecundity. Preliminary results also indicate that bugs selected for long wings fly for longer periods during a tethered flight assay (Palmer, unpublished). If this result is borne out when the flight test data and their analysis are complete, it suggests that it may indeed be appropriate to consider the genetic and phenotypic associations among traits a migration-life history syndrome adapted for colonization in this migrant population (Dingle 1984). In any event it is encouraging that the results of the sibling-analysis and of selection seem to be giving us some significant insights into the genetic architecture of life history structure in *O. fasciatus*.

The full results, analysis, and tests of significance for all traits examined in several population crosses, including Iowa, Maryland, and Puerto Rico, will be published elsewhere when complete (J. F. Leslie, unpublished). The preliminary data reported here, however, do suggest some interesting gene differences among bugs of different geographic origin. Females from Puerto Rico have quite different reproductive habits, apparently evolved in the approximately 450 years *O. fasciatus* has been present on the island (Dingle 1981). That period has been sufficient to evolve gene differences expressed as dominance deviations, overdominance, or intermediacy in the F_1 hybrids between populations. Our results also reveal that genome expression in the phenotype is environment specific. Another interesting result is that Iowa and Maryland populations seem to be similar genetically, we suspect both because they face similar selective regimes and because they exchange genes as a result of their strong migratory capabilities. A further question raised by our results is: at what point might the gene differences contributing to differences in life history phenotypes lead to speciation? The answer remains for future research to reveal.

Acknowledgments

Supported by grants from the U. S. National Science Foundation.

Literature Cited

Derr, J. A. 1980. The nature of variation in life history characters of *Dysdercus bimaculatus* (Heteroptera: Pyrrhocoridae), a colonizing species. Evolution 34: 548.

Dingle, H. 1974. Diapause in a migrant insect, the milkweed bug *Oncopeltus fasciatus* (Dallas) (Hemiptera: Lygaeidae). Oecologia 17: 1.

Dingle, H. 1978. Migration and diapause in tropical, temperate, and island milkweed bugs. *In* H. Dingle, ed. The Evolution of Insect Migration and Diapause. Springer-Verlag, New York.

Dingle, H. 1981. Geographical variation and behavioral flexibility in milkweed bug life histories. *In* R. F. Denno and H. Dingle, eds. Insect Life History Patterns: Habitat and Geographical Variation. Springer-Verlag, New York.

Dingle, H. 1982. Function of migration in the seasonal synchronization of insects. Entomol. Exp. Appl. 31: 36.

Dingle, H. 1984. Behavior, genes, and life histories: Complex adaptations in uncertain environments. *In* P. W. Price, C. N. Slobodchikoff, and W. S. Gaud, eds. New Ecology: Novel Approaches to Interactive Systems. Wiley, New York (in press).

Dingle, H., C. K. Brown, and **J. P. Hegmann.** 1977. The nature of genetic variance influencing photoperiodic diapause in a migrant insect, *Oncopeltus fasciatus*. Amer. Natur. 111: 1047.

Dingle, H., B. M. Alden, N. R. Blakley, D. Kopec, and **E. R. Miller. 1980a.** Variation in photoperiodic response within and among species of milkweed bugs (*Oncopeltus*). Evolution 34: 356.

Dingle, H., N. R. Blakley, and **E. R. Miller.** 1980b. Variation in body size and flight performance in milkweed bugs (*Oncopeltus*). Evolution 34: 371.

Dingle, H., W. S. Blau, C. K. Brown, and **J. P. Hegmann.** 1982. Population crosses and the genetic structure of milkweed bug life histories. *In* H. Dingle and J. P. Hegmann, eds. Evolution and Genetics of Life Histories. Springer-Verlag, New York.

Endler, J. A. 1977. Geographic Variation, Speciation, and Clines. Princeton University Press, Princeton, New Jersey.

Evans, K. E. 1982. The annual pattern of migration and reproduction in field populations of the milkweed bug, *Oncopeltus fasciatus*, in California. Ph.D. Thesis. University of California, Berkeley.

Falconer, D. S. 1977. Some results of the Edinburgh selection experiments with mice. *In* E. Pollack, O. Kempthorne, and T. O. Bailey, eds. Proceedings of an International Conference in Genetics. Iowa State University Press, Ames.

Falconer, D. S. 1981. Introduction to Quantitative Genetics. 2nd ed. Longman, London and New York.

Hegmann, J. P., and **H. Dingle** 1982. Phenotypic and genetic covariance structure in milkweed bug life history traits. *In* H. Dingle and J. P. Hegmann, eds. Evolution and Genetics of Life Histories. Springer-Verlag, New York.

Leslie, J. F., and **H. Dingle** 1983. A genetic basis of oviposition preference in the large milkweed bug, *Oncopeltus fasciatus* (Hemiptera: Lygaeidae). Entomol. Exp. Appl., 34:215.

Miller, E. R., and **H. Dingle** 1982. The effect of host plant phenology on reproduction of the milkweed bug, *Oncopeltus fasciatus*, in tropical Florida. Oecologia. 52: 97.

Rose, M. R., and **B. Charlesworth** 1981a. Genetics of life history in *Drosophila melanogaster*. I. Sib analysis of adult females. Genetics 97: 173.

Rose, M. R., and **B. Charlesworth.** 1981b. Genetics of life history in *Drosophila melanogaster*. II. Exploratory selection experiments. Genetics 97: 187.

Safriel, U. N., and **U. Ritte** 1980. Criteria for the identification of potential colonizers. Biol. J. Linn. Soc. 13: 287.

Chromosomal Polymorphism and Vagility in Natural Populations of *Drosophila pseudoobscura*

R. F. Rockwell

Louis Levine

Department of Biology
The City College of New York
Convent Avenue and 138th Street
New York, New York 10031

Introduction

If one desires to study the evolutionary genetics of a behavior, one must select a trait that is important in the natural setting of a species. However, given the differences in ecology that normally occur over the range of most species, one must anticipate that different populations of the species will vary in the level of expression of the behavior. In such a study, one wants to be able to identify ultimately the gene differences among populations that result in their observed behavioral differences. Unfortunately, for most behaviors we do not know which genes play major roles in the expression of the trait, and at least initially, we may have to look for associations between the behavior and chromosomes or even complexes of chromosomes. Dispersal is a behavior that is important in the natural setting of most animal species; *Drosophila pseudoobscura* Frolova is a species whose populations vary in their degree of chromosomal polymorphism. These polymorphisms have been found to be associated with certain behaviors and other factors affecting fitness, and it is therefore reasonable to look for their association with dispersal behavior.

Vagility is the freedom of movement that characterizes most organisms, being one of the mechanisms by which they respond to changes in the environment. It is highly developed in animals where it affects many complex behavioral repertoires, including those involved in searching for food, escaping from enemies, and finding a mate. It also forms the basis of migratory behavior, thus contributing to the spread of groups into new environments.

The movement of members of a population from one geographic locality to another is an important factor in the evolutionary process. It is clear that not all members of a given population emigrate. Of those members that do, the action of some is a reflection of their genotypes, for others it is a response to environmental factors, and for most it is probably caused by the interactive effects of both genotype and environment. Successful emigrants are those capable of surviving and reproducing in their new localities, thus making possible fine level adaptations to the new area. The possibility of extending the range of a species, therefore, depends both on the capacity of the genetic compositions of the emigrants to respond to a new environment and on their tendency to move from the old one.

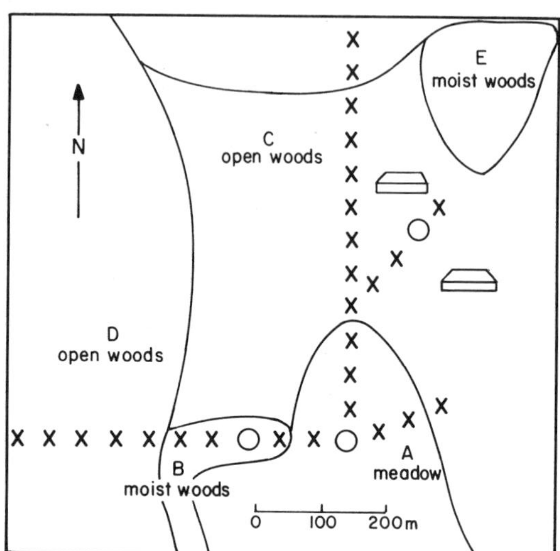

Fig. 1. Schematic map of the Mather, California, area showing adjoining areas of open woods, moist woods, and meadow. Locations of traps are indicated by X's, release sites by circles. (Source: Powell and Taylor 1979).

There are sufficient examples of the importance of vagility for both field and laboratory populations of *Drosophila*. In a study of habitat selection of populations of *D. persimilis* Dobzhansky and Epling (sibling species of *D. pseudoobscura*), Powell and Taylor (1979) chose an area with several distinct ecological niches (Fig. 1). A study of chromosome-inversion types and protein alleles revealed that even over these fairly short distances, inversion and gene frequencies differed from habitat to habitat, implying that the microgeographic genetic variation of these populations reflected adaptations to the respective habitats. The two areas of greatest ecological contrast were those designated B and C, which were, respectively, a dense, moist, dark woods and an open, dry, sunny woods. A capture-release-recapture experiment was conducted, using flies captured in areas B and C. The flies were marked with UV-fluorescent dusts of different colors so that each fly was identifiable both as to area of origin and point of release. The flies were released at points shown in Figure 1, in areas A, B, and C, and recaptured the following day in areas B and C. It was found that there was a distinct tendency for flies to return to the area in which they were captured initially and, hence, in which they were more highly adapted.

A most interesting discovery of a change in vagile behavior, which represented an evolutionary adaptation to a new environment, was made in population cage experiments involving the species *D. willistoni* Sturtevant (DeSouza et al. 1970). In one set of experiments, it was observed that a large number of larvae left the food cups and crawled onto the bottom of the population cage where most of them died from dehydration and lack of food. Only a small number of larvae that left the food cups formed pupae and produced adults. However, with time an increasing number of larvae that left the cups were observed to pupate on the cage floor and complete their life cycle successfully . At this point selection was started to obtain strains that pupated either inside or outside the cup. Pupae from inside the cups and those from outside were collected and kept separately. The adults produced by each type were introduced subsequently into different cages. The same separation of pupae and the adults derived from them was repeated in the following generations. The response to selection was very rapid and after six generations over 98% of the larvae pupated on the type of site for which they had been selected. As a result of this selection experiment, an "inside" and an "outside" strain were developed. In order to determine the genetic basis of pupation-site preference, reciprocal crosses were made between members of the two strains to obtain an F_1, and further crosses were made to obtain F_2 and backcross progeny. The results from the various crosses showed clearly that the difference between inside and outside strains is due mainly to a single pair of major alleles, outside being dominant over inside (see Table 1). Further studies on these two selected strains revealed that members of the "outside" strain had a faster rate of development and required less food than members of the "inside" strain. In this case, we have the colonization of a new ecological niche as a result of a change in larval vagility.

An experiment linking geotactic behavior to chromosomal polymorphism in *D. pseudoobscura* was reported by Dobzhansky and Spassky (1962). In a selection experiment for positive and negative geotaxis, flies heterozygous for the third chromosome gene arrangements Arrowhead (AR) and Chiricahua (CH) were used as the initial population. As shown in Figure 2, selection in both directions was successful in the case of females, and although not shown in Figure 2, almost identical results were obtained for males. After eight

Table 1. Percentage of pupation outside and inside the cups in different crosses with outside and inside strains.

Crossing	Percentage of Pupation		Total Number of Pupae
	Outside the Cups	Inside the Cups	
In X Out	98.5	1.5	1,000
Out X In	98.8	1.2	860
Out/In X Out/In	73.4	26.6	1,130
In/Out X In/Out	76.6	23.4	1,754
Out X Out/In	98.6	1.4	728
Out/In X Out	98.9	1.1	726
In X In/Out	50.3	49.7	1,369
Out/In X In	45.0	55.0	724
Out X In/Out	98.7	1.3	758
Out/In X Out	98.9	1.1	809
In X Out/In	46.6	53.4	749
In/Out X In	50.6	49.4	1,214

(Source: De Souza et al. 1970)

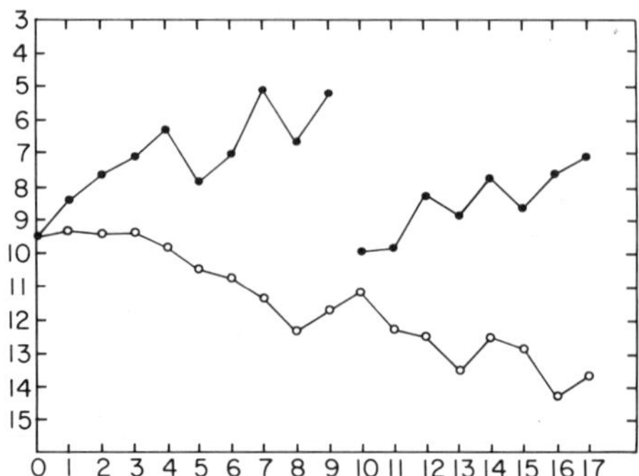

Fig. 2. Selection for positive and negative geotaxis in *D. pseudoobscura* females. Ordinate, the geotactic score; abscissa, generations of selection. Black circles, negative selection; white circles, positive selection. (Source: Dobzhansky and Spassky 1962).

generations of selection, the salivary gland chromosomes of larvae from each population were examined. The population selected for negative geotaxis proved to have only AR chromosomes while that selected for positive geotaxis maintained its polymorphism. At generation ten, the selection experiment for negative geotaxis was repeated, again using AR/CH flies as the initial population. Here too the response to selection was successful, and after seven generations of selection, a cytological examination showed that this population had also lost its CH chromosome and had become monomorphic for AR. This was the first report of an association of a chromosomal polymorphism and a behavior unconnected with sex. (Earlier, Spiess and Langer [1961] had reported differences in mating propensity among karyotypes of *D. persimilis*.)

Evolutionary genetic studies of a behavior should include an evaluation of changes in the expression of that trait both among populations and across time. In a fashion analogous to studies of morphological traits like industrial melanism, conclusions could then be drawn on such phenomena as allele replacement, enhancement of dominance and the establishment of balanced polymorphism. A project that aims to provide both temporal and spatial information on genetic and behavioral characteristics of natural populations of *D. pseudoobscura* has been in progress for the past two years. This is a binational effort involving American and Mexican geneticists, initiated some eight years ago under the direction of the late Theodosius Dobzhansky. During the first six years of the project, the basic thrust centered on an analysis of chromosomal polymorphism in three natural populations of the species over the various seasons of each year (Guzman et al. 1975, Powell et al., unpublished).

Two years ago, two additional goals were incorporated into the group's efforts: study of (1) behavior and (2) morphological and physiological characteristics of these populations. To date, two seasonal samples of flies from each of three collecting sites have been evaluated for vagility (Rockwell et al. 1983, Rosa et al., 1984). In addition, studies are in progress on photobehavior, digging activity, and pupation site preference. We plan to expand our efforts in the near future to include a variety of sexual and several other nonsexual behaviors. Thus far the investigation of morphological and physiological traits has involved gathering material for a wing-length study and collecting data on dessication resistance. Quite clearly, it is the total phenotype of the individual which confers fitness. By integrating temporal and spatial information on a variety of potentially interrelated traits, we expect to be able to evaluate more effectively the evolutionary genetics of specific behaviors and thereby gain clearer insight into microevolution in these populations.

In this paper we present our findings on vagility, relate them to levels of third chromosome inversion polymorphism, and discuss the potential importance of vagility to these populations.

Materials and Methods

The isofemale lines used in these experiments were each derived from randomly sampled, single, gravid females captured at Zirahuen, in the state of Michoacan, on May 18-19, 1981 and February 15, 1982; at Amecameca, in the state of Mexico, on May 28-30, 1981 and January 12-14, 1982; and at Tulancingo, in the state of Hidalgo, on May 24-25, 1981 and January 15-17, 1982. The locations of these sites are shown in Figure 3. The techniques involved in the collection of the flies and the subsequent determination of their third chromosome gene arrangements are described elsewhere (Levine et al. 1980). Our test system for vagility was devised by Sakai and his coworkers (1958) and consists of four interconnected vials through which the individual flies can move (Fig. 4). Twelve isofemale lines from each temporal sample of each locality were used in the experiments. The flies tested were taken from the fifth, sixth and seventh generation offspring of the females captured originally. The population size of each line in each generation was no less than 200 individuals.

Fifty male flies of a given isofemale line were collected fom a culture bottle, placed together in 8-dram food vials (20 per vial) and stored at 20 °C for 24 hours. Humidified carbon dioxide was used as the anesthesia for this procedure and the males collected were between 4 and 7 days old. Toward the end of the 24 hour storage period, approximately one hour before the beginning of the experiment, four vials were connected with clear tape to form a vagility test system. The unused connecting tubes were plugged with corks. Carolina Instant *Drosophila* medium was used as the food source in the vials. Each assembled system was placed in an open plastic tray measuring 28 x 28 x 16 cm high. The inside of each tray had been painted white and the outside black, to provide diffuse, uniform illumination. The trays containing the vagility test systems were placed in a 20 °C incubator, beneath a fluorescent light fixture consisting of two 40 W tubes suspended 61 cm above the trays.

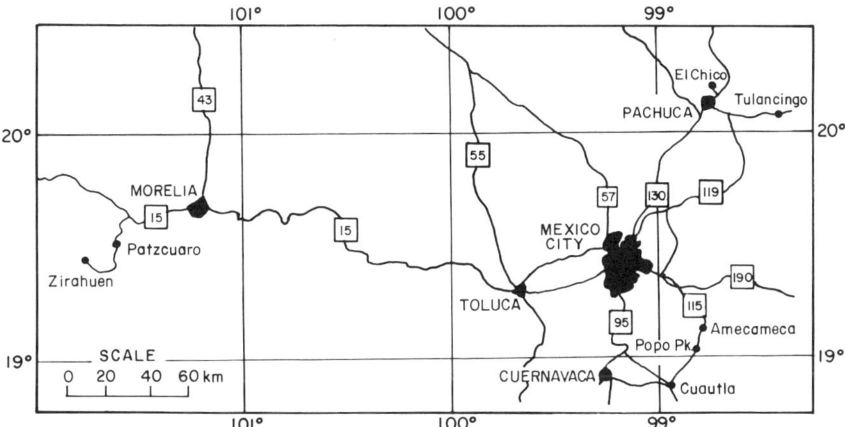

Fig. 3. Locations of the three geographic populations of *D. pseudoobscura*.

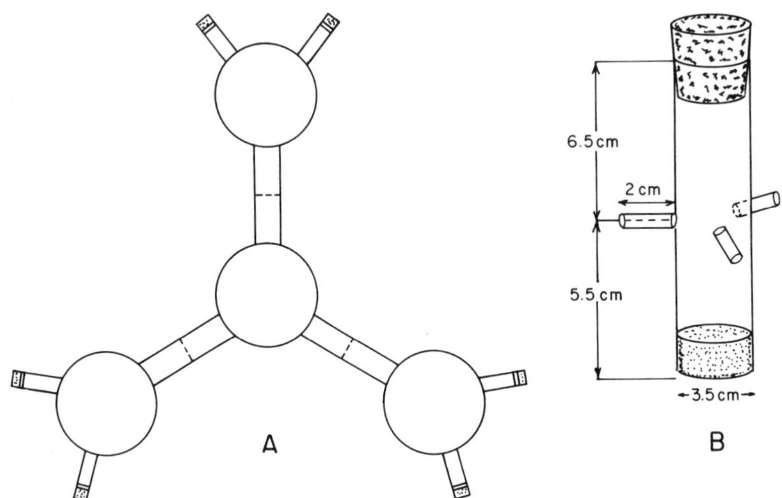

Fig. 4. The vagility test apparatus modeled after Sakai et al. (1958): (A) the arrangement of vials; (B) single vial showing connecting tubes and giving dimensions.

At the start of the experiment, between 11:30 AM and noon, a given set of 50 males was aspirated into the central vial of the system. The system was left undisturbed for 24 hours. At that time the number of flies in the three peripheral tubes was determined. The total number of flies in the three peripheral tubes is used as the measure of vagility and can range from 0 to 50. Three independent replicate trials were performed on all the isofemale lines of each geographical sample of the species in a replicate blocks fashion (Winer 1971). This design, in which replicate (i) of all lines is completed before replicate (i + 1) of any line is begun, was used for both temporal samples. The individuals for each replicate of a given isofemale line were drawn randomly from a different culture bottle of the line. This procedure maximized the likelihood that any "culture bottle" differences between lines were represented within lines, and, thus, minimized the potential effect of common environmental variation on similarities within isofemale lines (Falconer 1981). Since preliminary analyses demonstrated no significant effect of the replicate blocks, this source of variation was pooled with the error within each geographic and temporal sample of the species (Winer 1971, Rockwell et al. 1975).

Results

Inversion Polymorphism

The types and relative frequencies of the gene arrangements in both temporal samples of each geographic population are shown in Figure 5. The populations at Amecameca and Tulancingo are very similar in having the CU and TL inversions in high frequency (defined as greater than 5%), although the relative predominance of each of these arrangements varies with the population and through time (Tulancingo also has OL at a frequency in excess of 5%). The population at Zirahuen, on the other hand, is distinctly more polymorphic with CU, TL, SC, EP, and OA above or near 5% (in the May 1981 sample, the frequency of OL also exceeds 5%). These results are consistent with earlier estimates (Guzman et al. 1975) indicating that the overall degree of inversion polymorphism is reasonably stable in these populations.

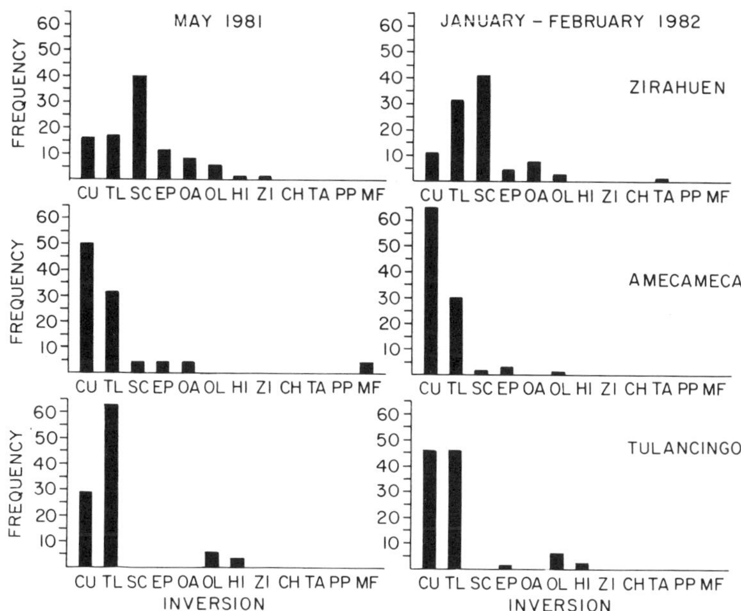

Fig. 5. Distribution of third chromosome inversion frequencies in three populations of Mexican *D. pseudoobscura* sampled at two different times. (CU = Cuernavaca; TL = Tree Line; SC = Santa Cruz; EP = Estes Park; OA = Oaxaca; OL = Olympic; HI = Hidalgo; ZI = Zirahuen; CH = Chiricahua; TA = Tarasco; PP = Pikes Peak; MF = Miraflores). (Source: Rosa et al. 1984).

Vagility—Intrapopulation Estimates

The vagility scores for the two temporal samples of the populations of *D. pseudoobscura* from Zirahuen, Amecameca and Tulancingo are depicted in Figure 6. For each sample of each population, a frequency distribution of the mean responses of the 12 isofemale lines is presented. For ease of comparison, the same scale, and class interval of 5 are used. Arithmetic means for both samples of each population were calculated from these isofemale line means. These are listed above the histograms and their positions are indicated by arrows.

Differentiation among the isofemale lines within each of the space-time populations was examined with one-way analysis of variance (Table 2). The actual level of isofemale line variation was estimated as the added component of variance due to isofemale lines (S^2_{IFL} in Table 2), partitioned from these ANOVAs (Rockwell 1980). The partition was formed as $S^2_{IFL} = (1/n)(MS_A - MS_W)$, where n = 3 is the number of replicates (Sokal and Rohlf 1981). It is interesting to note that only for the Zirahuen population are the mean-squares-among significant when tested over their respective mean-squares-within. This indicates that the estimated level of isofemale line variation is significantly greater than zero only for the Zirahuen population.

When the parametric value of isofemale line variation (as opposed to its estimate) is really zero, the mean-square-among and the mean-square-within are independent estimates of the parametric error variance and chance alone determines which of the two is larger. When the mean-square-within is larger than the mean-square-among, the partitioned, added component of variance, which estimates the parametric value of isofemale line variation, is negative. For some purposes, such estimates are set equal to zero. Here, however, the relative relationships among a set of variance components are of primary interest and the arbitrary setting of only

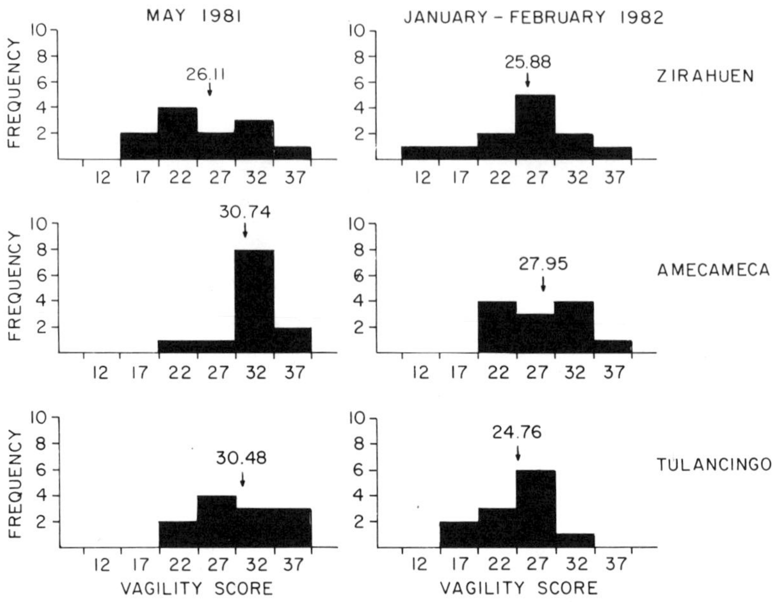

Fig. 6. Distributions of isofemale line mean vagility scores for the three Mexican populations of *D. pseudoobscura* sampled at two different times. Arrows locate the population means (numeric values) on the histograms. (Source: Rosa et al. 1984).

Table 2. Analyses of variance of the vagility scores of three Mexican populations of *D. pseudoobscura* sampled at two different times.

Source	df	Mean Squares					
		May 1981			January-February 1982		
		Z	A	T	Z	A	T
Among isofemale Lines	11	117.60*	43.17	71.42	123.52*	60.22	69.37
Within Isofemale Lines	24	43.31	39.53	87.92	42.92	50.92	69.43
Mean	—	26.11	30.74	30.48	25.88	27.95	24.76
Added Component of variance (S^2_{IFL})	—	24.77	1.22	−5.5	26.87	3.1	−0.02

NB: Z = Zirahuen: A = Amecameca; T = Tulancingo;
* = F is significant at $p < .05$. See text for details on S^2_{IFL}. (Source: Rosa et al. 1984)

the negative estimates to zero would clearly distort such comparisons. Thus, when they were encountered, the precise negative estimates were used for comparisons of variance components (see Rockwell 1980 for a discussion of these points). The statistical procedure used to evaluate such comparisons (see below) was designed to allow for negative estimates.

Isofemale lines of each space-time population were raised and tested in the same environment. Replicate behavioral measurements of each line used flies drawn from different replicate cultures of each line. Rockwell (1980) explained that with this experimental design, the component of variance due to differences among isofemale lines is a measure of expressed genotypic variance for the behavior of a given sample. That is, the component is a measure of the phenotypic variance resulting from genotypic rather than environmental differences.

Given the origin and maintenance of the isofemale lines, the expressed genotypic variance measured here is similar (but not identical) to the between-families component of variance evaluated in Falconer's (1981) analysis of full sib families. Like the latter measure, the expressed genotypic variance contains elements of the additive, dominance and epistatic components of genetic variance and, as such, is a broad measure of genotypic variation. Since the isofemale lines of each space/time sample were collected randomly from nature and exposed to a minimal amount of laboratory selection and inbreeding, the expressed genotypic variance reflects the broad genotypic structure of vagility for the populations under study.

Vagility—Spatial and Temporal Comparisons

The mean vagility scores and levels of genotypic variation for this trait are depicted with respect to space and time in Figure 7. A primary aim of this project is to detect any significant spatial and/or temporal differences in means or levels of genotypic variation for vagility. With respect to mean vagility, we initially performed a factorial ANOVA (Population by Time) on the isofemale line means. (As a test for main effects and first order interaction, this analysis is equivalent to a cross-nested ANOVA in which populations and time are factorially related, and the isofemale lines and their replicates are hierarchically nested at each intersect. Before the ANOVA was performed, heterogeneity of variance was evaluated with Cochran's test (Winer 1971) and found to be nonsignificant). The results of the analysis are summarized in Table 3. It is clear from the significant interaction term that spatial and temporal differences in mean vagility are not independent. As such, we must proceed by performing separate spatial and temporal analyses on the mean vagility responses. For convenience of presentation, we also consider the effects of space and time on the levels of genotypic variation in the same order.

Differences among the three populations in their mean vagility scores were examined for each temporal sample with one-way analysis of variance of isofemale line means (Rockwell 1980). Since a major thrust of this work centers on behavioral differentiation between populations displaying high *vs.* low degrees of chromosomal polymorphism, we also contrasted the vagility responses of Zirahuen to the mean of the two less polymorphic populations, using orthogonal decomposition (Sokal and Rohlf 1981). (Before these analyses were performed, heterogeneity of variance was evaluated with Cochran's test (Winer 1971) and found to be nonsignificant). The overall analyses and decompositions are summarized in Table 4. For the May 1981 sample, the overall variation in mean vagility score among the three populations is marginally significant. More importantly, the mean vagility score for the highly polymorphic Zirahuen population is significantly lower than that for the two less polymorphic populations. There is no evidence that the mean responses of these latter populations differ from each other. For the January-February 1982 sample, there is no significant overall variation among the three populations, nor are there any specific differences between the means of the populations contrasted, as above, on the basis of chromosomal polymorphism.

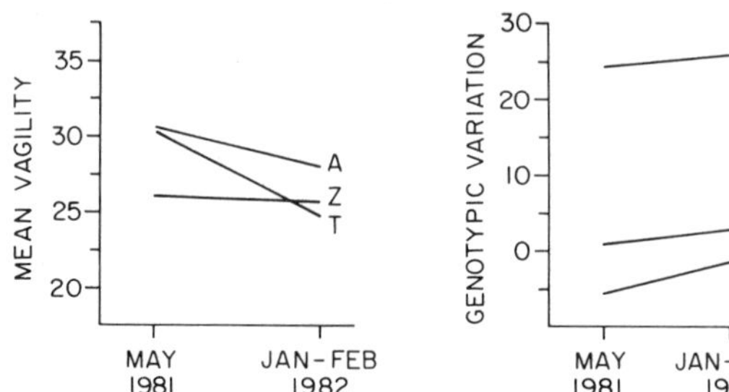

Fig. 7. Spatial and temporal patterns in the vagility of Mexican *D. pseudoobscura* in terms of both mean response (A) and genotypic variation (B). Spatial differences are depicted vertically while temporal differences are depicted horizontally. (Source: Rosa et al. 1984).

Table 3. Factorial analysis of variance of the vagility scores of three Mexican populations of *D. pseudoobscura* sampled at two different times.

Source	df	Mean squares
Among populations	2	144.36*
Between times	1	66.41
Populations by time	2	111.19*
Residual	66	34.27

NB: All sources tested over Residual;
* $p < .05$.

The levels of genotypic variation were compared between the populations in a pairwise fashion for each temporal sample using a Monte Carlo method described in Rockwell (1980). The method assumes that the two variance components being compared are the same and, as such, have an expected ratio of 1. The probability that the observed ratio departs from this expectation due to chance alone is evaluated by computer simulation (Edgington 1969). This statistical evaluation is used since the ratio of two variance components does not follow an F-distribution. For both temporal samples, the level of genotypic variation is significantly greater in the highly polymorphic Zirahuen population than in either of the less polymorphic populations. The latter two do not differ significantly in terms of this aspect of vagility.

Temporal differences in mean vagility were evaluated for each population with one-way ANOVA's of isofemale line means. (Before the analyses were performed, heterogeneity of variance was evaluated with Cochran's test (Winer 1971) and found to be nonsignificant). While there is a consistent decrease in mean vagility through time for all three populations, the effect is statistically significant only for the Tulancingo population.

The levels of genotypic variation were compared between the two temporal samples of each population using the Monte Carlo method. While there is a consistent increase in the level of genotypic variation through time for all three populations, the effect is not statistically significant.

Table 4. Analyses of variance of the vagility scores for spatial differences between three populations of Mexican *Drosophila pseudoobscura*.

Source	df	Mean Squares	
		May 1981	January-February 1982
Among populations	2	81.34#	31.42
High vs. low polymorphism	1	162.28*	1.78
Between low polymorphism	1	0.39	61.06
Within populations	33	25.80	28.12

$p = 0.07$
* $p < .03$

NB: The overall sources and orthogonal decompositions based on degree of chromosomal polymorphism are given for both temporal samples.

Discussion

The relationship of vagility to fitness, using wild type and mutant laboratory strains of *D. melanogaster* has been examined by Narise (1968, 1974). In the first study, separate groups of hybrids between wild type and vestigial wing flies were placed either in population cages or in double-ring (10 vial) vagility test systems. The frequency of vestigial flies decreased generation after generation in both types of populations. However, the pattern of selection against vestigial (vg) and its final frequency were significantly different in the two types of populations. These results were attributed to the differences in the two experimental systems. The space of the vagility test system is divided into ten compartments, allowing the vestigial flies opportunity to move about, thereby increasing their chances of avoiding severe competition with wild type flies. However, the space of a population cage is a single compartment with inescapable direct competition between the two types of flies. Thus the ability of vestigial flies to move about in a partitioned environment appears to have permitted the retention of the allele *vg* in the population.

In the second study, Narise (1974) examined the relationship between vagility and fitness of one wild type and five mutant strains of *D. melanogaster*. Vagility was tested with the single-ring (4 vial) Sakai test system. The components of fitness tested were: fecundity, viability and longevity. The data are shown in Table 5. From the results, it was concluded that in general both fecundity and viability are negatively related to vagility, while longevity is positively related. Although in no case was the relationship statistically significant, there does seem to be a general association of greater vagility with lower fitness.

Studies of vagility in natural populations of *Drosophila* are rather limited. Some of these have addressed themselves to the norm of reaction by investigating the interactive effects of genotypes and environments on this trait (Tantawy et al. 1975, Rockwell et al. 1978, Rockwell 1979, Mikasa and Narise 1983). While other studies have estimated the level of genotypic variation within natural populations (Rockwell and Levine, unpublished), they do not include either spatial or temporal dimensions. Sakai et al. (1958) and Mikasa and Narise (1979, 1983)

Table 5. Components of fitness (mean and standard deviation).

	Number of eggs layed	% of flies emerged	Longevity (days)	Vagility score
MS-1	351.44 ± 145.81	88.57 ± 7.18	21.93 ± 6.26	10.3
cn bw	290.76 ± 172.02	61.85 ± 16.35	41.65 ± 9.88	22.3
w^a	198.68 ± 112.82	61.92 ± 11.45	38.18 ± 12.07	12.3
se	217.90 ± 120.23	83.50 ± 14.54	38.01 ± 9.21	22.0
ss^a	300.74 ± 171.00	84.86 ± 18.57	24.32 ± 6.74	6.1
ss	408.68 ± 187.52	83.49 ± 19.19	20.31 ± 4.54	14.4

(Source: Narise 1974)

compared the vagility of several geographic strains of *D. melanogaster* and found substantial differences among them. Unfortunately, no attempt was made to identify other genetic differences among the six strains.

In the present study, we found that in the May 1981 sample, the highly polymorphic population at Zirahuen possessed a much lower mean vagility score than either of the lesser polymorphic populations. This interpopulation difference in mean response had disappeared in the January-February 1982 sample. In addition, the mean response of all three populations had decreased during the interim—that of Tulancingo markedly so. Whether this temporal change is a gradual time-related shift or a seasonal effect, and whether the change is cyclic or stochastic can only be ascertained with further temporal samples.

The differences between these populations in the level of genotypic variation for vagility is intriguing particularly given the observed pattern of inversion polymorphism. The level of genotypic variation and the expected level of inversion heterozygosity are highly correlated over the two time samples of the three populations ($r = 0.83$, $p < 0.05$). Recalling that each third chromosome inversion type in this species is resistant, when in a heterozygous association, to the dissolution of its particular allelic complex, one might suggest that the observed vagility patterns may be explained in terms of the expected relative frequencies of heterokaryotypes in the three populations. If one assumes that each heterokaryotype (and perhaps homokaryotype) displays a different level of vagility, then it is argued reasonably that the more polymorphic population should possess a higher level of genotypic variation for this behavior. Further temporal samples of these populations will test the precision and stability of this relationship.

It must, however, be made clear that neither vagility nor any behavior associated with it is influenced directly by gene arrangements as such, but rather by the gene contents of the chromosomes having certain gene sequences. As stated in the introduction of this paper, our ultimate goal is to be able to identify the gene differences among populations that result in their observed behavioral differences.

Literature Cited

De Souza, H. M. L., A. B. Da Cunha, and E. P. Dos Santos. 1970. Adaptive polymorphism of behavior evolved in laboratory populations of *Drosophila willistoni*. Amer. Natur. 104: 175.

Dobzhansky, T., and B. Spassky. 19962. Selection for geotaxis in monomorphic and polymorphic populations of *Drosophila pseudoobscura*. Proc. Natl. Acad. Sci. USA 48: 1704.

Edgington, E. S. 1969. Statistical Inference: The Distribution-free Approach. McGraw-Hill, Toronto.

Falconer, D.S. 1981. Introduction to Quantitative Genetics, 2nd ed. Longman, New York.

Guzman, J., L. Levine, O. Olvera, J. R. Powell, M. E. de la Rosa, V. M. Salceda, Th. Dobzhansky, and R. Felix. 1975. Genetica de poblaciones en *Drosophila* de Mexico III. Estudio preliminar sobre la variacion cromosomica de *D. pseudoobscura* en dos zonas geograficas del centro de Mexico. Anales Inst. Biol. Univ. Nal. Auton. Mexico 46 Ser. Zoologia (1): 75.

Levine, L., M. Asmussen, O. Olvera, M. E. de la Rosa, V. M. Salceda, M. I. Gaso, J. Guzman, and W. W. Anderson. 1980. Population genetics of Mexican *Drosophila* V. A high rate of multiple insemination in natural populations of *D. pseudoobscura*. Amer. Natur. 116: 493.

Mikasa, K. and T. Narise. 1979. The relation between dispersive behavior and temperature in *D. melanogaster* I. Dispersal patterns. Jap. J. Genet. 54: 217.

Mikasa, K., and T. Narise. 1983. Interactive effects of temperature and geography on emigration behavior of *Drosophila melanogaster*: Climatic and island factors. Behav. Genet. 13: 29.

Narise, T. 1968. Migration and competition in *Drosophila*. I. Competition between wild and vestigial strains of *Drosophila melanogaster* in a cage and migration-tube population. Evolution 22: 301.

Narise, T. 1974. Relation between dispersive behavior and fitness. Jap. J. Genet. 49: 131.

Powell, J. R., and C. E. Taylor. 1979. Genetic variation in ecologically diverse environments. Amer. Sci. 67: 590.

Rockwell, R. F. 1979. Emigration response behavior II. The responses of *Drosophila busckii*. Pan-Pac. Entomol. 55: 117.

Rockwell, R. F. 1980. Photobehavioral differentiation in natural populations of *Drosophila*: Changes of photoresponse over time in *D. pseudoobscura* and *D. persimilis*. Behav. Genet. 10: 521.

Rockwell, R. F., R. Harmsen, F. Cooke, and J. Grossfield. 1975. Photobehavioral differentiation within and between the sexes of *Drosophila pseudoobscura*. Can. J. Zool. 53: 1395.

Rockwell, R. F., J. Grossfield, and L. Levine, 1978. Emigration response behavior I: The effects of height and light on the Oregon-R and norp-A strains of *Drosophila melanogaster*. Egypt. J. Genet. 7: 123.

Rockwell, R. F., M. E. de la Rosa, E. Akin, M. I. Gaso, F. Gonzalez, J. Guzman, L. Levine, and O. Olvera. 1983. Chromosomal and behavioral studies of Mexican *Drosophila* I. Vagility characteristics of three populations of *D. pseudoobscura*. Behav. Genet. 13: 197.

Rosa, M. E. de la, M. I. Gaso, F. Gonzalez, J. Guzman, L. Levine, O. Olvera, and R. F. Rockwell. 1984. Chromosomal and behavioral studies of Mexican *Drosophila* II. Temporal changes of vagility characteristics of three populations of *D. pseudoobscura*. J. Heredity 75: 281-284.

Sakai, K. I., T. Narise, Y. Hiraizumi, and S. Iyama. 1958. Studies on competition in plants and animals IX. Experimental studies on migration in *Drosophila melanogaster*. Evolution 12: 93.

Sokal, R., and F. J. Rohlf. 1981. Biometry, 2nd ed. Freeman, San Francisco.

Spiess, E. B., and B. Langer. 1961. Chromosomal adaptive polymorphisms in *Drosophila persimilis*. III. Mating propensity of homokaryotypes. Evolution 15: 535.

Tantawy, A., A. Mourad, and A. Abou-Youssef. 1975. Studies on natural populations of *Drosophila* XVII: Migration in *D. melanogaster* in relation to genotype, temperature and population density. Egypt. J. Genet. 4: 263.

Winer, B. J. 1971. Statistical Principles in Experimental Design. McGraw-Hill, New York.

Between Population Variation in Spider Territorial Behavior: Hybrid-Pure Population Line Comparisons

Susan E. Riechert

Department of Zoology
University of Tennessee
Knoxville, Tennessee 37916

Introduction

When we consider the subject of behavioral genetics, we immediately think of courtship and habitat selection — traits that are involved with the differentiation of species and hence assumed to be under strong genetic influence. Less consideration has been given to those behavioral traits that represent adaptations to local environments or contexts (e.g., cooperative, competitive, thermoregulatory and foraging behaviors), because these are thought to be determined largely by experiential effects associated with temporally variable environmental contexts. No one would argue that such traits are not influenced by natural selection. However, the extent to which heritable variation exists in the genetic program is often open to question.

Territorial behavior is one of these environmentally linked behavioral traits. It is simply defined as the defense of space by fighting or display and is basically an economic problem. The exhibition of territorial behavior itself will be favored by selection when the benefits of achieving sole access to a resource outweigh the cost associated with its defense (Brown 1964, Davies and Houston 1981). Territory size likewise reflects the balancing of needs with costs (Pyke 1979, Myers et al. 1981). The extent to which territoriality and territory size are fixed characteristics of local populations is also an economic problem. If, for instance, the cost of increasing territory size by the removal of neighboring territory owners is higher than the continuous defense of a territory that is larger than required by needs at any given time, then we would expect territory size to be fixed at an area that meets the energetic needs of the members of a local population under the worst conditions experienced in that habitat. If adjacent territory owners, however, are removed as easily as territories are defended against floating individuals (those lacking territories), then we would expect the trait to be variable with individuals adjusting the sizes of their territories over time as needs demand. The nature of the environmental fluctuation (how variable is it and how predictable is the variation) also should be an important determinant of the genetic programing of this trait.

Territoriality is a trait that lends itself to genetic study, since both territory size and the conflict behavior used in defending the space can be quantified. This paper deals with the genetic basis of territory size in the spider *Agelenopsis aperta* (Gertsch). The vast majority of the approximately 30,000 species of spiders compete rather than cooperate with conspecifics for limiting resources. Further, this competition takes the form of active defense and cannibalism rather than direct exploitation of the contested resource. Territorial behavior is thus common in the Araneae with individual spiders competing for sites of sufficient area to afford survival to reproduction (See Riechert 1982 for a review of spider territorial behavior). Considerable evidence also exists in the ecological and agricultural literature for a spider territorial system that is adjusted to the lows in prey availability viewed by local populations over time (Riechert and Lockley 1984). I hypothesize that territory size in spiders occupying temperate habitats is fixed because: (1) prey availability at local sites is highly variable and difficult to monitor and (2) because, as one's area needs increase with decreased encounter with prey, the costs and risks of removing a neighboring territory owner are sufficiently higher than those required to defend a site against intrusion by floating individuals (Riechert 1981, 1982).

The territorial system exhibited by temperate spiders is best documented for *Agelenopsis aperta*, the subject of this paper. This is an arid land representative of the funnel web spider family Agelenidae. The spider builds a non-sticky web-sheet with an attached funnel extending into some feature of the surrounding habitat. *Agelenopsis* maintains a territory in excess of its web-trap throughout its life. Territory size corresponds to the spider's optimum energy needs during periods of low prey availability (worst times; Riechert 1981). Thus, it varies only with the age of the individual spider and with populations located in disjunct habitats (Riechert 1978a). In this study, I present the results of experiments that were designed to determine the extent to which genetic *vs.* experiential factors underlie territory size in this species. The genetic basis of the behavior exhibited during the course of territorial disputes is discussed elsewhere (Maynard Smith and Riechert 1984).

Methods

Study Areas

Two populations of *Agelenopsis aperta* occupying habitats affording individuals different feeding levels were used in this investigation: desert grassland, south-central New Mexico, elev. = 1,636 m; and desert riparian, Chiricahua Mountains, southeastern Arizona, elev. = 1,616 m. The desert grassland habitat imposes both stringent thermal and prey availability conditions on this spider. Temperatures permit foraging activity by adults for an average of only 525 min/day and prey availability averages 27.2 mg ± 7.3 mg/day (Riechert 1978a). The habitat is characterized by a sparse sod of the dropseed grass *Sporobolus flexuosus* (Thurb.) Rydb. that surrounds numerous holes or depressions created by the eroding effects of major summer rains. The median for distances between webs of adult spiders in this habitat is 109 ± 11 cm (May—July).

The riparian habitat is more favorable, affording *A. aperta* an average foraging time of 660 min/day and 79.6 mg ± 9.8 mg/dry weight of prey per day (Riechert 1978a). This study area consists of a tree canopy of *Acer negundo* L. (boxelder), *Juniperus osteosperma* (Torr.) (alligator juniper), and *Platanus wrightii* Wats (Arizona sycamore) and an herb layer dominated by grasses. The study area is bissected by a permanent stream. The mean nearest neighbor adult distance in this habitat is 42 ± 3 cm (mid-June—August).

Territory Size

The following experiment was completed on groups of both feral and laboratory reared individuals representing each of the two populations. In the studies of laboratory spiders, a third category was added: crossbred AZNM (Arizona riparian X New Mexico grassland). The design of the experiment varied to some extent depending on whether feral or lab reared individuals were used. Each will be considered separately herein.

Feral. Groups of sixteen adult female spiders were collected from their web-sites in the respective habitats. The spiders were then housed individually in plastic containers (dimensions ± 20 cm long, 10 cm wide, 8 cm deep) and fed daily according to the feeding regime each was assigned to: deprivation ± 20 mg of prey/day (live weight) or; satiation ± 250 mg of live prey/day. At the end of up to one month of pretrial exposure to given prey levels, the 16 individuals belonging to a given feeding group were introduced into the experimental enclosure shown in Figure 1. Each enclosure consisted of a 4 m^2 frame and canopy within which 16 screen cones (for the web-funnels) were buried flush with the substrate at regular intervals. Hence, each cone was 33 cm from its nearest neighbor and the enclosure wall. Note that this distance is closer than the nearest neighbor distances observed in the field for either population. Poultry netting ("chicken wire") was laid over the natural substrate within the confines of the enclosure to ensure that adequate substrate for web-sheet attachment was equally available at all potential web-sites demarcated by the positions of the screen cones. Finally, a cheesecloth canopy restricted the movements of spiders and released prey to the enclosure confines while providing a thermal environment similar to that impinging on this spider at its natural web-sites in the respective habitats.

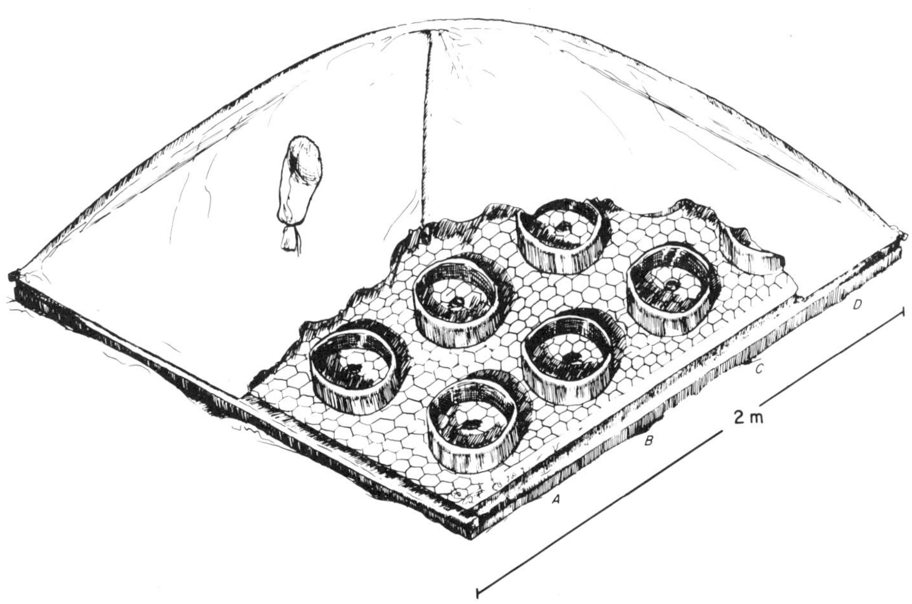

Fig. 1. Experimental enclosure used to test for territory size. A, B. C, and D denote rows of 4 evenly spaced web-sites established by the release of spiders into buried screen funnels and metal cylinders. Cylinders removed following web construction by all spiders (16) in the enclosure. Cheese cloth canopy removed once each day for censusing.

In preparation for introduction to the enclosure, each individual was weighed and paint-marked individually. It was then released into one of the screen cones and the opening plugged for a 12-h period to encourage the spider inside to initiate the building of its funnel retreat. Following the removal of the cotton plug, spider movement from the vicinity was restricted by the presence of a metal cylinder (20 cm in radius and 18 cm in height). These cylinders were removed after all 16 individuals in the enclosure had established webs within their confines (usually within two days). Thus, at the beginning of a test for territorial behavior, all spiders occupied webs.

Following the removal of the cylinders, the spiders within a given enclosure were allowed to expand their webs, and to interact freely with one another. During the accommodation period and for two subsequent weeks of free interactions, the two feeding levels initiated in the laboratory were maintained within the confines of the enclosure by offering each spider prey by forceps and in the case of the satiation runs by the additional release of prey through the sleeve in the top of the enclosure (Fig. 1). The top of each enclosure was removed once a day and the spiders checked for displacement of web owners, for changes in nearest neighbor distances (funnel to funnel) and for changes in web area. Spiders displaced to the cheesecloth canopy were removed.

Laboratory. The experiment described above also was completed on first generation (G_1) individuals resulting from pureline and mixed Arizona riparian and New Mexico grassland matings. Twelve matings were completed for each of the three sample groups (Arizona riparian, New Mexico grassland, and Arizona-New Mexico cross). There was an equal representation of Arizona and New Mexico mothers (6 and 6, respectively) in the cross population matings. The individuals used in the matings were collected from the respective populations as penultimates to ensure no prior mating. Following emergence from the egg sacs, the spiderlings resulting from the breedings were separated into individual plastic boxes and reared *ad libitum* on a mixed diet to maturity. Grow lights on a 12-h on-off cycle provided vitamin D, which is required for successful molting, and temperatures ranged between 25 °C (night) and 30 °C (day) in the rearing environment. Survivorship was high with 99.6% of the individuals reaching maturity. At maturity, groups of 16 females were selected at random from each of the three sample groups for introduction into the experimental enclosure. The same experimental design was used in this experiment as already described for the satiation runs completed on feral individuals. Since these experiments were completed in a controlled laboratory environment rather than the field, however, the enclosure design was altered to facilitate censusing and observations of territorial interactions. In this latter case, metal flashing was used to prevent spiders from leaving the enclosure and no cheesecloth cover was present. Crickets and clipped moths were used as prey. Spiders found along the edge of the flashing border were removed once each day.

Electrophoretic Analyses

Allozyme analyses were applied to representatives of the Arizona riparian and New Mexico grassland populations to provide insight into the questions of potential inbreeding and wide cross effects on the results of the pure line F_1-hybrid AZNM enclosure experiments. Twenty-five adult individuals were collected from six local Arizona populations and from seven local New Mexico populations. Allele frequency determinations were made for the following scorable loci: phosphoglucomutase, glutamate-oxaloacetate transaminase I and II, alcohol dehydrogenase, four esterases, 6-phosphogluconate dehydrogenase, lactate dehydrogenase, hexokinase I and II, isocitrate dehydrogenase, malate dehydrogenase I and II, α-glycerophosphate dehydrogenase, indophenol oxidase, glucose-6-phosphate dehydrogenase, general protein, octonol dehydrogenase, fumerase, and phosphoglucose isomerase.

Results

The results of the runs completed on feral spiders maintained at different feeding regimes for a one-month period prior to initiation of the test for territory size are shown in Table 1. The data are expressed in terms of mean nearest neighbor distances. There were no significant differences between treatments for either population (Mann Whitney Test: grassland $p < .50$; riparian $p < .50$). Short term feeding history, then, does not appear to influence the distance at which individual spiders will tolerate neighboring web owners. Significant differences were observed, however, in the nearest neighbor distances exhibited with respect to population. Riparian spiders demonstrated significantly smaller nearest neighbor distances in the experiments than did grassland spiders (Mann Whitney Test (treatments combined): $p < .001$). These experimental differences are consistent with those observed in the spacing of spiders in their respective habitats. Riparian spiders apparently tolerate neighboring conspecific web owners at closer distances than do grassland spiders. The result is a smaller area of occupation (territory size) for members of this population and higher densities of individuals exhibited per unit area (Mean number remaining in the enclosure at the end of the experiment for the riparian spiders = 8.0 ± .06; for the grassland = 1.8 ± .07).

Despite the fact that spacing in the experiments reflects the spacing patterns of individuals in the respective populations, there does appear to be a trend toward larger nearest neighbor distances in the experiment than that recorded in the field censuses (Table 1). Terrain is probably the key factor here. In the field, shrubs, rocks, etc. act as barriers preventing individuals from detecting the presence of conspecific web owners within distances usually cuing agonistic interactions. The resulting shorter nearest neighbor distances bias field estimates of territory size. This appears to be true especially of the riparian population where webs displaced vertically on shrubs are closer in proximity than webs displaced only horizontally (Riechert 1981). Barriers and vertical stratification were not included in the habitat design of the experimental enclosures.

Table 1. Comparisons of nearest neighbor distances between natural population estimates (based on data presented in Riechert (1978a) and outdoor experimental contexts, for two feral populations of *Agelelopsis aperta*: desert grassland, southcentral New Mexico; desert riparian, southeastern Arizona. Treatments represent one month of pretrial feeding of field collected individuals. Spider density at start of trials equaled 16 individuals. Distances (in cm) represent measurements completed at end of two week trial period.

	Arizona: desert riparian			New Mexico: desert grassland		
	Mean	Standard error	No. of replicates	Mean	Standard error	No. of replicates
Natural estimates	42	3	—	109	11	—
Experiments:						
Food deprivation	53	1	3	134	3	5
Food satiation	54	1	3	149	5	6

The results of the runs completed on first generation (G_1) lab reared spiders representing both pure population lines and crossbred AZNM individuals are presented in Table 2. The territory sizes of the G_1 pure population lines appear to shift toward one another: in both cases the shifts are significant when comparing feral with G_1 results ($p < .05$ in both cases) (Mann Whitney Test: Tables 1 and 2). Nevertheless, the behavior of individuals belonging to the respective populations still differ significantly from one another following lab rearing with no experience and abundant food (Mann Whitney Test grassland G_1 *vs.* Riparian G_1 results: $p < .001$). The crossbred individuals are outside the bounds of both pure populations, exhibiting even larger distances between neighbors than the New Mexico grassland population. The differences between these hybrids and the New Mexico G_1 trial results, however, is not significant ($p < .10$).

Table 3 shows one additional aspect of spider behavior that was detected in the trials completed on laboratory reared individuals: between population differences in the incidence of mortality suffered during the territorial disputes (Table 3). Mann Whitney tests comparing the numbers of individuals killed showed higher incidences of mortality in grassland G_1 trials than riparian G_1 trials ($p < .07$) and of significantly higher rates of mortality in the crossbred runs (grassland G_1 *vs.* crossbred $p < .025$). The crossbred individuals, then, in addition to exhibiting larger territories, apparently are more aggressive in their territorial interactions.

Electrophoretic Analyses

Heterosis in hybrids is considered to reflect new combinations of alleles resulting either from the crossing of inbred population lines with high degrees of homozygosity or from crossing populations with divergent allelic structure. Two indices of inbreeding applied to the Arizona and New Mexico electrophoretic data are presented in Table 4. Both the proportions of polymorphic loci and the frequencies of heterozygotes in the populations are consistent with outcrossing breeding systems. There is no pattern of either heterozygote deficiency or excess in either population.

Allele frequencies for four of the most polymorphic loci identified in the study, however, indicate that the populations are widely divergent (Table 5). Nei's (1972) index of genetic

Table 2. Comparisons of nearest neighbor distances for laboratory enclosure experiments involving lab reared individuals representing New Mexico grassland, Arizona riparian populations and (G_1s) between population crossed individuals. Treatments represent entire life of *ad libidum* food and no experience with conspecifics past the spiderling stage. Spider density at the start of the trials equaled 16 individuals. Distances (in cm) represent measurements completed at end of two week trial period.

	Mean nearest neighbor distance	Standard error	Number of replicates
Arizona riparian	60	5	10
New Mexico grassland	104	9	10
Crossbred AZNM[1]	160	176	10

Table 3. Comparison of mortality suffered in different G_1 experimental groups for trials of the enclosure experiment.

	Mean number of individuals killed/trial	Standard error	Number of replicates
Arizona riparian	2.0	.2	10
New Mexico grassland	2.6	.2	10
Crossbred AZNM	3.2	.2 5	10

Table 4. Test for evidence of inbreeding in Arizona riparian and New Mexico grassland populations. (Arizona values from single population used in experiments: New Mexico values averaged over four local populations used.)

	Arizona riparian	New Mexico grassland	Outcrossing invertebrates*
Proportion of polymorphic loci (P)	.31	.40	.18-.59
Representation of heterozygotes in populations sampled (H_0)	.15	.06	.06-.15

*Information presented in Selander (1976)

similarity (Identity) based on 22 scorable loci, shows the two sets of local populations, in fact, to exhibit an average genetic similarity of .87—a value that is in the range of described subspecies for other invertebrates. (Mean local population similarity for a variety of invertebrates = .98, sub- and semispecies = .82, species = .54: from Ayala 1974, Avise 1976). The hybrid AZNM G_1 individuals then are potentially the offspring of what is termed a wide cross — a cross between two populations that shows a high degree of genetic differentiation (Falconer 1981).

Discussion

Most local populations of the spider *Agelenopsis aperta* occupy relatively stringent environments — desert areas where prey are present only in abundance following highly localized rains. Perhaps, then, it is not surprising that territory size is an inherited characteristic of local populations. Prey availability is unpredictable and territorial expansion (by the elimination of established neighboring territory owners) required as prey levels decline would occur at great risk and cost to the individual spider. In other words, an ownership bias would no longer be a determinant of individual success in a territorial dispute (Riechert 1978b, 1982). *Agelenopsis aperta* does not adjust its territory size according to local conditions of prey

Table 5. Evidence of degree of genetic differentiation existing between Arizona riparian and New Mexico grassland populations: Allele frequency determination for four highly polymorphic loci. (Arizona values from single population used in experiments: New Mexico values averaged over four local populations used.)

Locus	Alleles	Arizona riparian	New Mexico grassland
Phosphoglucomutase (PGM)	a	.0	.03
	b	.0	.25
	c	.68	.71
	d	.32	.02
Phosphoglucose isomerase (PGI)	a	.0	.01
	b	.31	.99
	c	.39	.0
	d	.31	.01
Fumerase	a	.08	.0
	b	.33	.04
	c	.58	.92
	d	.31	.01
Esterase (ESTL)	a	.0	.01
	b	.0	.09
	c	.95	.65
	d	.05	.0
	e	.0	.25

*Underlined scores represent fixed alleles.

availability. Rather, it adjusts its foraging behavior, taking a broader range of prey when these are in short supply and a narrower range when they are abundant numerically (Riechert 1981).

To what extent is genetically determined territorial behavior characteristic of the Araneae (spiders in general)? Although Uetz et al. 1982 and in this symposium give a fine example of spacing in a colonial spider that is to a large extent facultative, (temporally flexible) there is little evidence to suggest that most temperate spiders exhibit the kinds of population fluctuations facultative territorial behavior would produce (see Riechert and Lockley 1984 for a review). Spiders fail to show changes in densities corresponding to changes in local prey abundance. This is despite the fact that they are capable of aggregating in areas where prey are prominent numerically, and produce offspring to some extent proportionally to prey intake. Spiders appear to fit the group of organisms that are said to follow the maximum strategy which is set to minimize losses in unfavorable times (Levins 1968, Pitelka et al. 1975, MacLean and Seastedt 1979). This phenomenon also has been referred to as the principle of stringency (Wilson 1975).

Territorial behavior then ensures individual spiders' survival to reproduction in times of low prey abundance. Its consequence on population size is to set it below average carrying capacity. Hence, spider populations do not track the densities of associated prey. We might, however, be able to augment the predatory effect of this ubiquitous predator on pest arthropods, if we were to circumvent the self-damping effects of their territorial behavior. The results of the crossbred trials indicate that sufficient genetic variability exists such that directional selection might be applied artificially to achieve individuals that are tolerant of nearest neighbors at closer distances, hence permitting realization of the already inherent aggregational and reproductive numerical response tendencies. These genetic programs might be applied to common spider species in important crop systems for the improved biological control of certain insect pests.

Continued investigation of territorial spacing in *Agelenopsis aperta* demonstrated that the high variance noted for F^1 hybrids between Arizona and New Mexico populations of this species reflects the operation of a cytoplasmic effect. The results of data collected from reciprocal backcrosses to the respective parent lines further show that there is interaction between nuclear and cytoplasmic components of the trait. Investigation of the agonistic behavior used in settling territorial disputes indicates a similar genetic mechanism is involved and a two gene complex model is proposed to explain this behavior.

Acknowledgments

I would like to thank Gary McCracken for this assistance in the electrophoretic analyses and Joe Hegmann, Milton Huettel and George Uetz for their suggestions. The work was supported by NSF Grant No. DEB 8002882.

Literature Cited

Avise, J. C. 1976. Genetic differentiation during speciation. *In* F. J. Ayala, ed. Molecular Evolution. Sinauer Associates, Sunderland, Massachusetts.

Ayala, F. J. 1974. Genetic differentiation within and between species of the *Drosophila willistoni* group. Proc. Nat. Acad. Sci. USA 71: 999.

Brown, J. L. 1964. The evolution of diversity in avian territorial systems. Wilson Bull. 76: 160.

Davies, N. B., and A. I. Houston. 1981. Owners and satellites: the economics of territory defense in the pied wagtail, *Motacilla alba*. J. Anim. Ecol. 50: 157.

Falconer, D. S. 1981. Introduction to Quantitative Genetics, 2nd ed. Longman, New York.

Levins, R. 1968. Evolution in Changing Environments: Some Theoretical Explorations, Princeton University Press, Princeton.

MacLean, S. F., Jr., and T. R. Seastedt. 1979. Avian territoriality: sufficient resources or interference competion. Amer. Natur. 114: 308.

Maynard Smith, J., and S. E. Reichert. 1984. A conflicting tendency model of spider agonistic behavior: hybrid-pure population line comparisons. Anim. Behav. 32:564-578.

Myers, J. P., P. G. Connors, and F. A. Piteika. 1981. Optimal territory size and the sanderling = compromises in a variable environment. *In* A. C. Kamil and T. D. Sargent, eds. Foraging Behavior: Ecological, Ethological and Psychological Approaches, Garland STPM Press, New York.

Nei, M. 1972. Genetic distance between populations. Amer. Natur. 106: 283.

Pitelka, F. A., R. T. Holmes, and S. F. MacLean, Jr. 1974. Ecology and evolution of social organization in arctic sandpipers. Am. Zool. 14: 185.

Pyke, G. H. 1979. The economics of territory size and time budget in the golden-winged sunbird. Amer. Natur. 114: 131.

Riechert, S. E. 1978a. Energy-based territoriality in populations of the desert spider, *Agelenopsis aperta* (Gertsch). Symp. Zool. Soc. London 42: 211.

Riechert, S. E. 1978b. Games spiders play I: Behavioral variability in territorial disputes. Behav. Ecol. and Sociobiol. 3: 135.

Riechert, S. E. 1981. The consequences of being territorial: spiders, a case study. Amer. Natur. 117: 871.

Riechert, S. E. 1982. Spider interaction strategies: communication *vs.* coercion. *In* P. N. Witt and J. Rovner, eds. Spider Communication: Mechanisms and Ecological Significance. Princeton University Press, Princeton.

Riechert, S. E., and T. Lockley. 1984. Spiders as biological control agents? Ann. Rev. Entomol. (in press).

Selander, R. K. 1976. Genetic variation in natural populations. *In* Molecular Evolution. Sinauer Associates, Sunderland, Massachusetts.

Uetz, G. W., T. C. Kane, and G. E. Stratton. 1982. Variation in the social grouping tendency of a communal web-building spider. Science 217: 547.

Wilson, E. O. 1975. Sociobiology. Belknap Press of Harvard University Press, Cambridge.

Environmental and Genetic Influences on the Social Grouping Tendency of a Communal Spider

George W. Uetz

Thomas C. Kane

Gail E. Stratton*

Michael J. Benton*

Department of Biological Sciences
University of Cincinnati
Cincinnati, Ohio 45221

Introduction

Within the study of animal social behavior, attention has been focused primarily on vertebrates or eusocial insects, and most models generated to explain the genetics of social grouping or the evolution of social behavior in general apply to those systems (Wilson 1971, 1975, Brown 1975, Oster and Wilson 1978, Wittenberger 1981, Hermann 1982).

A distinction is often made, particularly in reference to vertebrate animals, between social groups and colonies. Social groups of vertebrates, such as fish schools, bird flocks, whale pods, ungulate herds or primate troops, tend to be mobile and organized internally into subgroups, dominance heirarchies, mating consortships, and the like (Wittenberger 1981). Social groups are based in interattraction and communication among individuals, and appear to have evolved either because they provide protection from predators or because they enable more efficient use of resources. Colonies, on the other hand, are spatially organized groups associated with a central place (i.e., a nest or roost) and/or a fixed spatial organization among members (Wittenberger 1981). In general, cooperation and competition among members of groups are conflicting forces in vertebrate social groups and colonies and the benefits of group living must be high to overcome the high costs of living in close proximity to others.

Invertebrate sociality is usually thought of as coloniality. Colonial invertebrates (e.g., hydrozoans, bryozoans and sponges) tend to be sessile animals, whose groups proliferate by asexual reproduction and exist in fixed spatial locations relative to each other. Wilson (1975) refers to colonial hydrozoans as "perfect societies," where group members are organized like a superorganism, with individuals demonstrating the ultimate level of cooperation — func-

* Present address: Gail Stratton, Department of Biology, Albion College, Albion, Michigan 49224; Michael J. Benton, Department of Biology, University of Calgary, Alberta, Canada T2N IN4.

tioning as feeding appendages, digestive organs, reproductive organs, etc. Eusocial insects, by contrast, are not fixed in space, but center their activities around a colonial nest. They show a high degree of cooperation, including division of labor among individuals, with reproduction and maintenance assigned, respectively, to reproductive individuals (queens, drones, etc.) and sterile workers (workers, soldiers, etc.) (Wilson 1971). Models of evolution in colonial invertebrates and eusocial insects stress either cooperation based on genetic relatedness (i.e., genetically identical individuals in hydrozoans; kin selection in haplodiploid insects), or parental domination of offspring through biochemical means as the driving force behind evolution (Wilson 1971, Wittenberger 1981).

However, it may be possible that there are other groups of animals exhibiting social behavior that do not fit well into many current evolutionary schemes. Spiders, for example, are generally assumed to be solitary, exhibiting aggressive behavior toward other animals, and even conspecifics. However, social interactions play an important role in the lives of many spiders, and take the form of maternal care (Rovner et al. 1973), territorial contests over webs (Riechert 1982), interactions of recently hatched spiderlings on communal webbing (Christenson and Goist 1979), and even establishment of social hierarchies among males (Aspey 1977). Even though it is extremely unusual for spiders to live together, there are a number of group-living spider species (Shear 1970, Kullman 1972, Burgess 1978, Buskirk 1981, Krafft 1982). Within the Araneae, there exist many levels of sociality, ranging from temporary, fortuitous aggregations and maternal care of young to large, communal groups exhibiting cooperative prey capture and brood care. The variety of social phenomena in spiders suggests that arachnid sociality has multiple evolutionary origins, and that many selective regimes can account for its occurrence (Buskirk 1981). Comparison of social spiders with eusocial insects, however, suggests that they have little in common (Buskirk 1981, Smith-Trail 1983). Despite the fact that most social spiders have been studied incompletely, it is clear that the major features of insect sociality — reproductive division of labor and morphological castes — are absent in social spiders.

Groups of social spiders have other characteristics that make them difficult to place in the terminology of sociobiology, which has resulted in much confusion (Buskirk 1981, Smith-Trail 1982). The social spiders are all web builders. They are sedentary predators, analogous in many ways to filter feeders, and their activities would seem to be limited to a central location, or "colony." Many of the social spiders maintain fixed distances between individuals within "colonies" (Burgess and Uetz 1982). However, spider webs are not only habitations, but prey capture devices, so a group of spiders also might be considered a "foraging flock" (Rypstra 1978). Moreover, some of the group-living spiders exhibit cooperative prey capture behavior, while others are strictly noncooperative and capture prey individually. Thus, social spiders may be considered colonial invertebrates; social arthropods that behave in some ways like social mammals or flocking birds.

Social spiders provide an opportunity to test some ideas about the evolution of sociality in an alternative invertebrate. It is likely that some subset of models generated for vertebrates may apply to spider sociality, since competition and cooperation are surely conflicting evolutionary forces for spiders as well (Smith-Trail 1982). This conflict may be even greater for spiders than for vertebrates, since their highly territorial and aggressive nature usually prevents them from living together. Since the behavior of spiders (and invertebrates in general) appears to be under much tighter genetic control than that of vertebrates (Witt et al. 1968, Risch 1977, Riechert 1982, Maynard Smith and Riechert 1983), the study of within- and between-population variation in social phenomena in spiders should allow a more precise assessment of the relative contributions of environment and heredity to social grouping behavior.

An orb web building spider found in central Mexico, *Metepeira spinipes* O. Pickard-Cambridge (Araneidae), occurs solitarily, but is found more frequently in aggregations of 5-150 or more individuals (Uetz and Burgess 1979). Aggregations of *M. spinipes* remain together for long periods of time, and individuals taken from different colonies will build webs together, even when the colonies are from areas hundreds of miles apart (Burgess and Witt 1976). Although communal, individuals maintain webs and retreats within the colony, and capture their own prey. Aspects of the behavior and ecology of this species have been studied during several trips to Mexico (Uetz and Burgess 1979, Burgess and Uetz 1982, Uetz et al. 1982, Uetz 1983).

There is considerable geographic variation in the social grouping tendency of *M. spinipes*, which appears to be related to the availability of insect prey. Moreover, the social spacing patterns of this species are flexible, and respond quickly to changes in prey abundance. The wide range of spacing exhibited by the populations observed suggest that *M. spinipes* may represent an intermediate level in the evolutionary spectrum of social behavior in spiders: a "communal-territorial" living arrangement.

Geographic Variation in Social Grouping

Methods

Patterns of variation in social grouping of *Metepeira spinipes* were studied in a wide variety of habitats in central Mexico, ranging from severe climates with low prey availability to benign climates with high prey availability. At each site, data were collected on colony size, nearest-neighbor distance within colonies, and insect prey activity. (The latter was estimated by visual observation of the number of insects flying into a colony web over a specified period of time, then correcting this estimate by the number of hours of weather conditions favoring activity and feeding in a day).

To examine further the relationship between climate, prey availability, spider group size and spacing, we attempted a series of field experiments. Colonies were taken from a site in the agricultural valley of Toluca, where prey were abundant and the climate was typically moderate, to a high elevation mountainside in a nearby state park (Parque Sierra Morelos) where high winds and wide temperature and moisture fluctuations make the environment rigorous for these spiders. Prey abundance on the mountain is much lower than in the valley below. Twenty colonies of 10 individuals each were measured for nearest-neighbor distance, collected, and then released in agaves at Parque Sierra Morelos at an elevation of 3000 m. Spiders occurring naturally at this latter site are usually solitary. Ten days later, measurements were made of nearest-neighbor distances within the transplanted colonies.

A second experiment, which provided the opportunity to test the influence of prey availability on social grouping and spacing independent of differences in climate was conducted in Tepotzotlan, an agricultural area north of Mexico City. In this experiment, colonies were moved from a site of high insect activity (due to dumping of feedlot waste) to an area of low insect activity nearby. We assumed that there were no differences in microclimate between the areas. Nearest-neighbor distances were measured, spiders were collected in groups of 20, and "transplanted" as above. Five of the new colonies had prey supplemented by piling mounds of cow dung about the bases of the agaves in which they were located. Nearest-neighbor distances were measured after 10 days.

Results and Discussion

Group size in *M. spinipes* varies over the gradient of habitats studied (Fig. 1). In severe habitats where prey availability is low and environmental conditions are extreme (PSM;SMA),

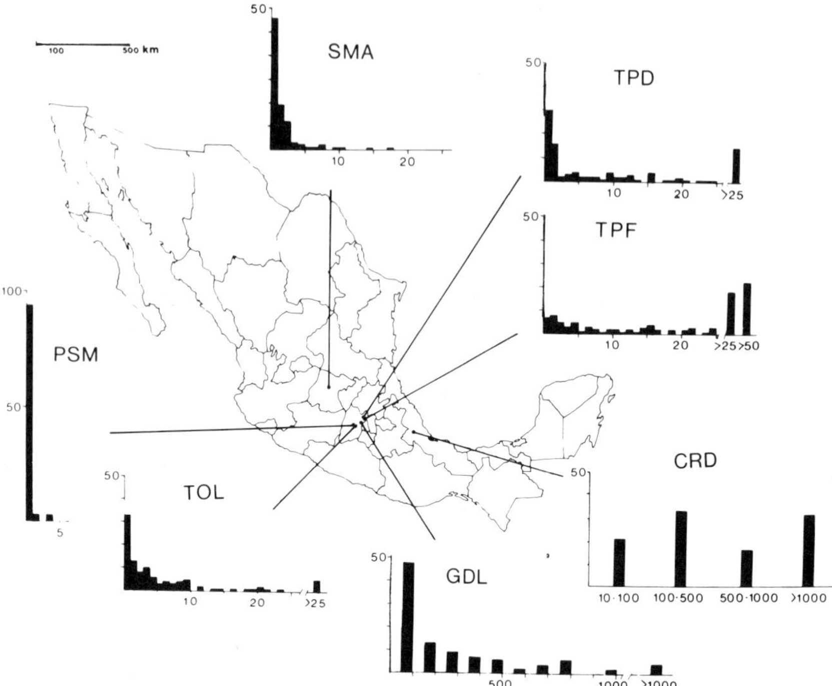

Fig. 1. Frequency distributions of group size for *M. spinipes* in various sites in Mexico. (SMA—San Miquel de Allende, desert grassland; TPD—Tepotzotlan, agricultural roadside; TPF—Tepotzotlan, feedlot waste disposal area; CRD—Cordoba/Fortin, tropical site; GDL—Guadalupe Lake; TOL—Toluca, agricultural site; PSM—Parque Sierra Morelos, high elevation mountainside). Copyright 1982 American Association for the Advancement of Science, used with permission.

individuals are predominantly solitary or in small groups. In intermediate sites (TOL, TPD, TPF, GDL), spiders occur more frequently in groups. Differences in group size between these sites would appear to be attributable to prey availability. In the moist tropical site (CRD), where climate is benign and favorable all year and insect abundance is great, colony size is very large. Group size distributions for all sites except PSM were significantly different ($p < .05$) from a Poisson distribution truncated at zero, indicating nonrandom clumping at larger colony sizes (Uetz and Burgess 1979, Burgess and Uetz 1982).

Within-colony nearest-neighbor distance decreases over the habitat gradient as well and, in general, appears to be inversely related to prey availability (Fig. 2). Analysis of variance of nearest-neighbor distances (Table 1) shows a significant between-site effect, as well as some between-colony, within-site effect. Despite within-site variation, multiple range tests (Duncan's; $p < .05$) reveal that the six sites may be sorted into five nonoverlapping ranges of nearest-neighbor distances (Fig. 2).

Field experiments testing the effect of climate and prey availability shed some light on the within- and between-site variance in social spacing (Uetz et al. 1982). In PSM, where colonies were moved to a site with an extreme climate, nearest-neighbor distances increased twofold (Table 2). In Tepotzotlan, differences were seen between transplanted colonies with and without prey supplementation. Nearest-neighbor distance increased in colonies after reloca-

Table 1. Analysis of variance for *M. spinipes* nearest neighbor distances within-colonies and between sites.

Source of Variation	df	Mean Square	F
Site	5	2769.05	22.19*
Colony (Within-Site)	142	124.78	7.21*
Error	2136	17.29	

* $p < .05$

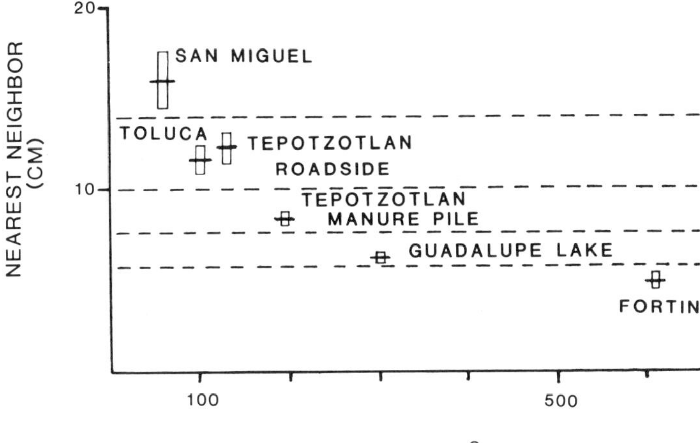

Fig. 2. Mean nearest neighbor distance (± 95% conf. limits) for colonies of *M. spinipes* in several locations in Mexico, with estimated prey availability. Dashed lines indicate means significantly different ($p < .05$) by Duncan's Multiple Range Test.

tion to a prey-poor site, except when colonies had prey supplemented (by addition of cow dung to their new site).

The results of both these experiments suggest that differences in group size and nearest-neighbor distance seen in different regions of Mexico are largely due to varying prey availability. *Metepeira spinipes* can tolerate conspecifics at closer distances in areas where prey are more abundant. Reichert (1978) found a similar reduction of interindividual distance over a latitudinal gradient in a solitary desert funnel-web spider, *Agelenopsis aperta* (Agelenidae), which was related to increased prey availability in lower latitudes. Her studies have suggested that *Agelenopsis* has a minimum territory size set at whatever area will provide a spider with sufficient biomass. In contrast to Riechert's finding that minimum territory size in these spiders is genetically fixed at the population level (Riechert 1981, and this volume), *M. spinipes* shows a rapid change in spacing when food availability changes. This may be explained by the fact that orb weavers, unlike funnel web builders, renew their web on a daily basis.

Table 2. Nearest neighbor distances of *M. spinipes* colonies before and after relocation to areas of lower prey density (Means + 95% conf. limits).

Site	Before	After
Toluca—relocated in Parque Sierra Morelos at high elevation (severe environment)	11.73 ±1.24 n=26	19.63* ±4.38 n=11
Tepotzotlan—relocated across road; half had prey supplemented w/cow dung	8.22 ±.704 n=37	22.15* ±4.95 n=5
Supplemented		11.21 ±1.62 n±5

* Before and after means significantly different (*t*-test, $p < .05$)

In a communal orb weaver, rising hunger and aggression levels associated with food deprivation may result in greater nearest-neighbor distances when the webs are rebuilt each day. Although these experiments have shown that *Metepeira* spacing is flexible, and capable of increasing as food supply decreases, they do not fully explain the differences between populations.

Laboratory Studies of Interindividual Spacing

Methods

For the first experiment, spiders brought back to the laboratory from Tepotzotlan after the 1979 field season were housed in laboratory cages (100 x 50 x 50 cm) and maintained at different levels of prey. Groups of 20 spiders/cage were subjected to three levels of availability of *Drosophila*: low — 3/spider/day; maintenance — 20/spider/day; and satiation — 35/spider/-day. Three cages per treatment level were maintained. Nearest-neighbor distances were recorded and occurrences of cannibalism and/or aggressive behavior were noted over a period of two months.

For the second experiment, representative field sites were chosen in areas along the gradient of habitats studied previously (San Miguel — desert grassland; Tepotzotlan — agricultural valley; Fortin — moist tropics). From each of these sites, spiders and egg sacs were collected for laboratory rearing in Fall 1981 and Spring 1982. Spiders were housed in the laboratory in cages, in a room with conditions of controlled climate (27 °C: 65% RH, 12 L:12 D cycle, and satiation levels of prey availability. Spiders were raised under these conditions until maturity, when nearest-neighbor distances were measured. (During the course of the experimental rearing, accidental spraying of insecticides by physical plant personnel resulted in large scale mortality of spiders. Only enough survived to permit one treatment cage per source population to be reared to adulthood.)

Results and Discussion

Observation of behavior and spacing patterns of the Tepotzotlan spiders in laboratory cages confirm the findings of field studies: spacing varies with food availability. Spiders in cages at low levels of prey availability were spaced at significantly greater distances from each other than those in cages at maintenance or satiation levels of prey (Duncan's Multiple Range Test; $p < .05$) (Fig. 3). Instances of cannibalism were highest in the low prey availability cages ($n = 9$), and very low (1 each) in the other treatments. Interestingly, there were no significant differences in nearest-neighbor distances of spiders at maintenance and satiation levels of prey. This suggests, as Riechert (1981) has shown for *Agelenopsis*, that there is an intrinsic lower limit of distance at which conspecifics can be tolerated (analogous to territory size).

Genetic differences in the degree of tolerance of conspecifics were suggested by laboratory observations of differences in the behavior of the 1981-82 field-collected spiders from various localities. In particular, spiders from the tropical population appeared more tolerant of each other, even at close distances. During transport, many adult and juvenile individuals could be put together in small containers with water but no food, and survive for up to two weeks without cannibalism. In contrast, spiders from the other populations could not exist under these conditions for more than a day or two. When released into new web sites (in laboratory cages or in the field), behavioral differences in construction of the communal web are apparent. Individuals from the tropical population all cluster together at first in a small

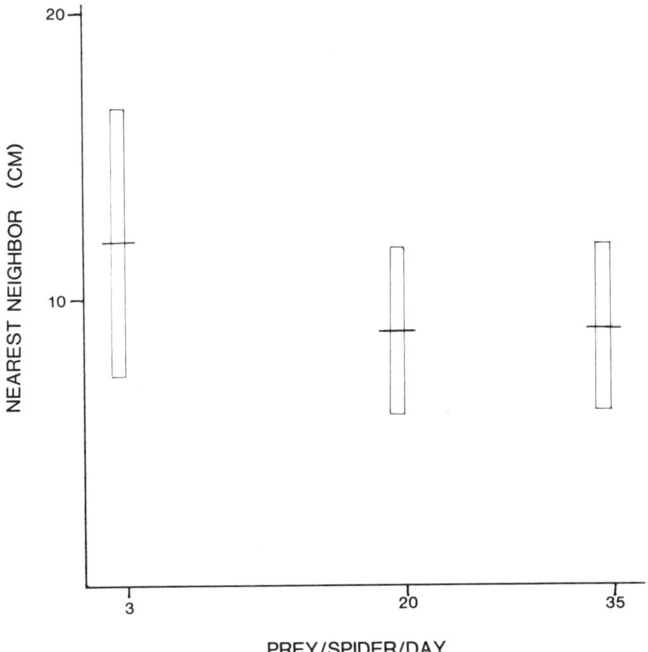

Fig. 3. Mean nearest neighbor distance (± 95% conf. limits) of colonies of *M. spinipes* from 3 different source populations, reared in the laboratory under identical conditions.

amount of webbing, with individuals 1-2 cm apart. Then, a scaffolding is built by the simultaneous efforts of many individuals, and spiders begin to space themselves out at greater distances within this space web, although few webs are built. By the third or fourth day, the communal scaffolding is expanded to its ultimate size, and individuals all have webs spaced at their previous distances within a single, central group. When individuals from Tepotzotlan or San Miguel undergo this process of rebuilding the communal web, the pattern is somewhat different. Individuals disperse about the cage or web site at greater distances (10-20 cm) from each other than those of the tropical population. Web construction begins with each individual laying silk lines between supports, and elaborating on silk laid by others. Within a day, webs are renewed, and individuals appear uniformly spaced within the colony, or in small clusters of two to five individuals. The presence of distinct behaviors in individuals from different populations placed in essentially identical conditions, suggests that these behaviors are under genetic control.

Spiderlings hatched from egg sacs collected in 1981-82 from the three source populations were reared in cages under identical conditions, and at satiation levels of prey. Each population cage had an equal representation of siblings from each of several egg sacs to equalize maternal effects. In this manner, any genetic differences in the degree of tolerance, or minimum distances set at the popuation level should become apparent. Spiderlings hatched from egg sacs collected in the tropical site remain closer together for longer periods of time after hatching than do spiderlings from other sites reared under identical conditions. The data that could be salvaged from this experiment (because of the insecticide induced mortality) suggest that genetic differences in tolerance exist, reflected in the spacing patterns of spiders from separate source populations. One way analysis of variance of the nearest-neighbor distances of spiders in the three remaining cages shows a significant between-population effect ($F = 3.97$: $p < .05$). The tropical population from Fortin had a significantly lower mean nearest-neighbor distance (Duncan's Multiple Range Test; $p < .05$) than the populations from Tepotzotlan and San Miguel, which were not significantly different from each other (Fig. 4). The mean spacing

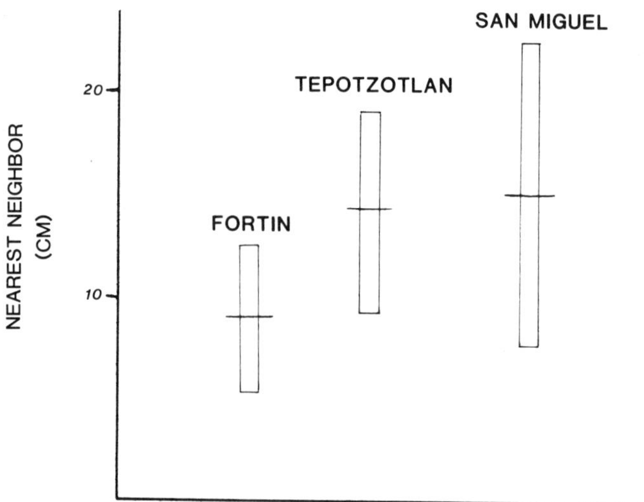

Fig. 4. Mean nearest neighbor distance (± 95% conf. limits) of colonies of *M. spinipes* from three different source populations, reared in the laboratory under identical conditions.

level for laboratory-reared tropical individuals at adulthood (8.9 cm ± 3.44) is similar to field data for adult nearest-neighbor distances in those populations (field colony means 7.64-11.43 cm). Measurements of nearest-neighbor distances between adult spiders of the other populations were much higher in the field than in the laboratory (Tepotzotlan lab ± 14.4, field ± 19.2-18.3; San Miguel lab ± 14.6, field ± 18.7-39.38 cm). Comparison of data from laboratory and field for these three populations support the notion that a lower limit of nearest-neighbor distance (or "territory size") exists, even at satiation levels of prey, and that this lower limit is set genetically in different populations. (Previously reported field and laboratory data (Figs. 2,3) are not really comparable, since they were taken from subadult spiders). These findings, although tentative and based on a small sample size, are similar to those from Riechert's more extensive work with *Agelenopsis* (Riechert 1981).

Preliminary Survey of Genetic Structure of *Metepeira* Populations

Methods

We have attempted to quantify the degree of geographic differentiation among *M. spinipes* populations using gel electrophoresis. All electrophoresis studies have been done on polyacrylamide slab gels, using a Hoefer SE—600 vertical slab gel system. To date, we have preliminary electrophoretic data on five enzymatic systems encoded by six loci in samples of 15 individuals from several colonies in each population. The enzyme systems include an esterase, malate dehydrogenase (two loci), phosphoglucose isomerase, phosphoglucomutase, and superoxide dismutase (SOD).

Of the six loci examined, all except SOD are variable in some or all populations. Although the esterase is variable within and among populations, we have not been able to determine the inheritance pattern, and, therefore, it is not included in any of the calculations. Genetic similarity between locations was determined using Nei's index (*I*) (Nei 1972).

Results and Discussion

Heterozygosity values (Nei 1972) suggest that the tropical populations may be less variable (H = 0.09) than those of the central valley (Toluca, H = 0.19; Tepotzotlan, H = 0.20) or the northern desert region (San Miguel, H = 0.21). Despite these apparent differences in variability, there is little genetic differentiation among the four geographic sites. Genetic similarity values (Nei 1972) between all populations were > 0.9 (Table 3), which are not different from values often reported for continuously distributed conspecific populations of other animals (Ayala 1976, Selander and Johnson 1973). Even though the number of loci we used is very small, it would appear that the amount of biochemical divergence we have observed between spider populations in different geographic areas is much less than the divergence in their behavioral characteristics.

The high degree of genetic similarity between *M. spinipes* populations has at least three possible explanations. It is possible that colonization across presumed geographic barriers has been too recent for genetic differentiation to have occurred. Alternatively, it is possible that presumed geographic barriers do not act to prevent gene flow and, therefore, high genetic similarity is merely the result of continuous genetic exchange. (Spiders are known to have a high dispersal capacity as a result of "ballooning" on silk threads.) A third hypothesis would suggest that genetic similarity is maintained through similar selection pressures on these enzymatic loci over the entire species range. The present data are insufficient to distinguish between any of these hypotheses.

Table 3. Genetic similarity values (Nei's Index, *I*) for sampled populations of *M. spinipes*.

Site	San Miguel	Toluca	Tepotzotlan	Fortin
San Miguel	—	.9739	.9703	.9579
Toluca		—	.9694	.9497
Tepotzotlan			—	.9731
Fortin				—

Conclusion

The data presented here, although preliminary, allow some tentative conclusions about the genetic basis of social grouping in *M. spinipes*. Considering the high degree of genetic similarity between geographic areas, it is probable that differences in levels of tolerance, reflected in minimum nearest-neighbor distances, constitute ecotypic variation of behavioral traits within the species' population. This species would appear to demonstrate the characteristics that would be expected of intermediate stages in the evolution of sociality in spiders.

Literature Cited

Aspey, W. P. 1977. Wolf Spider Sociobiology. I: Agonistic display and dominance-subordinance relationships in adult male *Schizocosa crassipes*. Behaviour 62: 103.

Ayala, F. J. 1976. Molecular Evolution. Sinauer Associates, Sunderland, Massachusetts.

Brown, J. L. 1975. The Evolution of Behavior. W. W. Norton & Company, New York.

Burgess, J. W. 1978. Social behavior in group living spider species. Symp. Zool. Soc. London 42: 69.

Burgess, J. W., and G. W. Uetz. 1982. Social spacing strategies in spiders. *In* P. N. Witt and J. S. Rovner, eds. Spider Communication: Mechanisms and Ecological Significance. Princeton University Press, Princeton.

Burgess, J. W., and P. N. Witt. 1976. Spider webs: design and engineering. Interdiscip. Sci. Rev. 1: 322.

Buskirk, R. E. 1981. Sociality in the Arachnida. *In* H. R. Hermann, ed. Social Insects. Vol. II. Academic Press, New York.

Christenson, T. E., and K. C. Goist. 1979. Costs and benefits of male-male competition in the orb weaving spider, *Nephila clavipes*. Behav. Ecol. Sociobiol. 5: 87.

Hermann, H. R., ed. 1982. Social Insects. Vol. I. Academic Press, New York.

Krafft, B. 1982. The significance and complexity of communication in spiders. *In* P. N. Witt and J. S. Rovner, eds. Spider Communication: Mechanisms and Ecological Significance. Princeton University Press, Princeton.

Kullmann, E. 1972. Evolution of social behavior in spiders (Araneae; Eresidae and Theridiidae). Amer. Zool. 12: 419.

Maynard Smith, J., and S. E. Riechert. 1984. A conflicting tendency model of spider agonistic behavior: hybrid-pure population line comparisons. Anim. Behav. (in press).

Nei, M. 1972. Genetic distance between populations. Amer. Natur. 106: 283.

Oster, G. F., and E. O. Wilson. 1978. Caste and Ecology in the Social Insects. Princeton University Press. Biology Series, No. 12, Princeton.

Riechert, S. E. 1978. Energy-based territoriality in populations of the desert spider, *Agelenopsis aperta* (Gertsh). Symp. Zool. Soc. London 42: 211.

Riechert, S. E. 1981. The consequences of being territorial: spiders, a case study. Amer. Natur. 117: 871.

Riechert, S. E. 1982. Spider interaction strategies: communication *vs.* coercion. *In* Spider Communication: Mechanisms and Ecological Significance. Princeton University Press, Princeton.

Risch, P. 1977. Quantitative analysis of orb web patterns in four species of spiders. Behav. Genet. 7: 199.

Rovner, J. S., G. A. Higashi, and R. F. Foeliz. 1973. Maternal behavior in wolf spiders: the role of abdominal hairs. Science 182: 1153.

Rypstra, A. L. 1979. Foraging flocks of spiders. Behav. Ecol. Sociobiol. 5: 291.

Selander, R. K., and W. E. Johnson. 1973. Genetic variation among vertebrate species. Ann. Rev. Ecol. Syst. 4: 75.

Shear, W. A. 1970. The evolution of social phenomena in spiders. Bull. Brit. Arachn. Soc. 1: 65.

Smith-Trail, D. 1982. Evolution of sociality in spiders — the costs and benefits of communal behavior in *Philoponella Oweni* (Araneae: Uloboridae). Ph.D. Thesis. Cornell University.

Uetz, G. W., T. C. Kane, and G. E. Stratton. 1982. Variation in the social grouping tendency of a communal web-building spider. Science 217: 547.

Uetz, G. W., and J. W. Burgess. 1979. Habitat structure and colonial behavior in *Metepeira spinipes* (Araneae: Araneidae), an orb weaving spider from Mexico. Psyche 86: 79.

Uetz, G. W. 1983. Web-building and prey capture in communal orb weavers. *In* W. A. Shear, ed. Spider Webs and Spider Behavior. Stanford University Press, Stanford (in press).

Wilson, E. O. 1971. The Insect Societies. Belknap Press of Harvard University Press, Cambridge.

Wilson, E. O. 1975. Sociobiology: The New Synthesis. Belknap Press of Harvard University Press, Cambridge.

Witt, P. N., C. F. Reed, and D. B. Peakall. 1968. A Spider's Web. Problems in Regulatory Biology. Springer. Verlag, New York.

Wittenberger, J. F. 1981. Animal Social Behavior. Duxbury Press, Boston.

Genetic Constraints on the Evolution of Cannibalism in *Heliothis virescens*

F. Gould

Department of Entomology
North Carolina State University
Raleigh, North Carolina 27650

Introduction

Most theories about the evolution of behavior treat organisms as if they had infinite genetic flexibility. Given a set of ecological circumstances, they predict deterministic, optimal outcomes (Pyke et al. 1977, Maynard Smith 1978). Our knowledge of the complex interactive nature of many population genetic processes, however, predicts some departure from optimal behavior in the real world. Although this problem has often been debated at a theoretical level (Jacob 1977, Lewontin 1978, Gould and Lewontin 1979), detailed empirical studies that are needed to resolve it are lacking.

Although many field studies show a statistically significant fit between the predictions of an optimality model and what an animal does, there is rarely a perfect fit between the observed and predicted behavior. This lack of fit is dealt with in a variety of ways. One common approach is to add biologically reasonable complexity to the model and thereby obtain a better fit. Another is to reconsider poor measurements of environmental variables that serve as our parameters in the model, and a third is to modify our definition of fitness. Once we obtain the closest fit of model and data that seems legitimate, we are at a loss to explain whether the "noise" left in the system is due to our current inability to measure correctly the important variables or to the truly non-optimal behavior of the organism studied. When the noise in the system is large enough to prevent a significant match between the model and the data, it is probably difficult for a scientist to convince a journal that he or she has found something worth publishing since reviewers are more likely to question the individual scientist's accuracy than the optimality paradigm. Although it is technically easier and psychologically more appealing to find optimal behavior than to quantify the non-optimal component of behavior, it is important that we actively seek to measure this non-optimal component and determine the reasons for its existence. The more complex an ecological system is, the more difficult it is to partition the "noise" into experimental error and non-optimal behavior. We must therefore seek for study simple, well-defined models and ecological systems.

Using the logic of one optimization model, the evolutionarily stable strategy model (Maynard Smith 1972, Maynard Smith and Price 1973), we can make deterministic predictions about the evolution of some types of social behaviors. This model predicts clearly that an unstructured animal population composed of cannibalistic (i.e., selfish) and noncannibalistic (i.e., altruistic) individuals would evolve to a point where all of the remaining individuals were cannibalistic, if there were no cost associated with cannibalism (Dawkins 1976, Stenseth and Reed 1978). It therefore interested me to find a high degree of genetic variability in the propensity for cannibalism among larvae of the moth, *Heliothis virescens* (Fabricius) (Lepidoptera: Noctuidae). When 11 strains of this phytophagous species, originating in different geographic areas of the United States (Gould et al. 1980), were tested under identical laboratory conditions, results ranged from no significant cannibalism to ca. 95% cannibalism. Since *H. virescens* is presumed to be very mobile (Wolfenbarger et al. 1973, Riley 1983), and has large population sizes in the areas sampled, explanations of interpopulation variation relying on genetic drift or complete genetic isolation of populations did not seem very reasonable (but see Wade and Breden 1981 for effects of partial isolation within populations). We therefore developed a number of hypotheses explaining the genetic variation in cannibalism based on differences in the costs and benefits associated with cannibalism in the various agroecosystems in which *H. virescens* is found. For example, in Texas, where cannibalistic propensity was high, *H. virescens* coexists with *Heliothis zea* (Boddie) on cotton. The latter is highly cannibalistic and predacious, even at low densities (Farrar, unpublished, using radioactive tracers). Contest competition with *H. zea* in this environment could have selected for heightened predacious behavior, which is correlated with cannibalistic behavior (Gould et al. 1980). In contrast, most of the *H. virescens* population in North Carolina feeds on tobacco, where interactions with *H. zea* are less frequent. Here the cannibalistic propensity is low. Although laboratory experiments indicate that successful cannibals are rarely harmed physically, cannibals could risk ingesting pathogens or sublethal doses of pesticides carried by their victims. Additionally, Joyner (1982) found that, at least for *H. zea*, the nutritional benefit of cannibalism to the victor was dependent on its previous diet. Each of these factors could be used to explain some of the variation in cannibalism among *H. virescens* populations. However, such *ad hoc* hypotheses are not very satisfying, and given the complexity of this system, testing these hypotheses might lead to an explanation of only a small fraction of the genetic variation found.

Instead of pursuing these hypotheses, I took the reductionistic approach of testing the original optimality model in a controlled laboratory environment. I first set out to determine whether cannibalism was inherited as a single locus trait and whether it exhibited additive genetic variation. Then I tested the following specific hypothesis: given a randomly mating population of *H. virescens* with additive genetic variation for propensity to cannibalize (PTC), the population will increase in PTC due to genetic changes if maintained in a situation conductive to cannibalism.

Methods

Stock Cultures and Rearing

Two strains from the previous study (Gould et al. 1980) were chosen for use in the present study. One, the Brownsville, Texas strain (Br), showed the highest propensity to cannibalize (PTC) of all strains tested. The other, New Raleigh (NR) from North Carolina, exhibited a very low PTC. Larvae were reared on artificial diet (Burton 1970) in 25 ml containers, mimicking the confined state of *H. virescens* in cotton bolls, tobacco buds, and corn ears where they usually develop. Both stocks were reared at a density of two neonate larvae per rearing container. Ten to 15 ml of artificial diet were placed in each container. In this environment, no evidence was found of a risk of injury to the cannibal by its victim. As in Gould et al. (1980), PTC was measured as the percent deviation from the expected percent of containers with one survivor per container if there had been no cannibalism.

The Br + NR stocks required the same number of days to reach their last larval instar. At this point, Br larvae were ca. 5% smaller than NR larvae. Therefore, size and growth rate were not responsible for the heightened aggreessiveness of Br larvae. The within cohort variance in development time of the stocks and hybrids tested in these experiments varied slightly, but was not related to the amount of cannibalism that occurred.

Genetic Crosses

In previous work (Gould et al. 1980), reciprocal crosses of the Br and NR strains yielded progeny exhibiting a PTC that was approximately intermediate between the PTC of the two parental strains. This indicated that genes exhibiting dominance were not involved. Additionally, progeny showed no tendency to resemble the maternal parent more than the paternal parent. In this study, the inheritance of PTC was further elucidated by following an initial cross of the two strains with two generations of backcrossing to each parental stock. A single-pair mating design was used in all backcrosses, and all individuals involved in the crosses were reared singly to avoid the potential selection of more cannibalistic individuals. One hundred and fifty progeny of each female were tested for PTC by rearing them at an original density of two larvae per container and counting the number of pupae in each container after ca. 20 days. If there were no cannibalism, only random mortality, the number of containers at the end of the experiment with 0, 1, and 2 pupae would be expected to follow a binomial distribution. An excess of containers with one pupa would be indicative of cannibalism (Gould et al. 1980). Once the existence of cannibalism was established, the extent of cannibalism could be approximated. The algorithm developed in Gould et al. (1980) for approximating the extent of cannibalism was complex and was not powerful statistically. Fortunately, the independently estimated random mortality of larvae from all strains used in this study was low and similar (ca. 1.5-3.0%). Given these conditions, the degree of departure from the binomial expectation (measured as percent deviation of the one pupa per container class) offers a relative monotonic measure of propensity to cannibalize (PTC) which is only slightly conservative in its estimates of differences in PTC among strains (T. Emigh, pers. comm.).

Selection Experiments

1) In the first selection experiment, four populations were constructed that were genetically variable regarding PTC. Three of these populations (A,B,C) were each initiated with 65 pupae from the NR stock, 30 pupae from a backcross of the F_1 hybrid progeny to NR stock, and 3 pupae from F_2 progeny of the NRxBr cross. This combination was chosen to approximate a population in Hardy-Weinberg equilibrium containing 90% low PTC and 10% high PTC alleles. To control for frequency dependent selection, the fourth population (D) was initiated with 65 Br pupae, 30 pupae from a backcross of the F_1 hybrid progeny to Br stock, and 3 pupae from F_2 progeny of the NRxBr cross. This would approximate 90% high PTC and 10% low PTC alleles in Hardy-Weinberg equilibrium, if there were single locus inheritance with fixed parental stocks (see Results and Discussion).

The neonate larvae produced by the initial generation of adults in each of the four populations were divided into two subpopulations. One subpopulation from each initial population was maintained with one larva reared per 25-ml container (A_1, B_1, C_1, D_1); the other had two randomly chosen larvae reared per 25-ml container (A_2, B_2, C_2, D_2). With this design, cannibalism could occur in only one of each paired set of subpopulations. These populations were maintained for six generations with a neonate population size of at least 150. The size of the randomly mated adult population depended on extent of cannibalism since random mortality was low (ca. 3%) throughout the experiment. PTC was monitored each generation in the subpopulations reared at two neonate larvae per container, and was assayed once at the beginning and once at the end of the experiment in the subpopulations reared with no opportunity for cannibalism. According to the previously stated hypothesis, the subpopulations reared at two larvae per container were expected to increase in propensity to cannibalize due to genetic changes. Since random mortality in all subpopulations was low and similar, departure from the binomial expectation offered a good estimate of PTC.

2) A second selection experiment was conducted in which samples ($n < 40$) of three natural populations of *H. virescens* from North Carolina were brought into the laboratory and were reared at a density of two larvae per container for 12 generations. Any additive genetic variation within these populations would presumably allow selection for an increase in PTC over time. Cannibalism rates were monitored as described above in each generation.

Results and Discussion

Figure 1 indicates that there is a large additive component to the genetic variation between strains in PTC. These results lend no support to the hypothesis that PTC is controlled by a two-allele, single locus system, but they cannot definitely refute it. If expression of cannibalistic propensity were governed mostly by a simple one-locus system, we would have expected 1/2 of the initial NR backcross progeny to be heterozygous at the "cannibalism locus" and the other half to be homozygous for low propensity to cannibalize. Thus, when randomly selected progeny from the initial backcross were again backcrossed to NR individuals, we would expect 1/2 of the replicates to produce all homozygous progeny with propensity for cannibalism equal to that of the NR stock, and the other half of the replicates to produce a 1:1 mixture of heterozygous to homozygous low cannibalism progeny. This would generate bimodality among the replicates in departure from the number of cups with 2, 1, 0 pupae expected based on the binomial distribution. The second backcross to the Br stock would be expected to produce a similar bimodal distribution. Results (Fig. 1) indicate no significant bimodality in either backcross.

Given the uncomplicated quantitative inheritance of PTC, results of the first selection experiment were quite revealing. As the experiment progressed and the first five generations of data were analyzed (Fig. 2), it appeared that rearing with two larvae per cup was resulting in the pattern expected, based on the previous finding (Gould et al. 1980) that larvae from the highly cannibalistic line survived best in contest competition with less cannibalistic larvae. In three of the four populations, there was a tendency, over time, for an increase in departure from the expected frequency of 2, 1, and 0 pupae per cup, thereby indicating an increase in cannibalism. However, after the final generation, when the four "control" subpopulations also were tested, I found that they too had increased PTC (Fig. 2).

The original NR and Br stocks that were reared at two larvae per container had not undergone a similar change in PTC. Percent deviations from the binomial expectation before and after the experiment were, respectively, 11 and 8 for NR; 45 and 47 for Br. (Additionally, for each generation, there was no significant covariation among the experimental populations in the rate, and in some cases, direction of change in PTC. Therefore, the overall increase in PTC could not be explained based on a change in environmental conditions over time. The large variation among subpopulations in the extent of change in PTC may have been due to stochastic effects caused by small effective population size. Yet, had effective population size been very small, genetic drift would have been expected in two directions. As it is, none of the eight subpopulations showed any overall decrease in PTC. A sign test (Sokal and Rohlf 1969) indicates that significantly more subpopulations increased in PTC than expected by chance ($p < .01$). The most plausible remaining explanations for the change in the "controls" are that one or more loci involved in expression of cannibalism are linked to other loci responsible for differences in mating vigor and/or fecundity, or that alleles controlling PTC have pleiotropic effects on mating vigor and/or fecundity. An analysis conducted after these experiments were terminated indicated the mean number of fertile eggs produced per female of the NR stock

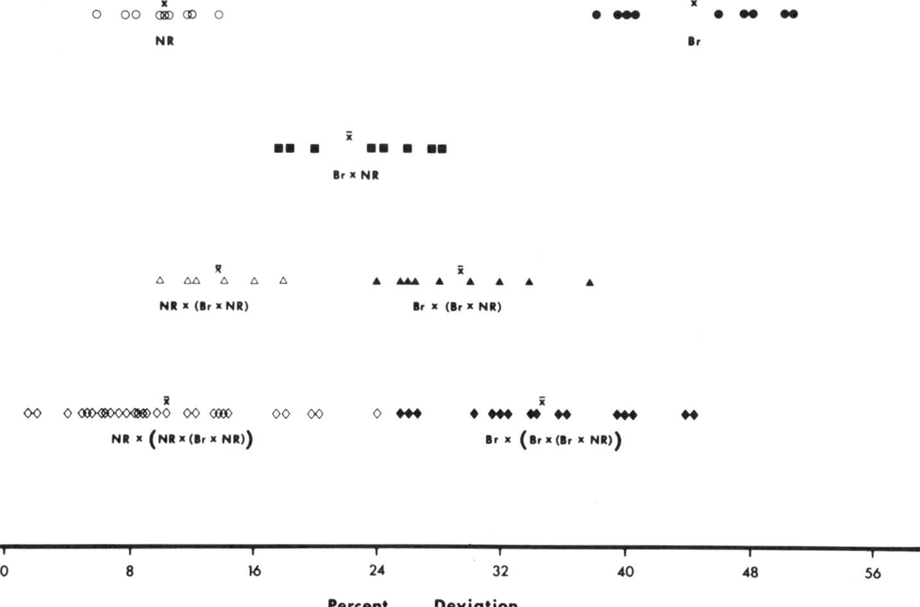

Fig. 1. Index of propensity to cannibalize (PTC) in progeny from a series of crosses. Percent deviation from expected number of containers with one survivor per container if there had been no cannibalism is used as a measure of PTC. Each symbol represents the deviation from expected in a single replicate, which consisted of 150 progeny from a single pair mating. Position of ˣ indicates mean deviation for all replicates of each type of cross.

(low PTC) was significantly lower than that produced by a cross of Br ♀ ♀ x NR ♂ ♂ in one of the two generations tested ($p < .05$).

Data gathered from the second set of selection experiments adds more evidence regarding the complexity of this system. In these experiments, three field-collected samples of *H. virescens* from North Carolina were maintained in the laboratory for 12 generations with larval population size of at least 150, using the two larvae per container rearing regime that allows cannibalism. Two of the populations initiated from the field samples did not change significantly in PTC. The third population, however, showed a significant decrease in PTC ($p < .05$), a result which was not at all predicted by the initial hypothesis.

It is interesting to note that the rate of change in cannibalism found in the first experiment for control and treatment groups was more rapid than would have been predicted, based on a simple one-locus, two-allele model (M. Wade, personal communication). A polygenic model would predict a small increase in cannibalism during the first two generations followed by an almost negligible rate of increase afterwards because the genetic component of variance between interacting larvae would be very small (F. Gould, unpublished data). These model predictions make the finding of a decrease in cannibalism in the second experiment somewhat less surprising, because the force countering this change may have been very weak.

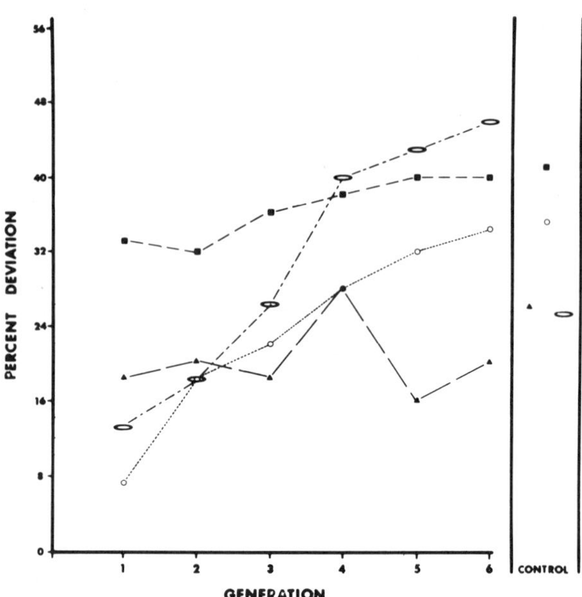

Fig. 2. Propensity to cannibalize in selected and control populations as indicated by percent deviation in number of containers with one survivor, where the expectation is based on a binomial distribution. The values labeled "control" are values obtained at the end of the experiment from subpopulations maintained at one larva per container. Linear regression analysis of PTC in subpopulations maintained at two larvae per container indicate a significant increase in PTC over time in subpopulations A_2, B_2, and D_2 ($p < .001$, $p < .001$, $p < .01$, respectively). In only one case, however, was the PTC in a "control" subpopulation less than that of its paired subpopulation where cannibalism had been allowed. A less powerful statistical test for change in PTC (Test for a Difference in Two Percentages) was necessary for comparing the percent deviations from expected in generations one and six. Subpopulations A_1 and B_1, reared at one larva per container and subpopulations A_2 and B_2, reared at two larvae per container, exhibited a significantly greater percent deviation at the end of the experiment than they did at the beginning ($p < .001$, $p < .001$, $p < .04$, respectively for A_1B_1, A_2,B_2). ○ = A, ⬭ = B, ▲ = C, ■ = D.

The obvious conclusion that can be drawn from this study is that even with basic information about the quantitative inheritance of a behavior (which is usually not available), it may be difficult to predict how evolutionary processes acting at the individual level will influence the frequency of alleles governing that behavior in a population. These results indicate that we cannot always rely upon simple, single-locus selection theory to explain why we see specific animal behaviors in nature. They also advise caution in using estimates of additive genetic variance to predict response to natural selection. Aside from the complexity of evolutionary outcomes generated by stochastic evolutionary events, we also must consider the fact that pleiotropy and interactions among loci may play an important part in determining the existence or extinction of a particular behavior regardless of whether that behavior, in and of itself, is of adaptive value to the organism (Maynard Smith and Haigh 1974, Rose 1982).

The present study was concerned with short-term evolutionary processes, but since these processes probably play an important role in long-term evolutionary changes, they certainly should not be ignored in our search to understand the behavior of organisms. This study was conducted in the laboratory, but there is no reason to believe that the population sizes used or the extent of change in environmental conditions imposed during the selection regimes are unlike those occurring naturally in many species.

I anticipate that the results described here will not be surprising to quantitative geneticists. I may even be accused of building a strawman only to knock it down. It is my feeling, however, that until enough empirical evidence is collected supporting the importance of genetic constraints in the evolution of behaviors, we will continue to witness the proliferation of optimality models devoid of detailed genetic considerations.

Acknowledgments

I thank L. Pearce, M. Ailor, and M. Smith for technical expertise, and A. Massey for discussion and encouragement of this endeavor. A. Massey, J. Ambrose, W. Blau, F. Breden, M. Huettel, G. Kennedy, C. Laurie-Ahlberg, R. Moll, S. Riley, R. Rough, M. Villani, M. Wade, and B. Wallace offered helpful comments on the manuscript. C. Satterwhite typed the manuscript. This work was supported by NSF grand DEB-7822738 and a NSF National Needs Postdoctoral Fellowship.

Literature Cited

Burton, R. L.. 1970. A low cost diet for the corn earworm. J. Econ. Entomol. 63: 1969.

Dawkins, R. 1976. The Selfish Gene. Oxford University Press, Oxford.

Gould, F., G. Holtzman, R. L. Rabb, and M. Smith. 1980. Genetic variation in predatory and cannibalistic tendencies of *Heliothis virescens* strains. Ann. Entomol. Soc. Amer. 73: 243.

Gould, S. J., and R. C. Lewontin. 1979. The spandrels of San Marco and the Panglossian paradigm: a critique of the adaptationist programme. Proc. Roy. Soc. London B 205: 581.

Jacob, F. 1977. Evolution and tinkering. Science 196: 1161.

Joyner, K. C. 1982. Developmental consequences of cannibalism in *Heliothis zea* (Lepidoptera: Noctuidae) on suboptimal diets. M.S. Thesis. North Carolina State University, Raleigh.

Lewontin, R. C. 1978. Fitness, survival and optimality. *In*: D. H. Horn, R. Mitchell and G. R. Stairs, eds. Analysis of Ecological Systems. Ohio State University Press, Columbus.

Maynard Smith, J. 1972. On Evolution. Edinburgh University Press, Edinburgh.

Maynard Smith, J. 1978. Optimization theory in evolution. Annu. Rev. Ecol. Syst. 9: 31.

Maynard Smith, J., and J. Haigh. 1974. The hitch-hiking effect of a favorable gene. Genet. Res. 23: 23.

Maynard Smith, J., and G. R. Price. 1973. The logic of animal conflict. Nature 246: 15.

Pyke, G. H., H. R. Pulliam, and E. L. Charnov. 1977. Optimal foraging: a selective review of theory and tests. Q. Rev. Biol. 52: 137.

Riley, S. 1983. The characterization of migratory flight ability in *Heliothis virescens*. Ph.D. Dissertation. North Carolina State University, Raleigh.

Rose, M. R. 1982. Antagonistic pleiotropy, dominance, and genetic variation. Heredity 48: 63.

Sokal, R. R., and F. J. Rohlf. 1969. Biometry. W. H. Freeman & Co., San Francisco.

Stenseth, N. C., and J. Reed. 1978. A comment on Bobisud's paper on evolution of cannibalism. Bull. Math. Biol. 40: 541.

Wade, M. J., and F. Breden. 1981. Effect of inbreeding on the evolution of altruistic behavior by kin selection. Evolution 35: 844.

Wolfenbarger, D. A., M. J. Lukefahr, and H. M. Graham. 1973. LD_{50} values of methyl parathion and Endrin to tobacco budworms and bollworms collected in the Americas and hypothesis on the spread of resistance in these lepidopterans to these pesticides. J. Econ. Entomol. 66: 211.

The Inheritance of Courtship Behavior in *Schizocosa* Wolf Spiders (Araneae; Lycosidae)

Gail E. Stratton*

George W. Uetz

Department of Biological Sciences
University of Cincinnati
Cincinnati, Ohio 45221

Introduction

The wolf spider genus *Schizocosa* is common throughout most of North America (Dondale and Redner 1978) and is a group that provides an example of how sexual behavior can function in reproductive isolation. Individuals of *Schizocosa ocreata* (Hentz) are found in deciduous forest leaf litter in the eastern United States and are nearly identical to individuals of *Schizocosa rovneri* Uetz and Dondale with respect to genital characters, body size, general morphology and coloration (Uetz and Dondale 1979). The two species overlap extensively in habitat, period of reproduction and geographic range (Uetz and Denterlein 1979, Stratton and Uetz 1981). Both species have ten autosomes (W. P. Maddison, pers. comm.) and both show identical patterns of maturation given a variety of temperature and diel light regimes in the laboratory (Stratton 1982). The females of the two species are indistinguishable on the basis of morphology, however the mature males of *S. ocreata* have prominent tufts of black bristles on the first pair of legs that are lacking in mature males of *S. rovneri*. These species were considered to be one (Uetz and Dondale 1979) until it was shown that their courtship behaviors are very different. Males of both species court females of either species, but females are receptive only to conspecific males (Uetz and Denterlein 1979, Stratton and Uetz 1981). It was demonstrated that sound production by males during courtship is critical for reproductive isolation as females are receptive only when the substratum-coupled acoustic signals are present (Stratton and Uetz 1981).

To test for mechanical barriers and for interfertility, a method of forcing copulation between the two species was devised that involved anesthetizing the female with CO_2, and allowing a heterospecific male to mate with her (Stratton and Uetz 1981). Typically, the female would stay anesthetized for 1-2 min while the male mounted and began copulation. When the female "awoke" she cooperated with the copulation by swiveling her abdomen to facilitate the insertion of the male palp, as is typical for wolf spiders (Rovner 1972). Offspring

*Present address: Department of Biology, Albion College, Albion Michigan 49224.

were produced in most interspecific crosses. There were no significant differences in either egg sac production or hatching success between conspecific and forced heterospecific matings (Stratton and Uetz 1983). Thus, there do not appear to be mechanical or postcopulatory barriers (such as gametic incompatibility or zygote mortality) to reproduction.

The F_1 hybrids produced from these matings matured and were apparently normal. They proved to be fertile when they were crossmated and backcrossed to both parental species (using the forced-copulation method). The ability of get F_1, F_2 and backcross progeny has allowed the opportunity to study the inheritance of courtship behaviors in a situation where it has been demonstrated experimentally that courtship functions to keep two species from interbreeding.

A comparison of the inheritance of courtship behaviors in other groups shows no consistent pattern of genetic control. Grula and Taylor (1980a,b) investigated the inheritance of courtship communication systems in *Colias* butterflies. In studying F_1, F_2 and backcross progeny, they found evidence for an X chromosome "supergene" which exerts a large influence on many of the distinguishing traits. Ewing (1969) found that in two sibling species of *Drosophila* (*D. pseudoobscura* Frolova and *D. persimilis* Dobzhansky and Epling) the genes that control the qualitative aspects of songs (such as the overall pattern of the song) are located on the X chromosome; while genes controlling quantitative differences (such as interpulse interval) are controlled autosomally. These data were supported in a study by Cowling and Burnet (1981). Wood et al. (1980) provided an analysis of the courtship patterns of hybrids between *Drosophila melanogaster* Meigen and *D. simulans* Sturtevant. They were able to demonstrate that several of the behavioral transitions in the courtship sequence were under partially independent genetic control. Manning (1961) studied the genetic basis of mating speed in *Drosophila melanogaster* by selecting for both fast and slow mating speeds. He found that there was no single factor controlling mating speed, but rather a complex mixture of behavioral elements and behavioral interactions between males and females that affected mating speed. The temperate fiddler crabs of the genus *Uca* use acoustic signals in their courtship. Salmon and Hyatt (1979) were able to cross two *Uca* species and get viable offspring that were reared subsequently in the lab. They found that the temporal patterns of the acoustic signals in the hybrids in some instances reflected the components of one or the other of the parental species, and in other instances were intermediate to what was seen in the parental species. Bentley and Hoy (1972) crossed *Teleogryllus commodus* (Walker) and *T. oceanius* (Orthoptera) and found that the calling songs of the F_1 hybrids were distinct from either of the parental species. Intrachirp and intratrill intervals were intermediate to those observed in the parental groups. The intermediacy of this parameter suggests polygenic inheritance. This was supported by the study of backcross individuals (Bentley 1971). It was found also that there were sex-linked factors affecting the calling song. Thus genetic control of the trait was distributed among both autosomes and the sex chromosome.

There are other examples where changes in single genes or gene complexes can cause significant behavioral changes in a species. If the differences in the alleles of a single gene can alter behavior, a simple mutation could lead to behavioral evolution. Behavioral traits may then be readily accessible to natural selection. Mayr (1963) suggests that a shift into a new niche or adaptive zone is almost without exception initiated by a change in behavior. Tauber and Tauber (1977) and Tauber et al. (1977) showed the importance of a very small number of genes affecting key factors controlling species isolation in the Neuroptera. Our study of *Schizocosa* wolf spiders provides an example of a change in a single gene, or a very few genes, that had profound consequences for the evolution of a group.

Methods

Subadult specimens from each of the parental populations were collected in the spring of 1979 and 1980. *Schizocosa rovneri* Uetz and Dondale was collected from floodplain forest leaf litter near the Ohio River, Boone County, Kentucky. *Schizocosa ocreata* was collected from upland deciduous forest leaf litter from the Cincinnati Nature Center, Clermont County, Ohio. All individuals were collected as immatures and housed individually in plastic boxes (7 x 7 x 13 cm) which both assured their virginity and shielded them visually from one another. Moisture was provided via a cotton stoppered glass vial and crickets (*Acheta domesticus* L.) were provided twice weekly as prey. Individuals were crossmated randomly in June 1980, using the "forced-copulation" method described by Stratton and Uetz (1981).

Courtship behaviors and copulations were observed in a circular glass culture dish (diam. 20 cm), with a piece of bond paper serving as substrate. A female's cage liner was placed on the floor of the test arena to provide pheromone and tactile stimuli to elicit courtship from the male (Tietjen 1978). All experimental pairings and observations were made between 0800 and 2100 h, with the exception of acoustic recordings which were made between 0400 and 0800 h. Room temperature was 20-22 °C.

Portions of the courtship of each group of spiders were filmed with a Nizo Super-8 movie camera (at 18fps and 54fps) or with a Bolex 160 Macrozoom camera. During courtship observations, each movement of the male during the first 3 min of courtship was recorded. Courtship of each individual spider was observed on a minimum of four different dates to ensure that the observer would see most of the individual's behaviors. All spiders were offered food within 2 days of any observations and spiders were not observed on consecutive days. In addition, each individual was scored at least once in a "blind" behavioral test.

Behavioral observations of *S. rovneri* and *S. ocreata* were done in the spring of 1980, and then concurrently with succeeding generations of spiders. Behavioral observations of the F_1 hybrids and the breeding experiments involving the F_1 generation were done in the fall and winter, 1980-1981. Offspring (F_2 and backcross progeny) from these crosses matured in the winter of 1982. Behavioral observations of this generation were made from January to April 1982. The number of individuals involved in each cross is indicated in Figure 3.

In instances where an individual displayed courtship on 2 or more days, the sequences of the courtship behaviors were coded numerically and analyzed using a computer program designed from descriptions of similar programs (Leonard and Ringo 1978, Aspey 1977). The output of the program consisted of a transition matrix in which each element in the matrix was represented by the number of times behavior *i* preceded behavior *j*. Pooled matrices of each group of spiders were prepared. For each group, at least 17 individuals were recorded and analyzed.

The pooled transition matrices were analyzed for nonrandom transitions using Chi square with $p < 0.05$ chosen as the level of significance. This was done by collapsing the complete table about the cell of interest (Slater 1973). For each group, transitions occurring more frequently than would be expected by a random model were placed together in a kinematic diagram. In each diagram, each behavior is enclosed in a box with the darkened area indicating its relative frequency of occurrence; the thickness of the arrow between behaviors is proportional to the frequency of transition between behaviors *i* and *j*, conditional upon the occurrence of *i* (Leonard and Ringo 1978). A standard form of the kinematic diagram was made to facilitate comparisons between all groups.

Sound and vibration recordings were made of males of each of the groups of spiders (i.e., *S. rovneri, S. ocreata,* F_1, F_2 and backcross progeny). Recordings were make by a Bruel and Kjaer accelerometer (Type 4366) high sensitivity vibration pickup leading to a Bruel and Kjaer sound level meter (Type 2203) whose output was recorded by a Teac tape recorder (Model 2300 Sx — frequency response range: 40 Hz to 24 kHz). Recordings were made at 19.05 cm/sec, between 0400 and 0800 h, with a room temperature of 20-22 °C. All recordings were made by placing the accelerometer on a piece of a female's cage liner that in turn was on a soapstone laboratory table top. The male was enclosed in a cage made by a clear acetate sheet that also surrounded the accelerometer. Equipment other than the accelerometer was placed on a separate table to minimize machine noise. Recordings of sounds judged to be typical were then played into a recording oscilloscope. Photographs were taken of the traces with a SLR camera.

Results

Behavior of Schizocosa ocreata

Typically, a male *S. ocreata* encounters female silk and shows "chemoexploratory" behavior. This involves rubbing the dorsum of the palp on the substrate where a female has laid her silk. This has been called the "searching phase" of courtship (predisplay) (Stratton and Uetz 1983). Interspersed with chemoexploratory behavior, the male walks and extends his forelegs and taps them in unison ("tapping"). The male may then go into the "jerky tapping" behavior, in which he produces stridulatory sounds by movement of the palp (Rovner 1975). In jerky tapping, the male also taps his forelegs, but will move his whole body forward and back as well. The movements involved in the jerky tap appear to occur in a series; that is, the spider does jerky tapping for several seconds, stops for 1-2 sec, then resumes jerky tapping. The jerky tap is seen particularly if the male is near the female or has been touched by her. The kinematic diagram for male *S. ocreata* (Fig. 1) shows that the predominant behaviors are the jerky tap, the double tap, walk, chemoexplore and leg I extend. The predominant transitions occur between walk and double tap, leg I extend to jerky tap, and jerky tap to jerky tap (i.e., the jerky tap occurs in a recognizable series). The sounds produced by the courting *S. ocreata* are complex without showing clear temporal patterning (Stratton and Uetz 1981). A single episode may last up to 10 sec (Fig. 1).

Behavior of Schizocosa rovneri

The courtship behaviors of *S. rovneri* have been described more extensively and more quantitatively than those of *S. ocreata* (Uetz and Denterlein 1979, Stratton and Uetz 1981, Stratton and Uetz 1983). Typically, a male *S. rovneri* shows chemoexploring interspersed with walking during the "predisplay" phase of courtship. The male spider then does a series of "bounces," each bounce producing an audible sound. During a bounce (which occurs in less than 1/60th of a sec), the male contracts all eight legs simultaneously and slams his body against the substrate. This occurs in even intervals (about one every 3-5 sec). During a bounçe, sounds are produced and are audible without amplifying equipment. The sounds produced form an even pulsed, pattern (Stratton and Uetz 1981) (Fig. 1). The kinematic diagram for *S. rovneri* shows that the predominant behaviors are bounce, chemoexplore, and walk (Fig. 1). The most important transitions occur from walk to leg I extend and bounce to bounce (i.e., the bounces occur in a series).

The courtship behavior of each of these species is clearly distinctive with very little overlap between the two in the predominant behaviors. Jerky tapping is typical of the courtship of *S. ocreata*; a series of bounces is typical of *S. rovneri*.

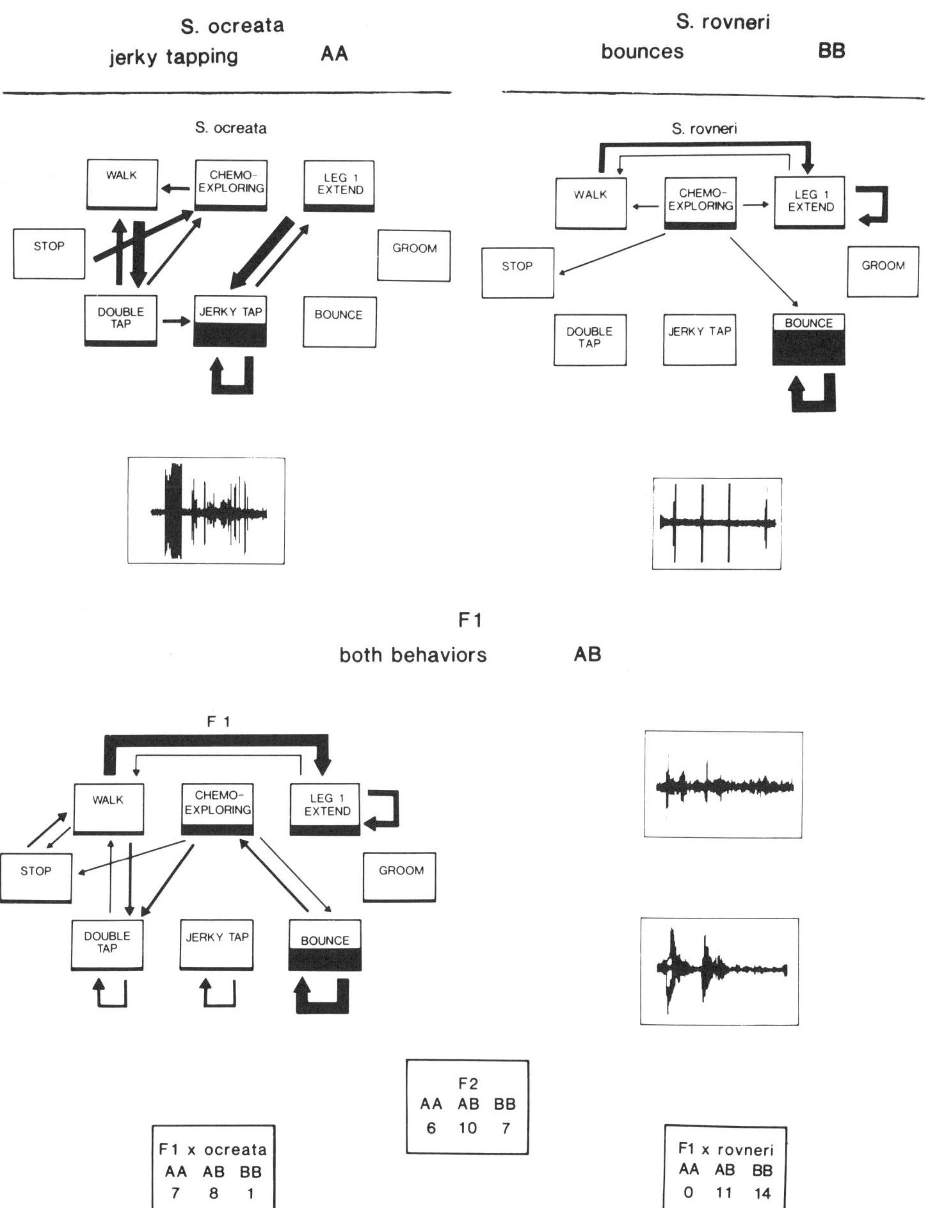

Fig. 1. Kinematic diagram for *Schizocosa ocreata* (n = 20), *Schizocosa rovneri* (n = 20) and F_1 hybrids (n = 20) males. Only nonrandom transitions are shown. Accompanying each kinematic diagram is a typical oscillograph of the sounds produced. Ratios of the typical pure species and phenotypes of F_2 and backcross progeny are shown in boxes below.

Morphology and Behavior of F_1 Interspecific Hybrids

The hybrids produced in the interspecific matings matured and were apparently normal although maturation was slower than seen in parental groups (Stratton 1982). The females were indistinguishable from the parental females. All of the mature males ($n = 50$) had tufts of bristles on the forelegs that were variable and intermediate between the tufts of *S. ocreata* and the complete absence of tufts in *S. rovneri.*

Most F_1 hybrids showed behaviors typical of both parental species, switching from sequences typical of one species to those typical of the other. The kinematic diagram of the F_1 males (Fig. 1) shows the complexity of the behaviors involved in switching between the behaviors of the two parental species. Bounce and leg I extend occur in series as in *S. rovneri*; jerky tapping occurs in series as in *S. ocreata.* The transition between walk and double tap reflects that which is seen in *S. ocreata*, while the transition between walk and leg I extend is similar to *S. rovneri.*

The sound produced by the F_1 hybrids likewise showed the switching between one parental type and the other. An oscillogram shows the courtship sequence of an F_1 hybrid, half of which is like the courtship song of *S. ocreata* and half of which is like the courtship song of *S. rovneri* (Fig. 1). Also typical of the F_1 hybrids was the presence of a "double bounce" or two bounces occurring immediately in succession (Fig. 1). This was never seen in *S. rovneri*, but occurred in a high proportion of the recordings of the F_1 hybrid males. It was noted also in the films made of the F_1 males.

There were no differences when the behaviors of the F_1 progeny of different parentage were compared (Fig. 2). F_1 individuals whose mothers were *S. rovneri* (H_{rov}) showed no qualitative or quantitative differences from individuals whose mothers were *S. ocreata* (H_{ocr}).

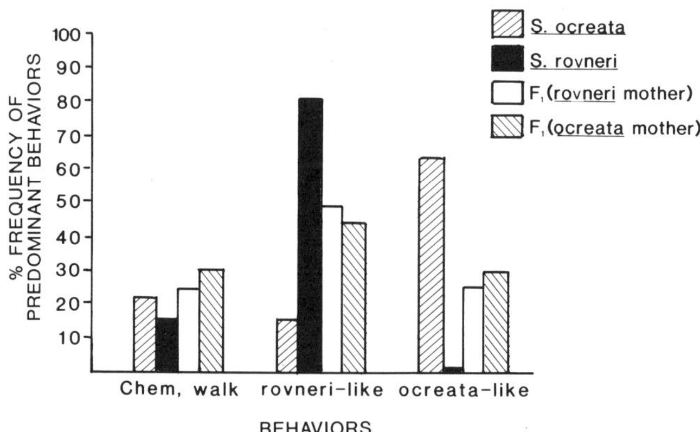

Fig. 2. Frequency of predominant courtship behaviors seen in parental *S. ocreata* (P_{ocr}), *S. rovneri* (P_{rov}) and their reciprocal hybrids. H_{rov} are hybrids whose maternal species is *S. rovneri,* H_{ocr} are hybrids whose maternal species is *S. ocreata.* Chemoexploratory and walking behaviors are seen in both species.

Morphology and Behavior of F_2 Hybrids and Backcross Progeny

The offspring of the crosses F_1 x F_1, F_1 x *S. ocreata*, and F_1 x *S. rovneri* all produced living offspring that matured and appeared to be normal although the rate of maturation was slower than was observed for the parental groups (Stratton 1982). While there were no differences between conspecific matings and heterospecific matings in either egg sacs produced or hatching success, there was a reduction in the mating success of the F_1's and the backcrosses (Fig. 3). To get the F_1 to breed, many more copulations had to be attempted. Although no mating of F_1 males with purebred females produced living offspring, the fertility of the F_1's was indicated by the successful production of F_2's.

The mature F_2 hybrid males showed a mixture of the patterns of tufts, with some showing full tufts, some no tufts, and some intermediate tufts. The relatively high number of individuals in the F_2 generation showing the parental patterning of bristles suggests a very small number of genes controlling this character. The data (Fig. 4) are consistent with a single gene hypothesis explaining the inheritance of the tufts of bristles (expected ratio = 1:2:1; χ^2 = 0.46, n.s.). Both test crosses (Fig. 4) are also consistent with this model (F_1 female x *ocreata* male, χ^2 = 0.22, n.s.; F_1 female x *rovneri* male, χ^2 = 2.77, n.s.).

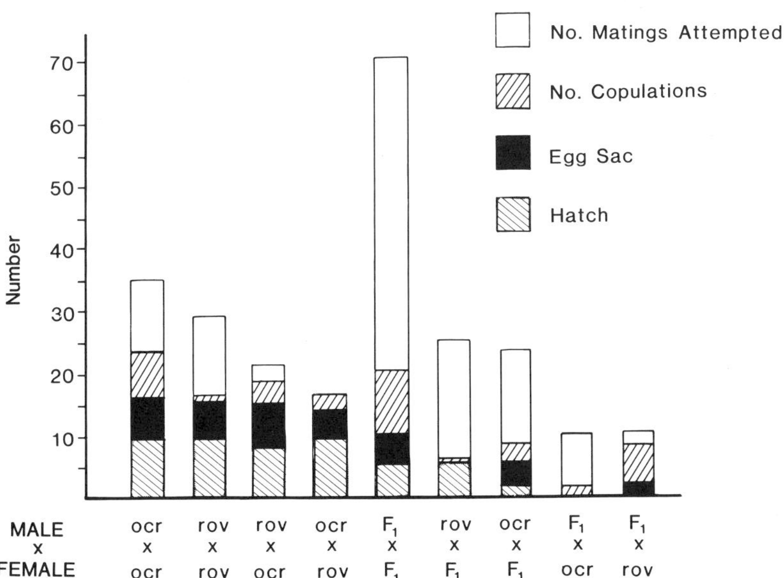

Fig. 3. Number of attempted matings, number of copulations and number of egg sacs produced, and number of egg sacs hatching for all crosses.

A relatively high proportion of the F_2 males also showed behaviors that were typical of one or the other of the parental species. Some males showed only a series of bounces (typical of *S. rovneri*); others showed the jerky tapping (typical of *S. ocreata*); while some switched between both types of behavior, as was seen in the F_1 males. The high number of parental types again suggests that a small number of genes are involved in controlling the inheritance of these characters. A single gene hypothesis would yield a 1:2:1 ratio in the F_2. The data (Fig. 1) for the F_2 and backcrosses are both consistent with this model (F_2, $\chi = 1$, n.s.; backcross $\chi^2 = 0.125$; $\chi^2 = 0.36$, n.s.).

The double tap behavior was examined independently of the other behaviors and was found to fit a single gene model with simple dominance. All F_1 hybrids showed tapping. In the F_2 hybrids, 18 of 23 individuals showed tapping. In a model with a single dominant gene, the expected ratio in the F_2 is 3:1 (phenotypic ratio). The data are consistent with this model ($\chi^2 = 0.134$; n.s.). In addition, backcrosses are consistent with this model ($\chi^2 = 0.04$; $\chi^2 = 0$; n.s.).

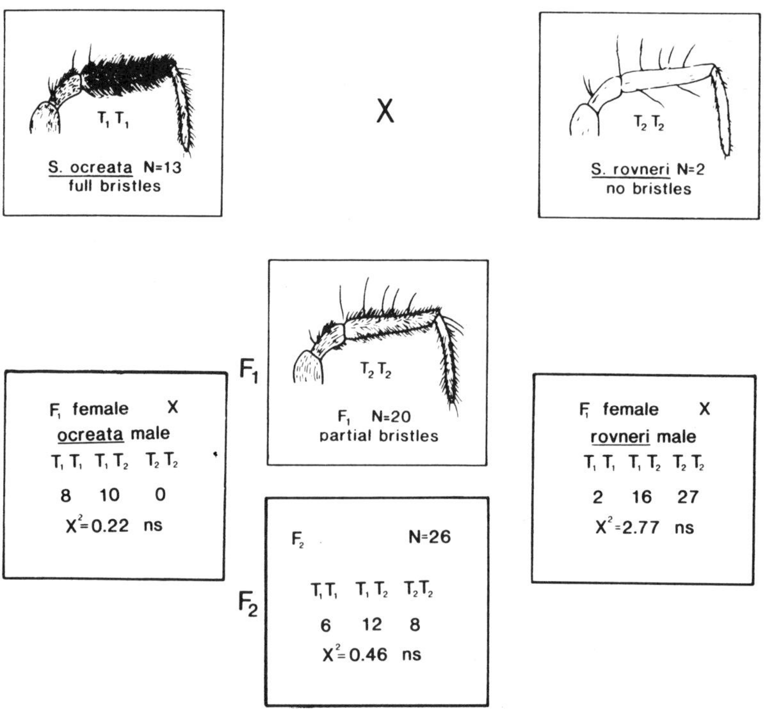

Fig. 4. Genetic model for the inheritance of the tufts of bristles in *S. ocreata* and *S. rovneri*.

There is evidence that suggests that the tufts of bristles and the major elements of courtship behavior do not assort independently ($\chi^2 = 25.8$; $p < 0.005$). In the F_2's, 5 of 23 showed recombination (i.e., individuals showing both full tufts and bounces, or no tufts and jerky tapping). This gives a cross over value of about 21%. Backcross progeny also indicate that there is a lack of independent assortment. In one cross, 12 of 45 were recombinant (24%) and in the other backcross, 4 of 16 (25%) were recombinant.

It would appear that the double tapping behavior and the jerky tapping behavior are either controlled by the same gene or they are very closely linked. It was clear that in the F_2 progeny, the double tapping behaviors and jerky tapping behaviors were not assorting independently ($x^2 = 33.3$; $p < 0.005$). In fact, in the F_2, there were no individuals that showed crossing over of genes controlling these traits. However, in the backcross progeny (F_1 x rovneri) 4 of 23 individuals showed a pattern of behavior that indicated that recombination was occurring (17% C.O.V.). In the F_1 x *ocreata*, 1 of 16 individuals showed recombination. This would suggest that these two behaviors are not controlled by the same gene but rather that the behaviors are controlled by linked genes.

The transitional frequency data from the F_2 generation were divided into three separate groups: those sequences that were like *S. ocreata* (labeled F_2 ocr-like in Fig. 5; n = 32), those that were like S. rovneri (F_2 rov-like; n = 22), and those that showed the mixture of behaviors as was seen in the F_1 generation (F_2-mixed; n = 27). Likewise, the backcrosses were similarly divided: the H-Pr cross (F_1 female x *rovneri* male) produced individuals that had either *S. rovneri*-like (rov-like) behaviors (n = 39) or a mixture of the behaviors seen in the parental groups (H-Pr mixed; n = 36). The H-Po backcross (F_1 female x *ocreata* male) produced individuals that had either *ocreata*-like behaviors (H-Po ocr-like; n = 16), or had the mixture of behaviors from both parental groups (H-po mixed; n = 31). This allowed pairwise comparison of transitional frequencies between groups of spiders that showed what appeared to be similar behavior.

Schizocosa ocreata and F_2 ocr-like (Fig. 5) were found to be very similar, with the predominant behaviors being jerky tap, double tap, walk, chemoexplore, and leg I extend. The only behavior that occurs in a series for *S. ocreata* is the jerky tap. For the H-Po ocr-like both leg I extend and grooming occurred in series as well. The group H-Po mixed was more similar to the F_1 than it was to the *S. ocreata* (Fig. 6). This was the only group that showed a nonrandom transition between double tap and bounce.

Both *S. rovneri* and F_2 rov-like (Fig. 5) show a significantly high proportion of walk to leg I extend transitions, a high relative frequency of bounces and chemoexplore, and a high probability that bounces will occur in a series. *Schizocosa rovneri* was then compared to the backcross H-Pr rov-like (Fig. 6). In both of these groups, bounce and chemoexplore were the predominant behaviors. The transition between walk and leg I extend occurred frequently, and both bounce and leg I extend occurred in series. The H-Pr mixed (Fig. 6) was clearly much more similar to the F_1 individuals than it was to the *S. rovneri* individuals. This was particularly due to the jerky tap and double tap occurring in series and due to the frequent transition between walk and double tap.

The F_2-mixed were then compared with the F_1 (Fig. 5). Both of these groups show bounces, jerky taps, and double taps occurring in a series. Also both show a relatively high frequency of bounces, jerky taps, chemoexplore, leg I extend and double tap. Both of these groups clearly show much greater complexity of behaviors than is seen in either of the parental groups. Curiously, the F_2-mixed lacks the transition from walk to leg I extend that is seen in the F_1.

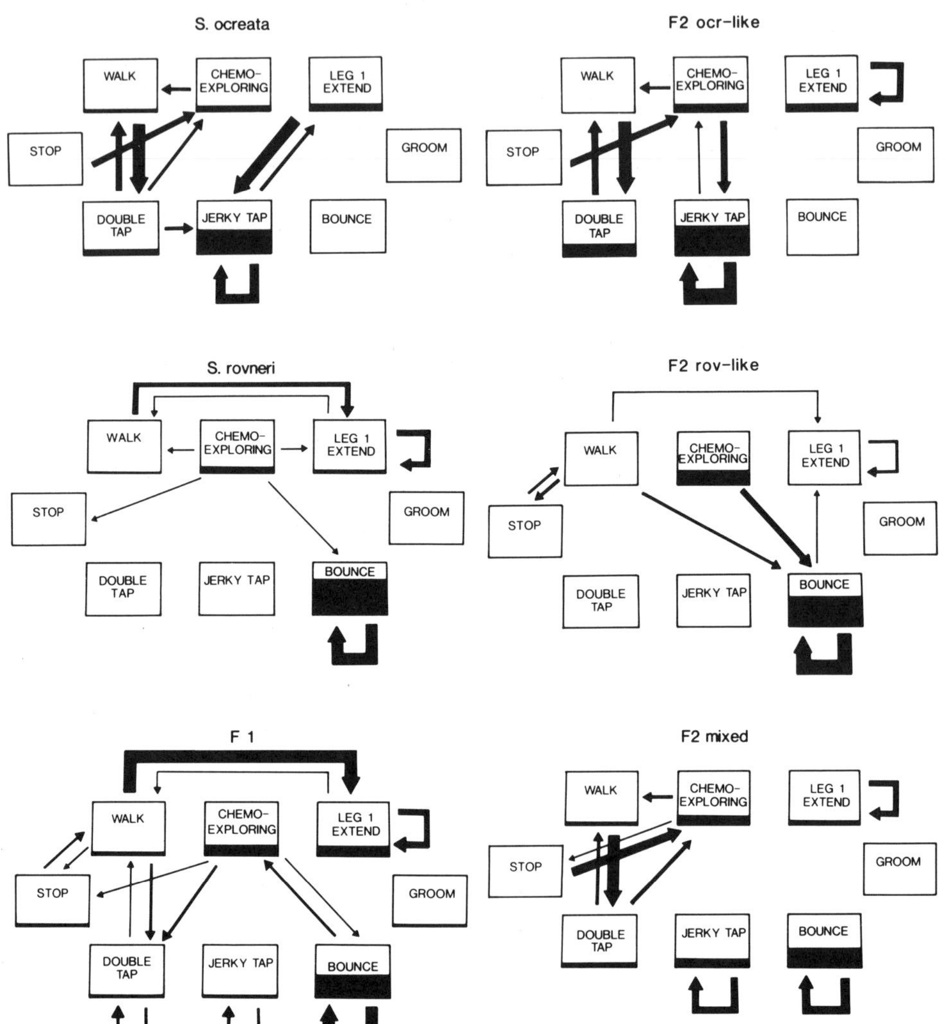

Fig. 5. A comparison of the kinematic diagrams of S. *ocreata*, S. *rovneri*, and F_1 individuals with the kinematic diagrams of the F_2's.

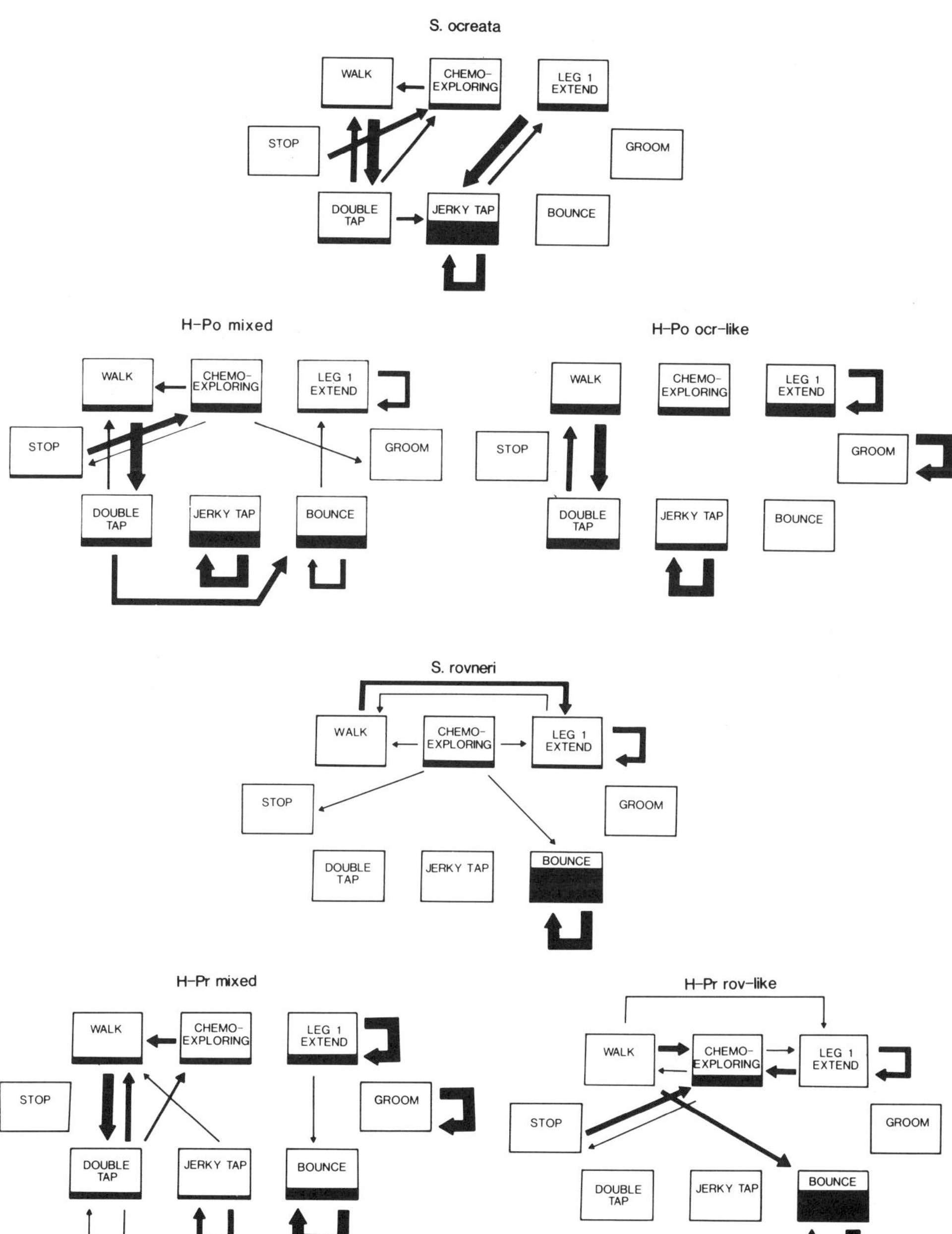

Fig. 6. A comparison of the kinematic diagrams of S. *ocreata*, S. *rovneri*, and backcross individuals.

Discussion

The courtship behaviors of the two species, *S. ocreata* and *S. rovneri*, are very clearly distinctive. Interbreedings and subsequent analysis of the courtship behaviors of the F_1, F_2 and backcross progeny demonstrated that the courtship behaviors are heritable and that they appear to be under the control primarily of very few genes located on the autosomes. The pair-wise comparisons of transition frequency matrices support the model of genetic simplicity. The overall patterns of behavior of the F_2 groups are either like one of the parental species or the other, or like the mixture of behaviors seen in the F_1 group. However, the complexity and variability seen in the kinematic diagrams suggest that the organization of behavior and the orchestration of sequences of movements are anything but simple. This is not surprising given that an individual can vary from day to day in being "primed" from courtship (Crane 1949), and differences can occur in the proximal stimuli, such as stimulus "strength." Other studies of behavioral genetics in arthropods have found similar mixtures of elements that can be isolated and explained by simple genetic models and other elements that are complex and cannot be explained so easily. Complexity is seen in the nonsexual behaviors such as walk and run, and in the transitions between these behaviors and those involved with the display movements. It is suggested that the overall pattern of display in courtship is under strong genetic influence. However, the variation seen in these other behaviors suggests that other factors, such as the proximal stimuli and situation of a male (i.e., strength of stimulus, or "priming" of male), may greatly affect the patterns of behavior.

Courtship behaviors and the tufts of bristles present in mature males of *S. ocreata* appear to be the main distinguishing features between these two species. They are sympatric and, while there appear to be slight differences in their habitat preferences, in several instances they have been collected in the same area. *Schizocosa ocreata* appears to prefer upland deciduous forest leaf litter, while *S. rovneri* has been collected primarily from floodplain leaf litter of major river systems. How and when these species diverged and how the premating isolating mechanisms originated are a matter of speculation.

The courtship behaviors of the two species appear to be most efficient in each species' preferred habitat. That is, the bounce of the *S. rovneri* courting male is louder than the stridulations of *S. ocreata*. The packed substrate of a floodplain could possibly carry sound better than the loose leaves normally found in upland litter typical of *S. ocreata* habitat. Thus, an *S. rovneri* male typically remains in one location while courting and avoids walking into a hungry and unsolicited female. The male *S. rovneri* will only approach a female when she turns, showing the receptive behavior (Stratton and Uetz 1983). A male of *S. ocreata* probably cannot project the sound (stridulation) very far and, thus, must move about more than *S. rovneri* to insure that he comes in contact with a female. The bristles on the forelegs of the mature *S. ocreata* male may help a female of *S. ocreata* make a quick visual assessment of the species of the courting male. This perhaps would lessen the risk to the courting male, although these bristles also may be important in agonistic interactions between males (Aspey 1977).

Two populations of an ancestral species could have been isolated geographically at some point, allowing subsequent adaptation to slightly different environments or habitats. If the two populations were separated by habitat (for example floodplain *vs.* upland leaflitter) as well as geography (preventing gene flow), each population could have evolved slightly different courtship signals that were more appropriate (efficient) in each habitat. For example, the jerky tap of *S. ocreata* has one element that is very similar to the bounce of *S. rovneri.* If by a genetic change, some individuals developed an accentuated "bounce-like" movement and could signal females at a greater distance as a result (and thereby not walk into a hungry unsolicited female), the bounce of the *S. rovneri* male could have been selected for. A genetic change in the female's preference could likewise have selected for certain elements in the courtship of males. This model assumes that *S. ocreata* is the ancestral species, or is more similar to the ancestral species than is *S. rovneri*. Their divergence also may have occurred sympatrically, but only if habitat isolation were very strong. This is similar to the "divergence-adaptation" models presented by Muller (1942) and Templeton (1981).

Lande (1981) suggests that rapid evolution can result from an interaction of natural and sexual selection with random genetic drift. He models genetic mechanisms and shows how rapid speciation can occur by sexual selection and divergence of secondary sexual characteristics. It seems possible that sexual selection could be operating in this system in a way similar to that described by the Muller model (or the Divergence-Adaptation model of Templeton). The female, selecting against the males that do not communicate effectively, could provide strong selection in this instance.

The Fisher (1930) and Dobzhansky (1940) model suggests that when isolated populations come into secondary contact, hybrid inviability or sterility results because of accumulated genetic differences. In this classical model of speciation, Fisher and Dobzhansky suggested that premating isolating mechanisms would arise to reinforce the postmating isolating mechanisms already in existence. The reinforcement of postmating mechanisms can be rejected for these wolf spiders, as the major differences between the species lie in the courtship behaviors.

Reproductive isolation could originate in a founder population, with differences subsequently being promoted by sexual selection (Kaneshiro 1980, Carson 1978). Could one of the species (for example *S. rovneri*) be a result of a founder group becoming isolated (for example, on islands in the floodplain of a major river)? A change in very few genes appears to be sufficient to explain the major observable differences between the males of the two species (the tufts of bristles and the courtship behaviors). The genetic basis of the receptivity of the female does not appear to be linked to the courtship behaviors of the male (Stratton 1982). This divergence between the species may be explained by a genetic change in what signals the female preferred and subsequent intense sexual selection for particular mates. Thus, with either founder effects, or adaptation-divergence, sexual selection appears to be theoretically important and possibly the means by which divergence in these two species has occurred.

In summary, the courtship behaviors of the two species, *S. ocreata* and *S. rovneri*, are distinctive and normally function to keep the species reproductively isolated. Interbreeding between the species was made possible by a "forced copulation" method. Analysis of the courtship behaviors of F_1, F_2 and backcross progeny showed that the inheritance of these behaviors is under the control of a single gene or gene complex. The origin of the premating isolating mechanisms may be explained by either an initial habitat separation between the two groups, or by a founding event, with each group subsequently diverging in slightly different habitats. It is suggested that the differences in the microhabitats, coupled with strong female preference for certain courtship signals, could account for the observed species differences.

Acknowledgments

This research is taken from a dissertation submitted in partial fulfillment of the requirements for the Ph.D. degree at the University of Cincinnati. This research was supported by funds from the University of Cincinnati Research Council, the University of Cincinnati Department of Biological Sciences, and the Society of Sigma Xi. We thank T. C. Kane, M. C. Miller, D. L. Meyer, D. Jackson, T. Jones, C. Meininger, N. Folino, T. Bultman, S. Hartsock, M. Benton and W. Maddison for various assistance during the research. Thanks are extended to Drs. M. Huettel, J. S. Rovner, S. Riecher, and C. Dondale for reading earlier drafts of the manuscript. Thanks also to T. Smith for redrawing the figures and to S. Bauerle for typing the manuscript.

Literature Cited

Aspey, W. P. 1977. Wolf spider sociobiology: I. Agonistic display and dominance-subordinance relations in adult male *Schizocosa crassipes*. Behaviour 62: 103.

Bentley, D. R. 1971. Genetic control of an insect network. Science 174: 1139.

Bentley, D. R. and R. R. Hoy. 1972. Genetic control of the neuronal network generating cricket (*Telegryllus gryllus*) song patterns. Anim. Behav. 20: 478.

Carson, H. L. 1978. Speciation and sexual selection in Hawaiian *Drosophila*. *In* P. F. Brussard, ed. Ecological Energetics: the Interface. Springer-Verlag, New York.

Cowling, D. E., and B. Burnet. 1981. Courtship songs and genetic control of their acoustic characteristics in sibling species of the *Drosophila melanogaster* subgroup. Anim. Behav. 29: 924.

Crane, J. C. 1949. Comparative biology of Salticid spiders at Rancho Grande, Venezuela. Part IV. An analysis of display. Zoologica 34: 159.

Dobzhansky, T. 1940. Speciation as a stage in evolutionary divergence. Amer. Natur. 84: 312.

Dondale, C. D. and J. H. Redner. 1978. Revision of the Nearctic wolf spider genus *Schizocosa* (Araneida: Lycosidae). Can. Entomol. 110: 143.

Ewing, A. W. 1969. The genetic basis of sound production in *Drosophila pseudoobscura* and *D. persimilis*. Anim. Behav. 17: 555.

Fisher, R. A. 1930. The Genetical Theory of Natural Selection. Clarendon Press, Oxford.

Grula, J. W. and O. R. Taylor, Jr. 1980a. Some characteristics of hybrids derived from the sulfur butterflies, *Colias eurytheme* and *C. philodice*: phenotypic effects of the X-chromosome. Evolution 34: 673.

Grula, J. W. and O. R. Taylor, Jr. 1980b. The effect of X-chromosome inheritance on mate-selection behavior in the sulfur butterflies, *Colias eurytheme* and *C. philodice*. Evolution 34: 688.

Kaneshiro, K. Y. 1980. Sexual isolation, speciation and the direction of evolution. Evolution 34: 347.

Lande, R. 1981. Models of speciation by sexual selection and polygenic traits. Proc. Natl. Acad. Sci. USA 78: 3721.

Leonard, S. H. and J. M. Ringo. 1978. Analysis of male courtship patterns and mating behavior in *Brachymeria intermedia*. Ann. Entomol. Soc. Amer. 71: 817.

Manning, A. 1961. The effects of artificial selection for mating speed in *Drosophila melanogaster*. Anim. Behav. 9: 82.

Mayr, E. 1963. Animal Species and Evolution. Harvard University Press, Cambridge, Massachusetts.

Muller, H. J. 1942. Isolating mechanisms, evolution and temperature. Biol. Symp. 6: 71.

Rovner, J. S. 1972. Copulation in the lycosid spider (*Lycosa rabida* Walckenaer): A quantitative study. Anim. Behav. 20: 133.

Rovner, J. S. 1975. Sound production by Nearactic wolf spiders: A substratum coupled stridulatory mechanism. Science 190: 1309.

Salmon, M. and G. W. Hyatt. 1979. The development of acoustic display in the fiddler crab *Uca pugilator* and its hybrid with *U. panacea*. Mar. Behav. Physiol. 6: 197.

Slater, P. J. B. 1973. Describing sequences of behavior. *In* Basteson, P. P. G. and P. H. Klopfer, eds. Perspectives in Ethology. Plenum Press, New York.

Stratton, G. E. 1982. Reproductive Behavior and Behavior Genetics in *Schizocosa* Wolf Spiders. Ph.D. Dissertation. University of Cincinnati.

Stratton, G. E. and G. W. Uetz. 1981. Acoustic communication and reproductive isolation in two species of wolf spiders. Science 214: 575.

Stratton, G. E. and G. W. Uetz. 1983. Communication via substratum-coupled stridulation and reproductive isolation in wolf spiders (Araneae: Lycosidae). Anim. Behav. 31: 164.

Tauber, C. A. and M. J. Tauber. 1977. A genetic model for sympatric speciation through habitat diversification and seasonal isolation. Nature 268: 702.

Tauber, C. A., M. J. Tauber, and J. R. Nechols. 1977. Two genes control seasonal isolation in sibling species. Science 197: 592.

Templeton, A. R. 1981. Mechanisms of speciation—a population genetics approach. Ann. Rev. Ecol. Syst. 12: 23.

Tietjen, W. J. 1978. Tests for olfactory communication in 4 species of wolf spiders (Araneae: Lycosidae). J. Arachnol. 6: 197.

Uetz, G. W. and G. Denterlein. 1979. Courtship behavior, habitat and reproductive isolation in *Schizocosa rovneri* Uetz and Dondale (Araneae: Lycosidae) J. Arachnol. 7: 121.

Uetz, G. W. and C. D. Dondale. 1979. A new wolf spider in the genus *Schizocosa* (Araneae: Lycosidae) from Illinois. J. Arachnol. 7: 86-87.

Wood, D., J. M. Ringo, and L. L. Johnson. 1980. Analysis of the courtship sequences of the hybrids between *Drosophila melanogaster* and *Drosophila simulans*. Behav. Genet. 10: 459.

The Effect of Successive Founder Events on Mating Propensity of *Drosophila*

J. M. Ringo

Department of Zoology
University of Maine
Orono, Maine 04469

Introduction

Nonrandom mating within and between experimental populations of *Drosophila* has been observed many times (Spieth and Ringo 1983). Deviations from randomness can take many forms, but in experiments testing pairs of populations under very simple conditions, three such deviations are measured most commonly. These are deviations from (1) equal mating frequencies of the two types of female, (2) equal mating frequencies of the two types of male, and (3) rates of intertype *vs.* intratype matings expected under a random model. The first two are equivalent to unequal female mating propensities and unequal male mating propensities; the third is either positive or negative assortative mating. The evolutionary implications of genetic variation for the three components of mating are profound. Assortative mating can affect the amount of inbreeding and level of heterozygosity within populations and the amount of gene flow between populations (perfectly positive assortative mating between populations is equivalent to complete sexual isolation). Variation in the mating success of different male genotypes within a population is the basis of sexual selection acting upon males. Although little considered by theoreticians or experimentalists, variation in the mating rates of different female genotypes would seem to play as important a role in intersexual selection as sexual responsiveness to particular male traits (e.g., sexual preferences).

Most of the experimental populations to which I just alluded have been either selected artificially (e.g., del Solar 1966, Burnet and Connolly 1974, Alahiotis and Kilias 1982) or inbred, e.g., Maynard Smith 1956, Averhoff and Richardson 1974. Powell (1978) measured mating frequencies among caged populations subjected periodically to bottlenecks and therefore to genetic drift. His experiments were designed specifically to test an hypothesis of speciation articulated by Carson (1970): species, particularly insular species, can undergo cladogenesis as a by-product of "flush-crash cycles" (exponential growth followed by a bottleneck). Lines carrying genetic markers or cytologically scoreable gene arrangements have been studied also (Kaul and Parsons 1965, Spiess et al. 1966). The most striking interpopulation differences in male mating propensity have occurred in inbred lines, outcrossed males usually outperforming inbred males (Spiess 1970). Variation in female mating propensity has been considered explicitly in few studies.

The experiments described here are part of a larger study designed to monitor the development of reproductive isolation between experimental populations. Motivated by Powell's (1978) work, I subjected one set of populations to successive flush-crash cycles. Using *Drosophila pseudoobscura* Frolova, Powell (1978) measured sexual isolation after the initial founder event, and then after three flush-crash cycles. I used *Drosophila simulans* Sturtevant and measured sexual isolation after each of six flush-crash cycles following the initial founder event. Unexpectedly, I observed marked, increasing variation among lines in female as well as male mating propensities. Here I describe the extent and nature of interline variation in mating propensities; results dealing with reproductive isolation will be reported elsewhere.

Materials and Methods

A genetically heterogeneous base population (BASE) was established by pooling 69 wild-caught stocks of *D. simulans* from recent worldwide field collections (Wood and Ringo 1982). The stocks were from Africa, North America, South America, Australia, England, Hawaii, Japan, and New Zealand. No chromosomal inversion polymorphism was detected in the stocks. Eight drift lines were derived from single pairs chosen randomly from BASE late in its exponential growth phase, three to four generations after interstock crosses. I use the term "drift" because these lines experienced population size bottlenecks and, consequently, inbreeding (Falconer 1981). Thereafter the drift lines and BASE were kept in large cages as in Powell (1978) except that every 6 months (every 10-12 generations) each drift line was passed through a bottleneck ($n = 2$) by choosing a breeding pair at random. The adult population size of BASE was fairly constant at about 8000, except for a reduction to about 1400 following a brief mite infestation. In the stationary phase, each drift line also was maintained at about $n = 8000$.

Two to six generations after each bottleneck, multiple choice mating tests (Spieth and Ringo 1983) were performed between each pairwise combination of the nine lines, as in Powell (1978) except that 24 3-day-old flies were used in each replicate test, six of each sex-line combination. In this way, six sets of data with 36 mating tests per data set were collected. Replicates with no matings were discarded. Flies were identified by clipping the distal margins of both wings. The effects of clipping were analyzed for each sex of each data set.

Indices of mating propensity and assortative mating (Spieth and Ringo 1983) were calculated for each of the 216 mating tests. I call the index of assortative mating Y_A after Yule who invented the statistic (Yule 1912, cited by Bishop et al. 1975). Let the number of matings between type i females and type j males be x_{ij}; then $a = (x_{11})(x_{22})/(x_{12})(x_{21})$ and $Y_A = (\sqrt{a} - 1)/(\sqrt{a} + 1)$; $-1 \leq Y_A \leq +1$. The null hypothesis $Y_A = 0$, may be tested conservatively by $x^2 = (\ln a)^2/\Sigma\Sigma(1/x_{ij})$, which is compared with χ^2 with 1 df (Bishop et al. 1975). Female and male mating propensity for a type were simply the percentage of matings for that type, significance testing being performed with the usual Pearson statistic. Finally, a subset of the mating tests containing all those utilizing BASE was analyzed by a log likelihood procedure, the data being cast into a four-way table (8 lines X 2 female types X 2 male types X 6 times [postbottleneck measurements]) and fitted to a saturated model (Bishop et al. 1975).

To illustrate the method of analyzing single mating tests, consider this example from Powell's (1978) study. He observed the following numbers of matings in a test between lines 1 and 2 (listing the female type first): 1x1, 43: 1x2, 5: 2x1, 12: 2x2, 11. The cross product ratio is $a = (43)(11)/(5)(12) = 7.88$. The index of assortative mating is $Y_A = (2.81 - 1)/(2.81 + 1) = .475$; $x^2 = [\ln(7.88)]^2/(1/43 + 1/5 + 1/12 + 1/11) = 10.73$, which is significant at better than the .005 level. Line-1 females had a significantly higher mating propensity (48/71 = .68) than line-2 females (0.32); $x^2 = 8.80$. Similarly, line-1 males had a higher mating propensity (55/71 = .78) than line-2 males (.22); $x^2 = 21.42$.

Results

The effects of wing clipping were, for the most part, negligible. For the six sets of data, i.e., following the six successive flush-crash cycles, only females in test 4 showed a significant effect (1046 matings by clipped females: 955 matings by unclipped females, $x^2 = 4.14$, $p < .05$). This deviation was ignored, since it was relatively small and since the counterbalanced experimental design overcame the problem of a skewed ratio.

Several significant interactions appeared in the log likelihood analysis (Table 1). The biggest interactions were between females and lines and between males and lines, owing to large variation among lines, both in female and male mating propensities. Significant FLT and MLT terms indicate that these effects changed from bottleneck to bottleneck. Overall, too, there was either positive or negative assortative mating (FM term), but its strength did not vary among lines or times. Time dependent differences between the drift lines and BASE for the three factors, female, male, and line, are measured by the FT, MT, and LT terms; of these, only the MT term was significant.

Table 1. Analysis of mating frequencies in multiple choice tests between BASE and eight drift lines after six successive bottlenecks. Likelihood ratio G^2 values measure partial association between variables. F= female, M= male, L= line, T = time (postbottleneck test).

Effect	df	G^2	p
FM	1	7.9	.005
FL	7	185.3	<.005
FT	5	9.4	.09
ML	7	33.0	<.005
MT	5	11.3	.046
LT	35	47.9	.07
FML	7	11.1	.14
FMT	5	1.8	.88
FLT	35	54.8	.02
MLT	35	51.4	.04
FMLT	35	51.1	.04

The significant four-way interaction urges one to subdivide the data set and to perform further analysis. Two biologically reasonable approaches would be to "slice" the four-way table by times into six subtables, or by lines into eight subtables. I chose the latter alternative to emphasize line differences, particularly differences in bottleneck-dependent change (Table 2).

Table 2. Further analysis of mating frequencies in tests between drift lines and BASE: drift lines compared. Abbreviations are the same as those in Table 1. Each drift line was named after a famous evolutionary biologist.

Line	Effect	df	$G\pm$	p
CHETVERIKOV	FM	1	1.8	.19
	FT	5	7.5	.19
	MT	5	0.9	.97
	FMT	5	0.6	.99
DOBZHANSKY	FM	1	2.4	.12
	FT	5	16.2	.01
	MT	5	4.2	.52
	FMT	5	2.4	.79
FISHER	FM	1	1.3	.26
	FT	5	7.9	.16
	MT	5	12.0	.04
	FMT	5	7.4	.19
HALDANE	FM	1	8.2	.01
	FT	5	5.0	.42
	MT	5	4.1	.53
	FMT	5	2.7	.74
LINNAEUS	FM	1	2.4	.12
	FT	5	10.3	.07
	MT	5	8.7	.12
	FMT	5	16.1	.01
MAYR	FM	1	0.3	.58
	FT	5	2.8	.73
	MT	5	16.4	.01
	FMT	5	3.8	.58
WALLACE	FM	1	1.4	.24
	FT	5	13.6	.02
	MT	5	11.1	.05
	FMT	5	11.7	.04
WRIGHT	FM	1	0.04	.84
	FT	5	1.9	.86
	MT	5	4.8	.44
	FMT	5	8.2	.14

Several new points emerge after subdivision of the data by lines. First, the overall FM interaction (assortative mating) stems mainly from mating tests using HALDANE and BASE, in which Y_A is consistently >1, significantly so after bottleneck 2. The overall assortative mating in the data set is reduced (but still significant) after removing crosses with HALDANE. Second, female mating propensity is significant in mating tests using LINNAEUS and WALLACE. This is a reflection of significant assortative mating with LINNAEUS after bottleneck 6 and with WALLACE after bottleneck 3, and nonsignificant assortative and disassortative mating at other times. Third, FT is significant in DOBZHANSKY and WALLACE, owing to a decline in female mating propensity of the two drift lines. Fourth and last, MT is significant in FISHER, MAYR, and WALLACE, owing to a decline in male mating propensity in the latter two lines and lower male mating propensity of FISHER after bottlenecks 2, 3, 5, and 6.

A similar analysis of Powell's (1978) data was performed, except that time (bottleneck number) was not a factor, only one set of mating tests having been reported (Table 3). The significant FM effect was caused by assortative mating with line 1. The significant FL interaction was in large measure due to the higher female mating propensity of line 7, relative to ORIGIN (Powell's equivalent of my BASE).

A second approach, simply to calculate the three mating indices for each test, yielded a similar amount of variation (Table 4) but made line differences clearer for female and male mating propensity. In Powell's (1978) experiment there was a higher proportion of nonrandom mating in tests not involving ORIGIN than in those using it, for the three components of mating considered here. He observed much more assortative mating and less variation among lines for female mating propensity than I, but the interline variation in male mating propensity was about the same in the two experiments.

Discussion

One usually thinks of genetic drift as random change in allele frequencies resulting from chronically small population size. However, the effect of bottlenecks on genetic variance are potentially very large, so that a population alternating between long periods of large size and short periods of small size must suffer inbreeding and, therefore, must lose genetic variation. The generations with the smallest numbers have the largest effect (Falconer 1981) because the mean rate of inbreeding is the harmonic mean of the rates for all generations (approximately $1/(2N + 1)$ for dioecious species). Other factors can reduce genetic variation in a population,

Table 3. Analysis of mating frequencies between drift lines and ORIGIN in an experiment utilizing *Drosophila pseudoobscura* (Powell 1978). Abbreviations are as in Table 1.

Effect	df	G^2	p
FM	1	5.4	.02
FL	7	14.8	.04
ML	7	4.9	.67
FML	7	7.1	.42

Table 4. The extent of nonrandom mating summarized for analyses of all mating tests. Data from Powell (1978) are included for comparison.

Tests after bottleneck no.	Number of nonrandom indices[1]		
	AM[2]	FMP[2]	MMP[2]
1	1+	17	7
2	1+,3−	21	10
3	2+	22	14
4	1+	26	14
5	3+	23	15
6	3+	27	21
POWELL (4)	11+	9	18

[1] χ^2 tests, $p < .05$; on a random basis, 1 or 2 out of each set of 36 to be "significant."

[2] AM = positive (+) or negative (−) assortative mating; FMP = female mating propensity; MMP = male mating propensity.

but considering only population size, the mean cumulative loss of genetic variation was theoretically 81% in the drift lines and nil in BASE.

Drosophila simulans was chosen for the experiment because it is monomorphic chromosomally, the life cycle is relatively short, it is easy to handle in mating experiments, and both sexes discriminate against the sibling species, *D. melanogaster* Meigen (Schilcher and Dow 1977, Wood and Ringo 1980). The first characteristic is important because a purely genic origin for any reproductive isolation among lines that might develop was of interest. The last factor seemed important because reproductive isolation might develop more easily in a species showing interspecific sexual discrimination in both sexes. Powell (1978), using *D. pseudoobscura*, established a chromosomally polymorphic ORIGIN (base population), leaving open the possibility that inversion-associated genes might have been involved in sexual isolation.

Powell's (1978) observation of strong positive assortative mating between many flush-crash lines of *D. pseudoobscura* differs strikingly from my results, but one can only speculate about the causes of the difference. It may be that there are basic differences between the two species in the genetic architecture of mate preference or discrimination (the presumed cause of assortative mating). Alternatively, there may be interspecific variation in rates of recombination. Templeton (1980), who posited such variation without explanation and apparently in the absence of evidence, pointed out that the greater the rate of recombination between interacting genes affecting fitness, the greater the chance that the frequency of a particular combination will change during a founder-flush-crash episode. He predicted a positive relationship between the number of chromosomes and the rate of recombination. *Drosophila melanogaster* and *D. simulans*, then, with only three major chromosomes, should develop reproductive isolation as a by-product of drift at a lower rate than *D. pseudoobscura*, with four major chromosomes, while *D. virilis* Sturtevant, with five major chromosomes, should develop reproductive isolation at an even faster rate, all other things being equal. Unless one of these explanations is correct, one is forced to attribute the different outcomes of the two experiments to trivial causes such as the density of flies used in the mating test, or some other environmental parameter which there is no *a priori* reason to suspect as being significant biologically.

The hypothesis that genetic drift has an important role in speciation remains evergreen, although experimental evidence accumulated to date in constitutively outcrossing species casts into serious question Carson's (1970) contention that reproductive isolation can develop quickly enough to be detected in laboratory or field experiments. Also, there are theoretical reasons for doubting the idea. Charlesworth and Smith (1982), modeling speciation via founder effects, determined that 50-100 generations per flush, few generations in the bottleneck, and moderate population size ($n \sim 6$) in the bottleneck were important in producing appreciable reproductive isolation; the rate of recombination had a significant but smaller effect.

Other models of speciation have been reported, but do not seem to be directly related to my experiment. Lande (1981) developed a mathematical model in which sexual selection and genetic drift acted in concert to produce reproductive isolation. Sved (1981) implemented a computer model of the Wallace effect — selection for assortative mating between populations whose hybrids have reduced viability or fertility. Felsenstein (1981) developed a three-gene model in which there are two subpopulations with different selection regimes for two genes that affect fitness and interact epistatically; a third, linked gene influences assortative mating between the two populations, which are in free contact. The latter two models, then, do not deal with the development of reproductive isolation allopatrically, as a by-product of some genetic process. On the other hand, the similarity of results with mating propensities in Powell's (1978) and my experiments leads one to suppose that random fixation of alleles may frequently lead to decreased mating rates. There was a correlation between the overall mating frequencies of males and females within lines for each postbottleneck set of mating tests (r ranged from .76 to .89, df $= 7$, $p < .05$), in line with Spieth's (1951) proposal that female and male mating propensities are "balanced" in natural populations. My results also were concordant with Sharp's (1982 and personal communication) experiment with *D. melanogaster*, in which the mean male mating propensity (really, competitive male mating ability as measured in a specially designed female choice experiment) of inbred lines declined linearly with the inbreeding coefficient, *F*. No experiment employing a similarly precise measure of female mating propensity has been performed.

There is reason to believe that male mating propensities would have varied much more under natural conditions, as might have female mating propensities. Sharp (1982) found that males of inbred lines mated as often as males from outcrossed lines when equal numbers of both sexes were present in a female choice experiment, but that outcrossed males outcompeted inbred males when 25 males of each type were placed with 25 females at the beginning of the experiment, another 25 females being added 2 h later. In nature, the ratio of sexually mature males to virgin females is even larger than 2 : 1 (Spieth 1968), so that sexual selection would be even more intense. There are other potentially important differences between nature and virtually all mating experiments using *Drosophila* reported in the literature. Drosophilids in nature typically mate at feeding and oviposition sites, where adults of all ages congregate; mature males engage in courtship persistently but copulations rarely occur (Spieth 1968). Also, under natural conditions males have an opportunity to learn, thus altering rates of courtship directed toward virgin females *vs.* males or mated females (Siegel and Hall 1979); flies are not disturbed before courtship (Robertson 1982); females mate when younger than in experiments; and there is an opportunity for territories to be established (Jacobs 1978, but see Ringo et al. 1983). What kind of biases in measurement these factors introduce into experiments deserves serious consideration in future research, even though the technical difficulties would be formidable.

Powell's (1978) and my experiments do not give any information about behavioral mechanisms. For example, male courtship in low male propensity lines may be less vigorous or less persistent, or males may be slow to initiate courtship; this appears not to be the case, but an experiment to test the idea is in progress. It is known, for example, that inbred males of *D. subobscura* Collin are less vigorous and persistent than outcrossed males (Maynard Smith 1956). Genetic causes of differences in male or female mating propensities also are unknown. Genetic analysis of the lines would be useful for shedding light on the genetic components of interline variation (Fulker 1966). It also might prove instructive to derive and analyze chromosome lines from natural populations of *D. melanogaster* and *D. pseudoobscura*, say, as Brittnacher (1981) has done for male "virility: (male mating propensity in female choice experiments).

Both experiments have an important implication for theories of sexual selection. Intrasexual selection (for example, occurring in male "scramble" for females) and intersexual selection (for instance, occurring when females discriminate between males) would be enhanced by a higher threshold of acceptance (Manning 1967), i.e., lower mating propensity. As Templeton (1980) points out, species undergoing genetic drift via founder-flush-crash cycles (Carson 1975) would retain large amounts of genetic variation. Genetic variation for male sexual behavior, that has been documented for natural populations (Wood and Ringo 1982) could then allow for sexual selection to act. This is a particularly attractive idea, since so many conspicuous examples of sexual selection occur in island species (Darwin 1871, Carson et al. 1970), which have often started from small founding populations. Specifically, I suggest that the triggering of the "runaway" process of sexual selection described by Fisher (1958), acting principally upon males by the agency of females, is enhanced by genetic drift. In this way, genetic drift may have an even greater role in evolution than has been thought previously.

Acknowledgments

The research was supported by NSF grant DEB 7903866, and by a grant from the University of Maine. Dave Wood was indispensable, tirelessly helping with data collection and stock maintenance. Rocky Rockwell ran the log likelihood program and helped immensely in the interpretation of the analyses. Jeff Hall graciously provided necessary facilities during the writing, while I was visiting his laboratory. Francisco Ayala, Don Gailey, Jeff Hall, Jeff Powell, Paul Sharp, Eliot Spiess, and Herman Spieth made helpful comments on the manuscript.

Literature Cited

Alahiotis, S. N., and G. Kilias. 1982. Mating propensities and variations in enzyme activities in long-term cage populations of *Drosophila melanogaster*. J. Hered. 73: 53.

Averhoff, W. W., and R. H. Richardson. 1974. Pheromonal control of mating patterns in *Drosophila melanogaster*. Behav. Genet. 4: 207.

Bishop, Y. M. M., S. E. Fienberg, and P. W. Holland. 1975. Discrete Multivariate Analysis: Theory and Practice. The MIT Press, Cambridge.

Brittnacher, J. G. 1981. Genetic variation and genetic load due to the male reproductive component of fitness in *Drosophila*. Genetics 97: 719.

Burnet, B., and K. Connolly. 1974. Activity and sexual behavior in *Drosophila melanogaster*. *In* J. H. F. van Abeelen, ed. The Genetics of Behavior. American Elsevier, New York.

Carson, H. L. 1970. Chromosome traces and the origin of species. Science 168: 1414.

Carson, H. L. 1975. Genetics of speciation at the diploid level. Amer. Natur. 109: 83.

Carson, H. L., D. E. Hardy, H. T. Spieth, and W. S. Stone. 1970. The evolutionary biology of the Hawaiian Drosophilidae. *In* M. K. Hecht and W. C. Steere, eds. Essays in Evolution and Genetics in Honor of Theodosius Dobzhansky. Appleton-Century-Crofts, New York.

Charlesworth, B., and D. B. Smith. 1982. A computer model of speciation by founder effects. Genet. Res. Camb. 39: 227.

Darwin, C. 1871. The Descent of Man and Selection in Relation to Sex. Burt, New York.

del Solar, E. 1966. Sexual isolation caused by selection for and against positive and negative phototaxis and geotaxis in *Drosophila pseudoobscura*. Proc. Nat. Acad. Sci. USA 56: 484.

Falconer, D. S. 1981. Introduction to Quantitative Genetics. 2nd ed. Longman, London.

Felsenstein, J. 1981. Skepticism towards Santa Rosalia, or why are there so few kinds of animals? Evolution 35: 124.

Fisher, R. A. 1958. The Genetical Theory of Natural Selection. Dover Publications, New York.

Fulker, D. W. 1966. Mating speed in male *Drosophila melanogaster*: A psychogenetic analysis. Science 153: 203.

Jacobs, M. E. 1978. The influence of beta-alanine on mating and territorialism in Drosophila melanogaster. Behav. Genet. 8: 487.

Kaul, D., and P. Parsons. 1965. The genotypic control of mating speed and duration of copulation on *Drosophila pseudoobscura*. Heredity 20: 381.

Lande, R. 1981. Models of speciation by sexual selection on polygenic traits. Proc. Nat. Acad. Sci. USA 78: 3721.

Manning, A. 1967. The control of sexual receptivity in female *Drosophila*. Anim. Behav. 15: 239.

Maynard Smith, J. 1956. Fertility, mating behaviour, and sexual selection in *Drosophila subobscura*. J. Genet. 54: 261.

Powell, J. R. 1978. The founder-flush speciation theory: An experimental approach. Evolution 32: 465.

Ringo, J., M. K. Kananen, and D. Wood. 1983. Aggression and mating success in three species of *Drosophila*. Z. Tierpsychol. 61: 341.

Robertson, H. M. 1982. Female courtship summation in *Drosophila melanogaster*. Anim. Behav. 30: 1105.

Schilcher, F., and M. Dow. 1977. Courtship behaviour in *Drosophila*: Sexual isolation or sexual selection. Z. Tierpsychol. 43: 304.

Sharp, P. M. 1982. Competitive mating in *Drosophila melanogaster*. Genet. Res. Camb. 40: 201.

Siegel, R. W., and J. C. Hall. 1979. Conditioned responses in courtship behavior of normal and mutant *Drosophila*. Proc. Nat. Acad. Sci. USA 76: 3430.

Spiess, E. B. 1970. Mating propensity and its genetic basis in *Drosophila. In* M. K. Hecht and W. C. Steere, eds. Essays in Evolution and Genetics in Honor of Theodosius Dobzhansky. Appleton-Century-Crofts, New York.

Spiess, E. B., B. Langer, and L. D. Spiess. 1966. Mating control by gene arrangements in *Drosophila pseudoobscura*. Genetics 54: 1139.

Spieth, H. T. 1951. Mating behaviour and sexual isolation in the *Drosophila virilis* species group. Behaviour 3: 105.

Spieth, H. T. 1968. The evolutionary implications of sexual behavior in *Drosophila*. Evol. Biol. 2: 157.

Spieth, H. T., and J. M. Ringo. 1983. Mating behavior and sexual isolation in *Drosophila. In* The Genetics and Biology of *Drosophila*. Vol. 3c. Academic Press, London.

Sved, J. A. 1981. A two-sex polygenic model for the evolution of premating isolation. I. Deterministic theory for natural populations. Genetics 97: 197.

Templeton, A. R. 1980. The theory of speciation via the founder principle. Genetics 94: 1011.

Wood, D., and J. M. Ringo. 1980. Male mating discrimination in *Drosophila melanogaster, D. simulans*, and their hybrids. Evolution 34: 320.

Wood, D., and J. M. Ringo. 1982. Artificial selection for altered wing display in *Drosophila simulans*. Behav. Genet. 12: 449.

Yule, G. U. 1912. On the methods of measuring association between two attributes. J. Royal Statist. Soc. 75: 579.

Genetic and Sensory Basis of Sexual Selection in *Drosophila*

Therese Ann Markow

Department of Zoology
Arizona State University
Tempe, Arizona 85287

Introduction

Years of research on mating behavior have revealed that matings rarely occur at random between individuals in the same population. Why are some individuals more successful than others when competing for mates? What factors account for the particular combinations of types we observe in mating pairs? Current evolutionary theory explains mating patterns on the basis of maximization of fitness on the part of each sex and most species that have been investigated provide at least indirect support for this hypothesis. With *Drosophila*, it is possible to design controlled experiments that can answer directly questions about the relationship between mating behavior and fitness, and discover those factors that produce the nonrandom associations we observe.

Recent investigations in our laboratory have uncovered a number of examples in which male mating advantage is associated with superior fertility. Markow et al. (1979) reported that among *Drosophila melanogaster* Meigen males of the same strain and age, individuals that were temporarily less fertile due to previous copulations were less successful than virgin males when competing for females. When enough time had elapsed to restore fertility, this disadvantage disappeared. Since both types of males appear to court with equal vigor, it could be argued that females perceive them to be different qualitatively and actually choose the virgin male, thereby increasing their own productivity from a given mating. In another study, 4-day-old *D. melanogaster* males were found to give more progeny on a single mating than 2-day-old males of the same strain (Long et al. 1980). Despite the apparent absence of differences in courtship vigor, 4-day-old males had an advantage over 2-day-old males when competing for females. Also, in this situation it is tempting to conclude that female choice is operating.

The explanation of female choice in the above examples assumes that the competing males are providing different sensory cues. It follows that in the absence of such information, it should be impossible for females to make their choices properly and observed mating patterns should approach randomness with respect to male fertility. On the other hand, if male courtship vigor, rather than female choice, is the determinant of success, the absence of any contrasting sensory cues should make little difference to the success of the more fertile males under competitive conditions.

These predictions have been tested in my laboratory with *D. melanogaster* using sensory deficient mutants and males that differ in fertility due to known and controlled factors. We have conducted a series of experiments in which males of contrasting fertility competed for mates. Females either were wild type or had various sensory deficiencies.

Methods

Fly strains

Two wild type strains, Oregon-R (OR) and Canton-S (CS), were obtained from the *Drosophila* Stock Center at the California Institute of Technology. Another wild type strain, TM4, was collected in Tempe, AZ in May 1980 by the author. The mutant *norpA*^{P24} was obtained from Dr. William Pak, Purdue University, *sbl* from Dr. Jeffrey Hall, Brandeis University, and the double mutant *al th* (aristaless, thread) was constructed from strains from the California Institute of Technology and Bowling Green State University stock center. All strains were reared in half pint milk bottles containing standard cornmeal molasses agar medium at 24 ± 1 °C on 12 h L/12 h D cycle.

Male competition experiments

A single female was placed in a vial with two males. The female was either TM4 wild type, *norpA*^{P24} (a blind mutant), *sbl* (smell blind, olfactory deficient), or *al th* (impairment of sound perception). The vials were observed and a record was kept of the first male to begin courting, courtship latency of each male, courtship intensity and successful male ("winner"). After mating, the thorax lengths of both the winner and loser were measured in order to compare body size. In competition experiments conducted with TM4 females, all females mating with winner males were saved, transferred individually to fresh vials daily, and their progeny counted. The unpreferred "loser" male was immediately mated with another TM4 female (no choice or competition involved) which was also saved to determine progeny numbers.

Four competition experiments were performed. Females used in the first three were TM4; in the last, mutant. In experiment one, 4-day-old males of the Canton-S strain and 4-day-old Oregon-R males were used. In the second, 4-day-old CS males were in competition with 2-day-old CS males. In experiment three, single females were placed with CS virgin males and CS males that had mated twice in the hour immediately preceeding the choice test. Both males were 4 days old. Finally, in the fourth experiment, two males were chosen at random from a pool of virgin 4-day-old TM4 males. Each of the four competition experiments was conducted with females that were blind (*norpA*^{P24}), olfactory deficient (*sbl*) or auditory defective (*al th*) as well as with wild type TM4 females.

Results

In Table 1, we report the outcome of the first three competition experiments in which TM4 females were used. Canton-S males were twice as successful as OR males. In similar experiments using females from other wild type strains, CS males consistently were more successful than OR (unpublished observation). The data on virgin *vs.* mated males and 4- *vs.* 2-day-old males confirmed earlier findings from this laboratory (Markow et al. 1979, Long et al. 1980). When the number of progeny was compared (Table 2), the winner males were seen to produce more offspring than the losers.

In view of the finding that successful males were more fertile than unsuccessful males, we determined whether the successful male of two taken at random from an outbred strain showed a higher fertility than the loser. Among competing TM4 males of the same age and mating status, the winner was significantly more fertile than the loser (Table 2).

In seeking the biological and behavioral basis for the apparent correlation between success and fertility, these competition experiments were repeated and the courtship latency of each male as well as his thorax length were measured. Table 3 reports the type of male that

Table 1. Results of three competition experiments. (1) Four-day-old males from the CS and OR wild type laboratory strains competed for TM4 wild type females. (2) Four-day-old CS virgin males and 4-day-old males having mated twice previously competed for TM4 virgin females. In all experiments, a single female was placed in a vial with the two males. (3) Four-day-old virgin and 2-day-old virgin males from the CS laboratory strain competed for TM4 females.

Kinds of males		Male mating		x^2	
A	B	A	B	(1:1)	p
Canton-S	Oregon-R	85	49	9.672	<0.01
Virgin	Mated (1 h)	115	65	13.88	<0.01
4-Day	2-Day	42	24	4.91	<0.05

Table 2. Average number of offspring produced by successful and unsuccessful males. In experiments 1, 2, and 3, average progeny produced by each class are compared. In experiment 4, average progeny of successful and unsuccessful individuals are compared.

Experiment	Males		Average Number of Progeny		
number	A	B	A	B	*p*
1	Canton-S	Oregon-R	342.11 ± 16.99	267.2 ± 18.5	<.01
2	4-Day	2-Day	202.85 ± 9.98	128.13 ± 5.23	<.001
3	Virgin	Mated (1 h)	288.0 ± 17.4	206.0 ± 21.0	<.01
4	TM4 (W)†	TM4 (L)†	438.0 ± 22.1	361.0 ± 19.2	<.01

†W = winner, L = loser.

courted first, size of males of both types and type of male that was successful. First to court (or shortest latency to courtship) and the amount of time spent courting were positively correlated in each case, therefore only first to court is reported in the table. When TM4 females were used, there was no difference between CS and OR males for being the first to initiate courtship. Thorax lengths of CS and OR males were similar and yet CS males were significantly more successful in gaining matings. Four-day-old males were more successful than 2-day-old males but there were no significant differences between them in which male was the first to court or for the size of thorax. The same was true of virgin and mated males. Virgin males were significantly more successful than mated males but were, as was expected, the same size and also were equal in courtship initiations. Only with TM4 males were there differences between the "winners" and "losers" in courtship latency and body size. Winners tended to court first and were larger. The data show that successful males in the first three competitions

Table 3. Four male competition experiments conducted with females from the wild type strain TM4 or with sensory deficient females having *norpA*^{P24} (blind), *sbl* (smell-blind), and *al th* (auditory interference). In experiment number 1, Canton-S and Oregon-R males competed for females. In experiment 2, 4- and 2-day-old males were in competition. Virgin and mated males competed for females in experiment 3 and in experiment 4, pairs of virgin males were selected at random from the TM4 wild type strain and competed for females. The data on thorax lengths are presented to give the average sizes of the actual winners, regardless of their strain or treatment. The thorax lengths for flies of each treatment (A or B) appear one time in parenthesis under the values for winners and losers in the experiments using TM4 females. The average thorax length for each strain and treatment did not change over the course of the experiments.

Exp. no.	Type male and *n* of each type mating per total A	B	First to court A		B	Thorax length W (A)		L (B)
			TM4 Females					
1	CS (34/46)	*OR (12/46)	22		24	26.41 (26.13)		26.32 (26.03)
2	4-Day (24/30)	*2-Day (6/30)	18		12	25.95 (26.17)		26.11 (25.90)
3	V (33/51)	*M (17/51)	23		29	25.78 (26.02)		25.93 (25.84)
4	TM4-W	TM4-L	41	*	8	28.69	*	27.83
			norpA Females					
1	CS (23/39)	OR (16/39)	23		16	25.56		25.97
2	4-Day (24/40)	2-Day (16/40)	22		18	26.43		26.01
3	V (29/43)	*M (14/43)	20		23	25.86	*	26.00
4	TM4	TM4	23	*	12	28.73	*	27.66
			sbl Females					
1	CS (16/32)	OR (16/32)	16		16	26.96	*	25.69
2	4-Day (19/41)	2-Day (22/41)	21		20	27.01	*	25.88
3	V (20/37)	M (17/37)	19		18	26.89	*	25.81
4	TM4	TM4	29	*	16	28.66	*	27.29
			al th Females					
1	CS (26/33)	*OR (7/33)	16		17	26.18		25.99
2	4-Day (29/38)	*2-Day (9/38)	21		17	26.19		26.07
3	V (28/40)	*M (12/40)	19		21	26.29		26.01
4	TM4-W (37)	TM4-L (37)	27	*	10	28.51	*	27.14

*$p < .05$; comparison of A and B

were no larger than unsuccessful males nor did they court sooner or more intensely. While body size and courtship intensity were associated with success for TM4 wild type, random-selected males, there is nothing in the experiments with TM4 females that points to the sensory basis of male success.

The hypothesis that differential success depends upon differences in the particular sensory information that males present to females was tested by depriving females of specific sensory information during competition experiments. Females homozygous for the mutant $norpA^{P24}$ are blind and are, therefore, deprived of visual information about courting males. Canton-S males were still more successful than OR males but the proportion of the time they ''won'' was no longer significant as it was with wild type females. The same was true for 2-and 4-day-old males but not for virgin *vs.* mated males. However, with TM4 males, the winners were still the first to court and were larger.

When competition experiments were conducted with olfactory deficient females (homozygous for *sbl*), a significant change was observed in the outcome of the first three competition experiments. The advantage that one type had over the other disappeared completely. There was still no difference between types for first to court but winners, regardless of type, were those males that courted soonest and courted most intensely. Winners also were larger when females were *sbl*. When females were deprived of olfactory information, differential success of particular types of males disappeared. In the absence of olfactory information, a new system of success determination occurs in which success is associated with short courtship latency, greater courtship intensity, and larger body size.

Finally, these experiments were repeated with *al th* females whose ability to perceive courtship songs was obstructed. Here the results of all four competition experiments resemble the findings when TM4 females were used. When females' song perception is altered, CS males are more successful than OR, older males more successful than young males, virgins more successful than mated males. In the case of TM4 males, the larger males are still the winners.

Discussion

In nature, courtship and mating occur in aggregates of flies found on rotting fruit. Females are exposed to a large number of males that vary in almost every way imaginable: genotype, age, mating status, environmental conditions during development, etc. Much of this variability is correlated with both fertility and the ability to obtain mates. In the experiments described here, males that because of their age, mating status, or genotype were more fertile were most successful also in gaining matings with wild-type females. An important question facing evolutionary biologists is whether sexual selection operates by female choice or simply by differences in male vigor. These experiments suggest that mechanisms of sexual selection may be heterogeneous and depend upon the nature of the differences between competing males in a given situation. In the first three competition experiments, male differences were highly contrived. Using wild type females, however, there was a strong and consistent advantage for males of a given treatment, whether it was strain, age or mating status. That these advantages were dependent upon different olfactory cues is suggested by the change that occurs when males of various treatments competed for *sbl* females. Canton-S males no longer had an advantage over OR, 4-day-old males were no longer more successful than 2-day-old males, and virgin males lost their advantage over nonvirgins. In the absence of the female's ability to detect olfactory cues, male success appears to be dependent upon two correlated traits, larger body size and greater courtship propensity. Therefore, CS and OR males must be providing different olfactory cues to females. The same can be said for old and young males and virgin and mated males. However, when females were blind, we saw also that

CS males lost their significant advantage and so did older males. Although CS and OR males court on the average with equal vigor, and so do 4- and 2-day-old males, a fine structure analysis of their courtship behavior (Markow and Hanson 1981) reveals differences in the sequence of courtship components, and in the stationary probability of occurrence of specific components, some of which have a visual basis.

It is impossible to determine whether wild type females are "preferring" CS males over OR males because of differences in their odors and courtship structure. The olfactory cues and other courtship signals of OR males simply may not be adequate to evoke a receptive state in females. In such a situation, it could appear as though females are choosing CS males when CS males may be winning by default. The same could be true of 4-day-old males and virgin males.

When no age or mating status difference exists among males of a genetically heterogeneous population, male success is correlated with body size and courtship propensity. The question about choice can be asked again here. Are females choosing larger males or are the large vigorous males winning the contests? These two alternatives may not be different when viewed from the standpoint of the females' central nervous system. Certain features of a courting male, whether they be specific sensory cues or overall courtship intensity, may bring a female into a receptive state. In certain contests described above, it was possible to distinguish between the importance of particular sensory modes, for example olfaction *vs.* overall courtship vigor, in producing differential success of certain types of males.

Several evolutionary consequences of these processes are obvious. A system in which female receptivity is stimulated only by a particular constellation of courtship components provides a mechanism for conserving a species specific mate recognition system. At the same time, the tendency for larger males to be more successful may reflect the action of balancing selection. There is some evidence suggesting that within the TM4 strain, larger males may be more heterozygous. When the number of sternopleural bristles on the right and left sides of body were counted, winner males (symmetry index, $L/R = 0.96 \pm 0.02$) showed a significantly higher degree of symmetry than did loser males (symmetry index, $L/R = 0.87 \pm 0.03$, $p < .001$). According to Lerner (1954) and Waddington (1957), bilateral asymmetry results from a poor capacity for developmental homeostasis associated with homozygosity at a large number of loci. If, in fact, larger males are more successful at inseminating females and are relatively more fertile than smaller males, and if both are a funtion of increased heterozygosity, success might act to maintain genetic variability in a population.

A question that appears to be more interesting than whether or not "female-choice" is operating is concerned with the apparent association between success and fertility. Is it simply a fortuitous observation in all four experiments that the males that win contests are also the most fertile? This laboratory is in the process of comparing the relative courtship success and fertility of males from additional inbred strains, varied rearing conditions and additional, more complex treatments.

Acknowledgment

This work was supported by NIH grants NS15263 and GM30638-01. The author thanks Ms. Rosa Tang for her excellent technical assistance.

Literature Cited

Lerner, I. M. 1954. Genetic Homeostasis. Oliver and Boyd, London.

Long, C., P. C. Yaeger, and T. A. Markow. 1980. Relative male age, fertility, and competitive mating success in *Drosophila melanogaster*. Behav. Genet. 10: 163.

Markow, T. A., and S. J. Hanson. 1981. Multivariate analysis of *Drosophila* courtship. Proc. Nat. Acad. Sci. USA 78: 430.

Markow, T. A., M. Quaid, and S. Kerr. 1979. Male mating experience and competitive courtship success in *Drosophila melanogaster*. Nature 276: 821.

Waddington, C. H. 1957. Strategy of the Genes. Allen and Unwin Ltd., London.

Reproductive Isolation in a Neotropical Insect: Behavior and Microbiology

Lee Ehrman

Division of Natural Sciences
State University of New York
Purchase, New York 10577

Norman L. Somerson

Department of Medical Microbiology and Immunology
Ohio State University
Columbus, Ohio 43210

Frederick J. Gottlieb

Department of Life Sciences
University of Pittsburgh
Pittsburgh, Pennsylvania 15260

Introduction

The *Drosophila paulistorum* Dobzhansky & Pavan complex offers an excellent opportunity for the study of the relationship between an insect host and its microbial symbiont. Electron microscopy reveals that all members of this insect group harbor endosymbionts which, by all appearances, are cell-wall deficient organisms (CWD). While it is not known if the CWDs of *D. paulistorum* are essential to their hosts, it is clear that a stable, benign relationship exists between the host and its own CWD. However, our evidence indicates that transfer of an endosymbiont from one semispecies of *D. paulistorum* to another semispecies results in male sterility. The specificity exhibited by the endosymbionts increases reproductive isolation between semispecies and may have contributed to the evolutionary and taxonomic complexity of *D. paulistorum* as a widespread superspecies.

As their names imply, the six known neotropical semispecies of *D. paulistorum*—Centroamerican, Amazonian, Orinocan, Transitional, Andean-Brazilian, and Interior—are identified by their geographical distribution, an extrinsic isolating mechanism in evolution. Three intrinsic isolating mechanisms also are operative: sexual (or behavioral or ethological) isolation, hybrid sterility, and hybrid inviability. Of these isolating mechanisms, the behavioral one is the most effective. This is seen in the few cases where the geographical distribution of two or three semispecies overlap. Even then, the semispecies do not interbreed. When hybrids are produced in the laboratory as a result of crosses between semispecies, regardless of the direction of the cross, the male hybrids are sterile and the female hybrids are fertile. It appears that sterility is a consequence of incompatibilities between cytoplasmically-transmitted infectious agents contributed maternally and genetic material contributed paternally.

Genetics of Sexual Isolation and Hybrid Sterility

Ehrman observed in 1960 that crosses between separate strains of *D. paulistorum* produced hybrid sterility, even though the strains were morphologically indistinguishable from one another. Techniques of using marked chromosomes in controls and crosses are described and referenced comprehensively in Ehrman and Parsons (1981). The genetic basis of the sexual isolation has been studied in the hybrids between the Centroamerican and Amazonian semispecies, among others. These are incipient species with overlapping geographic distributions. They are indistinguishable morphologically yet, when crossed, produce fertile female and sterile male hybrids. Crosses have been made in which the distribution of a certain pair of chromosomes was followed with the aid of mutant genes that served as genetic markers. The sexual preferences were studied in the F_1 hybrids between the semispecies and in a series of backcrosses to each of the parental semispecies. The evidence obtained shows that sexual isolation is controlled by factors distributed on all three pairs of chromosomes that the species possesses. The many genes that control behavior produce additive effects, the sum of which makes the bar to crossing nearly complete between the semispecies. The object was to transfer one marked chromosome into the nuclear and cytoplasmic background of an alien semispecies. Backcross progenies were obtained by crossing F_1 hybrid females carrying suitable marker genes in the third chromosome to males of one or the other of the parental semispecies. The backcrosses were repeated in each of three successive generations, always selecting as female parents the carriers of semispecies-foreign chromosomes. For each combination of parental semispecies, two series of recurrent backcrosses were made to each of these semispecies. The F_1 hybrid females between semispecies A and B obviously contain one A and one B chromosome in each pair. The F_1 hybrid males have the X of their mother, the Y of their father, and an A and a B autosome of each parent. In the backcrosses of the A/B hybrid females to A males, the B chromosomes, except the one with the genetic marker, tend to be replaced gradually because only the one with the mutant genetic marker is selected for: whereas, in the backcrosses to B males, the A chromosomes tend to be eliminated. In the progeny of the third backcross, most of the flies carry chromosomes of one semispecies only, except the foreign chromosome with the genetic marker (and sometimes also a foreign Y chromosome), as itemized in Table 1.

The control crosses involved the use of sisters of the same females used in the experimental series, but not containing the foreign chromosomes with the genetic markers. The progenies were, consequently, like the experimental ones, except that they did not contain the foreign marked chromosome.

In most instances, the foreign chromosome contained only a single mutant gene, which served as a marker. This, nevertheless, was deemed a satisfactory experimental technique for two reasons: (1) Whenever more than one marker was present, crossing over between homologous pairs was found to be suppressed in the intersemispecific crosses; (2) The semispecies involved in these experiments differed in at least one inversion in each of their five chromosome "arms," so that effective exchange between the homologues was reduced.

Though the role of each chromosome was analyzed in more than one interpopulational cross, only a single set of datum will be presented as an example. To test the effects of the foreign third chromosomes on sexual preferences, Amazonian females heterozygous for the dominant marker *Delta* (wing venation) in the third chromosome were crossed to Centroamerican males which carried on one of their third chromosomes the dominant *Minute*

Table 1. Fractional and percentage dilutions of alien chromosomes in repeated backcrosses between semispecies A and semispecies B, where semispecies B is the recurrent parent. See figures in Ehrman (1960) for outlines of all of these interpopulation crosses incorporating mutant marker genes.

	Chromosomes				% A in entire genome	% B in entire genome
	Marked		Unmarked			
	A	B	A	B		
F_1	½	½	½	½	50.0	50.0
BC_1	½	½	¼	¾	33.3	66.7
BC_2	½	½	1/8	7/8	25.0	75.0
BC_3	½	½	1/16	15/16	20.8	79.2
BC_n	½	½	$\frac{1}{2}^{n+1}$	$1-(\frac{1}{2}^{n+1})$	$(0.5+\frac{1}{2}^{n})/3$	$1-[(0.5+\frac{1}{2}^{n})/3]$
			Controls			
BC_2	0	1	1/8	7/8	8.3	91.7
BC_3	0	1	1/16	15/16	4.2	95.8
BC_n	0	1	$\frac{1}{2}^{n+1}$	$1-(\frac{1}{2}^{n+1})$	$(\frac{1}{2}^{n})/3$	$1-[(\frac{1}{2}^{n})/3]$

(bristles) and the recessive *ebony* (body color). In the F_1 generation, the *Delta/Minute-ebony* females were used as progenitors of the backcross progenies, and their sibs were used for tests of the mating preferences of the F_1 hybrids (*ebony* is approximately 50 crossover units from Minute and is used here as a check on the suppression of crossing over in the hybrids). The results are reported in Table 2. The F_1 hybrid females accepted Centroamerican males, while the F_1 hybrid males appeared to be neutral. The backcross progenies show that the sexual preference of the hybrids was for the semispecies of the recurrent parent. It therefore appears that the third chromosome does not by itself determine sexual preferences in these crosses. Indeed in Amazonian x Centroamerican hybrids, sexual preference is decided by which semispecies contributes more than half of the genome, with no one chromosome being clearly more important than the others. Therefore, the sexual isolation analyzed here, in which matings between females and males of the different *D. paulistorum* populations are much less likely to succeed than matings within a population, is apparently controlled by polygenes scattered in every one of the three pairs of chromosomes.

Extensions of these experiments illuminated a fascinating variation of hybrid behavior. F_1 hybrid females from a cross between Andean-Brazilian and Amazonian semispecies were observed to accept no males courting them. Most crosses between these two semispecies fail because of the powerful sexual isolation barrier. However, after repeated and lengthy attempts, viable male and female hybrids were obtained. It should be emphasized that these are normal males and normal females as far as external and internal anatomy are concerned. Yet the genic endowments contributed by the parents of these hybrids are so discordant that the hybrids are virtually unable to perform successfully the mating rituals normal in either parental semispecies.

A study of the behavior of these flies under a microscope in special observation chambers shows that the hybrid females will not accept any males that court them regardless of how vigorous or persistent the courtship is. They have been observed to reject consistently the

Table 2. Direct observation of matings testing the influence of the third chromosome in the genetic architecture of sexual isolation between two strains of *D. paulistorum* (from Ehrman 1961).

	No.	Copulate with CA	Copulate with AM	x^2	P
		Tests of Hybrid Females			
F_1	19	17	2	10.3	<0.01
		Backcrosses to Centroamerican Parent			
BC_1	20	18	2	11.3	<0.01
BC_2	20	19	1	14.5	<0.01
BC_3	20	19	1	14.5	<0.01
		Backcrosses to Amazonian Parent			
BC_1	20	2	18	11.3	<0.01
BC_2	20	1	19	14.5	<0.01
BC_3	20	1	19	14.5	<0.01
		. . . .			
		Test of Hybrid Males			
F_1	20	8	12	0.5	0.70-0.50
		Backcrosses to Centroamerican Parent			
BC_1	20	14	6	2.5	0.20-0.10
BC_2	20	19	1	14.5	<0.01
BC_3	20	19	1	14.5	<0.01
		Backcrosses to Amazonian Parent			
BC_1	20	2	18	11.3	<0.01
BC_2	20	2	18	11.3	<0.01
BC_3	20	2	18	11.3	<0.01

males of both parental semispecies, as well as their own hybrid brothers. They accomplish this by assuming the posture of rejection of the courtship that is characteristic of *D. paulistorum*: the female lowers her head and elevates the tip of her abdomen so that the vaginal orifice is inaccessible to an approaching male.

If populations have diverged genetically, developing different coadapted complexes as a result of having adapted to different environments, then gene exchange between these populations is likely to produce ill-adapted genotypes. Natural selection acts to build and reinforce the barriers to gene exchange between populations whose hybridization results in reproductive wastage. The appearance of hybrids with inferior fitness is, in this way, minimized or avoided altogether. Alternatively, one may assume that reproductive isolation arises as an accidental by-product of genetic divergence. As populations become adapted to different environments, they become different in progressively more and more genes. Reproductive isolation arises because the action of many genes is pleiotropic. Some gene differences selected for different reasons or resulting from random genetic drift may thus have isolating side effects.

However, evidence that natural selection can strengthen reproductive isolation in wild populations comes from multiple-choice experiments using Elens-Wattiaux mating chambers and scoring by direct observation. Joint isolation coefficients were computed for given pairs of semispecies which have been found to occur both sympatrically and allopatrically. In allopatric crosses the average isolation coefficient was +0.85 while in sympatric crosses it was 0.67 (Table 3). Thus, pairs occurring sympatrically exhibit more sexual isolation than the same pairs occurring allopatrically, or, semispecies coexisting geographically tend to be reproductively more isolated than those that do not, which is reasonable since the production of large numbers of hybrids would be very inefficient.

Almost all studies of sexual isolation have utilized young virgin females but work has been completed with aged, experienced females. *Drosophila paulistorum* provides rich material for a survey of the effects of early experience on later mate selection because this superspecies is composed of six semispecies or incipient species, among which many degrees of sexual isolation exist. Then too, females of this superspecies have been shown to mate repeatedly, unlike many insects.

Table 3. Numbers of matings observed and isolation coefficients calculated for sympatric and for allopatric crosses; 1,695 matings were observed (from Ehrman 1965).

Races	Origin	Matings	Coefficient
1. Amazonian x Andrean	Sym	108	0.86 ± 0.049
	Allo	100	0.66 ± 0.074
2 Amazonian x Guianan	Sym	104	0.94 ± 0.033
	Allo	109	0.76 ± 0.061
3. Amazonian x Orinocan	Sym	106	0.75 ± 0.065
	Allo	124	0.61 ± 0.070
4. Andean x Guianan	Sym	109	0.96 ± 0.026
	Allo	102	0.74 ± 0.066
5. Orinocan x Andean	Sym	100	0.94 ± 0.033
	Allo	111	0.46 ± 0.084
6. Orinocan x Guianan	Sym	104	0.85 ± 0.053
	Allo	100	0.72 ± 0.069
7. Centro-American x Amazonian	Sym	102	0.68 ± 0.072
	Allo	103	0.71 ± 0.070
8. Centro-American x Orinocan	Sym	110	0.85 ± 0.052
	Allo	103	0.73 ± 0.069
Average (sympatric) = 0.85			
Average (allopatric = 0.67			

In direct observation of the mating of *D. paulistorum* semispecies females with heterogamic (unlike) or homogamic (like) males, it was demonstrated that aged females' mate selection did not differ significantly from that of young females. Previous heterogamic copulatory experience did not change consistently the degree of sexual isolation; however, females with homogamic copulatory experience showed a significantly higher preference for homogamic males in subsequent tests. A test of the proportions of homogamic matings relative to total matings indicated significant differences between subjects with homogamic experiences and naive ones (9 days old) across all combinations. Hence, we have evidence for the effects of experience. (See Ehrman and Parsons 1981 for details.)

Identifying an Endosymbiont

Ehrman and Williamson (1965) showed that the sterility of intersemispecific *D. paulistorum* hybrid males was transmissible by injection of testicular extracts which gave the pattern for pedigrees seen in Fig. 1. Further studies by Williamson et al. (1971) proved the specificity of testicular extracts for each strain studied. Extracts of male flies of one semispecies injected into females of a different semispecies induced sterility in the male progeny even when these females are fertilized by males of the same semispecies. If the material injected were obtained from hybrid males that had a mother genetically different from the female recipients, the male progeny are, again, sterile, the female progeny fertile. Eight variations of this test are shown in Table 4. The flies used in this work are designated by their points of origin. Mesitas (M) is a strain of the Andean-Brazilian semispecies, and Santa Marta (S) is a strain of the Transitional semispecies.

There are two kinds of reciprocal hybrids, depending on which strain provided the maternal parent. If a homogenate of hybrid males were injected into females of the same semispecies as their mothers (Row 3) the sterility rate was low: 6.7%. If, however, the hybrid male donors were from the reciprocal cross (Row 4) sterility of males was greater than 50%. Material from one semispecies injected into the same semispecies resulted in low sterility rates for male progeny as shown in Rows 1 and 5. The specific sterility factor of a semispecies is shown to be transmitted through the females while affecting males exclusively. Whether the recipient semispecies was Mesitas (Rows 1-4) or Santa Marta (Rows 5-8) the same pattern of sterility was observed.

All *D. paulistorum* males and females examined contain cell wall deficient (L-form) microbial forms, Kernaghan and Ehrman (1970a,b) treated female flies with antibiotics and found that hybrid sterility could be partially and temporarily alleviated, but no *D. paulistorum*, male or female, has been found to be totally free of L-forms. The ubiquity of the endosymbionts and the lack of a symbionet-free semispecies hampers studies of the individual characteristics of each species-specific L-form. The L-forms are incorporated into embryonic germ primordia and are transmitted vertically through generations. While each semispecies possesses its own L-form in a benign relationship, there is a rapid proliferation of the L-forms in the testes of hybrid males and a concomitant breakdown of normal spermiogenesis (Ehrman and Kernaghan 1971, 1972). Daniels and Ehrman (1974) showed a direct relationship between the presence of numerous L-forms in the testes of males and the maternally transmitted sterility of hybrid males.

Attempts to culture the L-forms on artificial media failed, an alternate insect host was sought. Gottlieb et al. (1977) injected *D. paulistorum* extracts into larvae of *Ephestia (=Cadra) kuehniella* Zeller, the Mediterranean meal moth. The injections killed the larvae. Extracts of the dying larvae were pathogenic when injected into healthy *Ephestia* larvae or into adult *D. paulistorum*. In the two types of recipients, both time of onset and frequency of

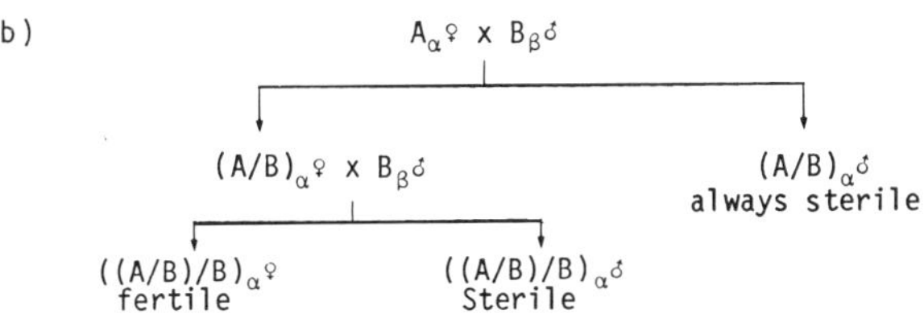

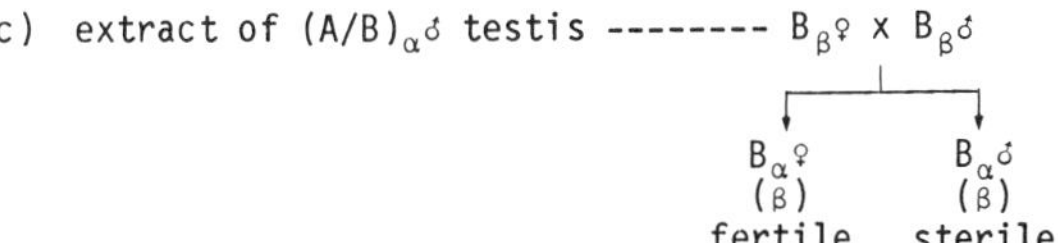

Fig. 1. Flow diagram of the passage of infectious agent in *D. paulistorum.* (a) Results of crosses within each of two semispecies; (b) results of a cross between semispecies and of tests of the resulting hybrids; (c) passage of infectious agent into nonhybrid females of the paternal semispecies (B), as a result of the injection of an extract of testes of sterile hybrid males. Key: "A" and "B" represent two different semispecies of *D. paulistorum;* "A/B" represents a hybrid from a cross between an "A" female and a "B" male; "and" represent the infectious agent resident in semispecies "A" and semispecies "B," respectively. From Gottlieb, Simmons, Ehrman, Inocencio, Kocka, and Somerson 1981.

Table 4. Percent sterility in male progeny of Mesitas (M) or Santa Marta (S) *Drosophila paulistorum* females injected with homogenates from males of the same strains or their hybrids (From Williamson et al. 1971).

	Recipient semispecies	Material injected	Sterility of males (%)
1.	M	Fertile M	3.3
2.	M	Fertile S	80.0
3.	M	Sterile F_1 (M ♀ x S ♂)	6.7
4.	M	Sterile F_1 (S ♀ x M ♂)	60.0
5.	S	Fertile S	6.7
6.	S	Fertile M	53.3
7.	S	Sterile F_1 (S ♀ x M ♂)	6.7
8.	M	Sterile F_1 (M ♀ x S ♂)	6.00

death are dose dependent. Larval *Drosophila* recipients of infected *Ephestia* extracts may survive to pupation but most of them die during the pupal stage. Surviving fertile adult females demonstrated the same pattern of sterility among their progeny as is found in hybrids of *D. paulistorum* semispecies and in *D. paulistorum* recipients of *D. paulistorum* testicular extracts. The L-forms retain their specificity for their host semispecies of *D. paulistorum* even when passed through *Ephestia*.

The use of *E. kuehniella* led to the first successful culturing of the L-forms outside of *D. paulistorum* and made possible the study of the sterility inducing agent in an alternative host. However, *Ephestia* appears to harbor its own symbiont. Gottlieb (1972), using electron microscopy, observed numerous inclusions in *Ephestia* larval forms, adult testis sheath, and other pigment cells. The *Ephestia* symbiont proved to be harmless to *Drosophila* and is not at all implicated in the induction of sterility in hybrids.

Symbiotic microbes are commonly found to be essential for the survival of their insect hosts and are often characterized by cell-wall deficiency, according to Mattman (1974). Therefore, a major effort was directed toward the isolation of Mollicutes (including mycoplasmas, acholeplasmas, and spiroplasmas). Our culture media were adjusted to isolate (L-forms). An SP4 formulation had been used successfully by Tully et al. (1977) for the isolation of spiroplasmas, and as a superior medium for culturing *Mycoplasma pneumoniae* from human throat specimens (1979). We used a semisolid SP4 medium (0.05% agar concentration) and a solid medium containing Noble agar at a final concentration of 1.0%. The Hayflick formulation was also employed. Primary cultivation of L-forms was attempted in media supplemented with 7% sorbitol, as shown in Figure 2.

We have isolated L-forms from three *D. paulistorum* semispecies. Using the procedure outlined in Fig. 2, we observed colonies after three weeks of incubation in an atmosphere of 10% CO^2 and 90% nitrogen. The colonies had a "fried egg" morphology, i.e., a translucent periphery and a denser central core. Their surfaces showed a "lacy" character with vacuolated peripheral areas. No cell walls were seen in electron microscopic study of the colonies. The individual colonies resembled L-forms more than mycoplasmas in their morphology and growth on medium of elevated osmolarity. This first isolate from Mesitas was designated FM1. A duplicate experiment was performed with an inoculum subjected to a freeze-thaw cycle as a bacterial decontamination measure. No L-phase isolates were obtained from that experiment.

Following the primary isolation, we had a great deal of difficulty subculturing the FM1 colonies and, for several months, were in danger of losing the isolate. Within 5 weeks of the first appearance of the FM1 colonies, we had tried over 60 subcultures on various combinations of liquid or agar media formutations adjusted for osmolarity, antimicrobials, sugar concentrations, and incubation atmosphere. In broth transfers or by the "push block" agar method, growth was increased when the sorbitol concentration was increased to 9%. The FM1 colonies will grow on trypticase soy agar containing 9% sorbitol, or even better, with 12% sucrose instead of sorbitol. Later, we found that omission of yeast extract, yeast autolysate, and fetal bovine serum improved growth. It became clear that FM1 was not a *Mycoplasma* species.

On electron microscopic examination of FM1, neither cell walls nor cell-wall material were seen. All of our evidence indicated that the colonies were L-forms and we had isolated a microbe that had no direct relationship to mycoplasmas. The FM1 isolate was subcultured for more than ten passages on antibiotic-free medium and maintained its colonial morphology. (No L-form has been isolated from the *Drosophila* food used in the routine handling and breeding of flies at SUNY-Purchase.) In experiments performed to obtain symbiont-free flies,

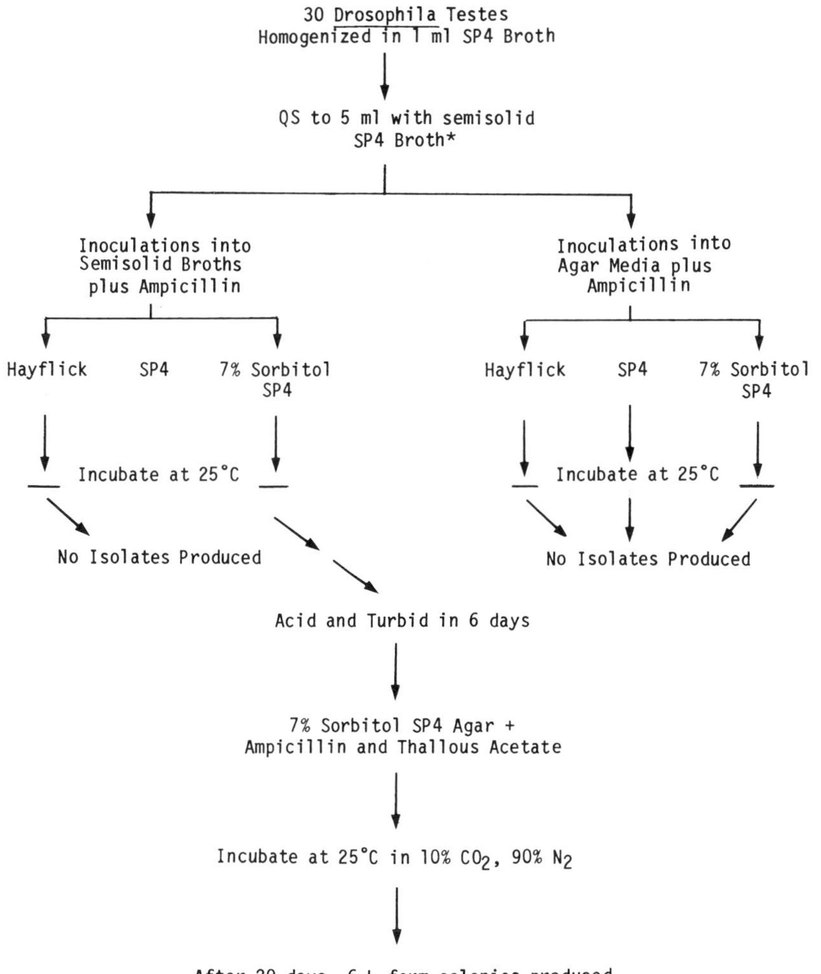

Fig 2. Isolation of an L-phase variant (FM1) from Mesitas testes.

we used Mesitas *D. paulistorum* eggs warmed to 28.5 °C and subjected to refrigerated tetracycline buffered with ascorbic acid. Of over 700 eggs so treated, about 4% survived. We crossed tetracycline-treated Mesitas females with Santa Marta males. In the preliminary results, it appeared that the symbionts had been eliminated successfully since the hybrid male progeny were fertile. However, we were able to isolate an L-form from these flies and labeled it TET1. Apparently, tetracycline, a bacteriostat, suppresses the symbiont, but does not eliminate it.

Other L-form Isolates

Santa Marta females injected with FM1 from Mesitas yielded sterile male offspring from which we recovered an L-form organism. Because of our experience in cultivating the original FM1 isolate, we were able to re-isolate colonies within 11 days. SP4 agar medium containing ampicillin, thallous acetate, and 9% sorbitol, but omitted yeast products and serum. This re-isolate was an attempt to verify the microorganism's role in hybrid sterility and to show retention of semi-species specificity. The re-isolate was labeled FM1-RA, and had the characteristics of the FM 1 isolates..

We also isolated L-forms from both the Centroamerican and Transitional semispecies. The L-forms named C2 were isolated from a preparation containing Centroamerican testes only; L-forms named SM were obtained from a homogenate of three whole Santa Marta flies.

An isolate was obtained from the moribund larvae of *Ephestia* that had been injected with *Drosophila* testicular extract. The microbe, labeled EK1, was cultured under circumstances similar to those of the other isolates. But unlike FM1 and FM1-RA, EK1 showed greater growth with serum and yeast products incorporated into the culture medium. Colonial morphology of this particular organism seemed distinct from the other two isolates.

Conclusion

One may represent the processes of race divergence and speciation by the following sequence of events.

1. Expansion of the distribution area occupied by a species, leading to formation of allopatric populations more or less isolated by distance from one another. Or else, secular changes taking place in the environments of the different parts of the geographic area which the species inhabits may make its populations face novel conditions.

2. Genetic divergence of the populations leading to formation of allopatric (geographic) races. This divergence occurs through natural selection and random genetic drift, making the populations different in the incidence of gene alleles, chromosomal variations, or both in their gene and chromosome pools. The genetic differences may be adaptive to the environmental conditions in which the population live.

3. Integration of the diverging corporate genotypes of the races (populations). Because of the epistatic interactions between the constituent genes in different populations, or simply because of accumulation of genetic differences which make the populations specialized to exploit different environments, mixed or compromise genotypes become disadvantageous. Gene exchange between the populations produces mostly or only genotypes which cause inferior fitness in their carriers.

4. Natural selection favors situations which minimize or eliminate the gene exchange between the divergent populations. These selection processes involve the appearance of rarely a single one, more commonly of a group of several, reproductive isolating mechanisms. Very often none of these isolating barriers alone is sufficient to make the isolation complete but they do so in combination.

5. Most of the processes of race divergence described here, and of formation of isolating mechanisms, take place while the populations are allopatric or only slightly overlapping in their geographic distributions. Reproductively isolated species may continue to be more or less strictly allopatric, but perhaps more usually these distribution regions will overlap, making them sympatric in more or less extensive territories. It should be noted that although sympatric coexistence does not necessarily require the reproductive isolation to be complete, it

must be strong enough to make the gene exchange between the species populations not exceed the rate at which the diffused genes can be eliminated by natural selection.

The *D. paulistorum* complex seems to be between stages four and five. The host-symbiont relationship, while not the primary isolating mechanism in the *D. paulistorum* complex, certainly fostered fragmentation of a relict, once unified species into semispecies. More importantly for our purposes, the presence of endosymbionts confers certain properties on their hosts. Possession of a L-form that shows the same semispecies specificity as observed with the endosymbionts, offers opportunities to characterize the microbe and, perhaps, to show its effects in *Drosophila* and other organisms. Patterns of association and the levels of integration of the partners provide valuable clues to (i) the history of the coevolution of the symbionts and (ii) the way their relationships have shaped metabolic and behavioral interactions.

Acknowledgments

The work described above was greatly aided by the professional expertise and moral support extended by Jack Kocka at Ohio State University; and Bertha Inocencio Green at SUNY-Purchase. We are most grateful for their contributions.

Literature Cited

Daniels, S., and L. Ehrman. 1974. Embryonic pole cells and mycoplasma-like symbionts in *Drosophila paulistorum*. J. Invert. Path. 24: 14.

Ehrman, L. 1960. The genetics of hybrid sterility in *Drosophila paulistorum*. Evolution 14: 212.

Ehrman, L. 1961. The genetics of sexual isolation in *Drosophila paulistorum*. Genetics 46: 1025.

Ehrman, L. 1965. Direct observation of sexual isolation between allopatric and between sympatric strains of the different *Drosophila paulistorum* races. Evolution 19: 459.

Ehrman, L., and R. P. Kernaghan. 1971. Microorganismal basis of infectious hybrid male sterility in *Drosophila paulistorum*. J. Hered. 62: 67.

Ehrman, L., and R. P. Kernaghan. 1972. Infectious heredity in *Drosophila paulistorum*. CIBA Foundation Symposium: "Pathogenic Mycoplasmas," Associated Scientific Publishers, Amsterdam.

Ehrman, L., and P. A. Parsons. 1981. Behavior Genetics and Evolution. McGraw-Hill, New York.

Ehrman, L., and D. L. Williamson. 1965. Transmission by injection of hybrid sterility to nonhybrid males in *Drosophila paulistorum*: Preliminary report. Proc. Natl. Acad. Sci. USA 54: 481.

Gottlieb, F., 1972. A cytoplasmic symbiont in *Ephestia (=Anagasta) kuehniella*: Location and morphology, J. Invert. Path. 20: 351.

Gottlieb, F. J., R. Goitein, L. Ehrman, and B. Inocencio. 1977. Interorder transfer of mycoplasma-like microorganisms between *Drosophila paulistorum* and *Ephestia kuehniella*: Tissues, dosages, and effects. J. Invert. Path. 30: 140.

Gottlieb, F. J., G. M. Simmons, L. Ehrman, B. Inocencio, J. Kocka, and N. Somerson. 1981. Characteristics of the *Drosophila paulistorum* male sterility agent in a secondary host, *Ephestia kuehniella*. Appl. Environ. Microbiol. 42: 838

Ignoffo, C. M. 1973. Effects of entomopathogens on vertebrates. Annals N.Y. Acad. Sci. 217: 141.

Kernaghan, R. P., and L. Ehrman. 1970a. An electron microscopic study of the etiology of hybrid sterility in *Drosophila paulistorum*. I. Mycoplasma-like inclusions in the testes of sterile males. Chromosoma 29: 291.

Kernaghan, R. P., and L. Ehrman. 1970b. Antimycoplasmal antibiotics and hybrid sterility in *Drosophila paulistorum*. Science 169: 63.

Mattman, L. H. 1974. Cell Wall Deficient Forms. CRC Press, Cleveland, Ohio.

Tully, J. G., D. L. Rose, R. Whitcomb, and R. P. Wenzel. 1979. Enhanced isolation of *Mycoplasma pneumoniae* from throat washings with a newly modified culture medium. J. Infect. Dis. 139: 478.

Tully, J. G., R. Whitcomb, H. F. Clark, and D. L. Williamson. 1977. Pathogenic mycoplasmas: Cultivation and vertebrate pathogenicity of a new spiroplasma. Science 195: 892.

Williamson, D. L., L. Ehrman, and R. P. Kernaghan. 1971. Induction of sterility in *Drosophila paulistorum*: Effect of cytoplasmic factors. Proc. Natl. Acad. Sci. USA 68: 2158.

Opportunities for Selection Derived from Variation in Mating Success in Milkweed Beetles (*Tetraopes tetraophthalmus:* Cerambycidae)

David E. McCauley

Department of General Biology
Vanderbilt University
Nashville, Tennessee 37235

Introduction

One approach to understanding the evolution and maintenance of the various mating systems found in invertebrates is to document the selective mechanisms generated by a given system. One would want to identify those behavioral and life-history characteristics correlated with mating success and then document the amount of phenotypic variation for those characters found to be important in a given population. Ideally, the estimates of individual mating success should be based on lifetime studies. In order to understand the evolutionary dynamics generated by such selection one would then need to estimate the heritability and genetic covariance of the various traits that contribute to the operation of selection. Heritability estimates are often difficult to obtain except in the laboratory for logistical reasons and many of those estimates are of limited applicability to field populations. It is well known that an estimate of heritability is only valid for the environment in which it is measured since the amount of environmentally derived phenotypic variance is important in determining heritability. Environmental effects on the phenotype can differ between laboratory and field and from field population to field population. Separation of the behavioral and ecological processes of selection from the genetical information required to predict a response to selection allows us to understand some important aspects of the dynamics of microevolution through field studies, even when logistical considerations prevent genetic analysis of the population. Wade and Arnold (1980) suggest the use of the concept of "opportunities for selection" when quantifying selection generated by variable mating success in males. This opportunity for selection is calculated as the variance in numbers of mates obtained by males divided by the square of the mean number of mates per male. This statistic allows for the quantification of the potential for selection generated by a given distribution of mating successes, even when the genetical information necessary for predicting a response to selection is lacking.

Studies of invertebrate, particularly insect, mating systems focus frequently on phenotypic variation that contributes to variation in mating success. Such attributes as body size (Mason 1964, McCauley and Wade 1978, Thornhill 1980), horn size (Eberhard 1982, Brown and Siegfried 1983) and behavioral repertoires (Cade 1981, Thornhill 1981) contribute to variation in male mating success. Typically, these studies follow individual male-male or male-female interactions, classify individuals as successful or not at obtaining a mate during that particular interaction, and relate the probability of success to the phenotype. Rarely do studies of insects quantify variation in lifetime mating success in natural populations. Both episodic and lifetime measures of selection pressures can be used to provide information on the opportunity for selection in a given insect system. Often these studies seek to differentiate the operation of intra- and intersexual selection or identify variable mating strategies. One question rarely addressed is the effect of variable survivorship by males during the reproductive season on variation in the number of mates accrued during a lifetime. Since life-history theory often assumes or predicts a trade-off between survivorship and reproductive effort this information should be of general interest.

This is a study of selection pressures generated within the mating system found in natural populations of the cerambycid milkweed beetle *Tetraopes tetraophthalmus* (Forster). It seeks to identify those behavioral, morphological, and life-history characteristics whose variation is responsible for variation in the lifetime mating success of adult male beetles. It is a study of opportunities for selection in natural insect populations based on variance in fitness as measured by mating success. In particular, it contrasts the variation among males in the ability to court and defend mates with variation in the survivorship of males over the reproductive season. Variation in both characteristics contributes to overall variation in lifetime male mating success. Selection resulting from variation in courtship and mate defense is often classified as sexual selection; selection generated by differential survivorship is often classified as natural selection (Arnold 1983).

The System

Tetraopes tetraophthalmus is a cerambycid beetle whose chief host plant is the common milkweed, *Ascelpias syriaca* (L.) Nearly the entire life cycle of the insect is tied to the one host plant species. There is one generation per year. Adults can be found feeding on the leaves and flower buds of the host plant from June to mid-August, depending on locality. Mating occurs daily at the feeding sites and is observed easily. Typically, about 35% of the adult population is found to be mating at any one time. Both sexes mate periodically throughout an adult lifespan that ranges up to 6 weeks in duration. Mason (1964) has shown that males of intermediate body size are more likely to be found in copula than odd sized males. Fighting between males, especially in the presence of females, can be observed. Females oviposit clutches of 10-20 eggs into the hollow stems of grasses that grow in association with milkweed. Oviposition also occurs periodically, perhaps daily, over the adult life of the female. The eggs hatch in about ten days and the larvae make their way into the ground where they locate the milkweed rhizomes on which they feed. Larval feeding lasts until late autumn at which time the larvae leave the rhizome and form a cell in the soil. Pupation occurs in this cell in the late spring.

Adult *T. tetraophthalmus* are rather poor fliers and disperse very little from their emergence sites (McCauley et al. 1981, Lawrence 1982, McCauley 1983). Their sedentary nature, local abundance, discrete generations, and frequent copulations make them ideal for the study of variation in mating success in natural populations. Individual beetles can be identified by a marking system at the beginning of the adult flight period and observed repeatedly throughout the season. Thus, there is the opportunity to document variation in the lifetime mating success of males and correlate this to various behavioral, morphological or life-history attributes of individuals.

Most field studies to be presented here were conducted in the vicinity of the Mountain Lake Biological Station in Giles County, southwest Virginia. In a typical study, a large milkweed field would be gridded off into 10 m x 10 m squares prior to the onset of adult emergence. As adults were found, they were given an individually identifying mark, based on a combination of colors and mark locations, of spots of enamel based paint. The sex, size and point where originally captured would be noted for each individual. Censuses of the study site would then be conducted daily until the end of the adult flight season. The location and behavior (mating status) were noted for all marked individuals located at each census.

Some laboratory studies will be presented to supplement the field data. Adult *T. tetraophthalmus* can be maintained easily in 305 x 305 x 610 mm wire cages available from BioQuip. They will feed and copulate readily when provided with cut milkweed stalks maintained in flasks of water. In a typical experiment, some number of marked field-caught adults were introduced into a cage at some experimentally determined combination of population density and sex ratio. The cages were then observed periodically and behaviors such as mating, fighting, and oviposition noted. Field and laboratory techniques are provided in greater detail in McCauley (1983).

Male Fighting Behavior

Mating success as determined by the aggressive interactions of competing males is a classical component of sexual selection. Adult male *T. tetraophthalmus* can be seen fighting with one another, particularly in the presence of females. In a typical fight, a single male will approach a mating pair and attempt to dislodge the mating male from the female. If successful, he then engages the newly single male in a pushing match that consists of head butting and locking of mandibles. Eventually, one male will back off and fall or fly from the plant. One hundred and two fights were observed in the field from initiation to conclusion. Of these, 77 involved single males confronting mating pairs. In 37 of the 77 fights (48%) the single male could not break up the mating pair and left eventually. In 47% he succeeded in dislodging the mating male from the female but the female left the area before either male could re-mate with her. In 5% of the fights the intruding male dislodged the mating male and actually went on to mate with the newly single female. Thus, since over 50% of fights resulted in a mating male being removed from a female, variation in fighting behavior could contribute to variation in lifetime mating success.

Fighting is important in this system but what are the sources of any variation in fighting behavior? Males vary in body size (Mason 1964, McCauley 1979) and there could be a correlation between body size and fighting ability. Additional fights were observed in the field and this time the participants were captured and measured with an ocular micrometer in the lab. Winners were declared to be those mating males that resisted being driven from the plant by aggressive single males or those single males that were successful at dislodging and driving away mating males. The relative sizes of the winners and losers of 28 fights are presented in Figure 1. A paired comparisons *t*-test reveals the mean size of the winners to be significantly greater than that of the losers ($t = 3.61$, 26 d.f., $P < 0.01$). Differential fighting ability results, in part, from variation in body size. McCauley (1982) presented this work in greater detail as well as some corroborating laboratory studies.

Lifetime Mating Success

One source of variable mating success would be variation among males in fighting ability. That fighting ability contributes to the probability that a male will mate at any given time does not document, however, the extent to which there is variation among males in lifetime mating success. In order to quantify this, one has to follow a cohort of males through their reproductive lives. In a species such as *T. tetraophthalmus*, where mating is possible at any time over an adult lifespan of several weeks, lifetime mating success could be influenced as much by adult longevity as by the ability of males to find, court, and defend females. This longevity or survivorship component to variable mating success would be described ordinarily as natural selection.

Lifetime mating success was assayed by taking 40 daily censuses of the behavior of over 350 males marked individually in a field population where the sex ratio of adults was approximately 50:50. Marked males were recaptured an average of five times with some males being observed on as many as 15 census days. Data on the number of times individuals were observed to be in copula were collected in the histogram presented in Figure 2. There was an average of 1.96 matings observed per individual with a standard deviation of 1.76. Since males are capable of copulating more than once a day and since not all males known to be alive were observed at each census, this is a measure of variation in relative lifetime mating success.

Information concerning the sources of this variation is available from these data. Assuming that males are equally capable of mating throughout their lives, mating success must be related directly to longevity. By graphing, in Figure 3, the percentage of males observed mating as a function of their age on the day of observation, we can see that there is no temporal trend in mating ability. Allowing male longevity to be scored as the number of days between marking and last observation, the data can be used to construct a survivorship curve as shown in Figure 4. Mean survivorship is 15.95 days, with a standard deviation of 9.51 days.

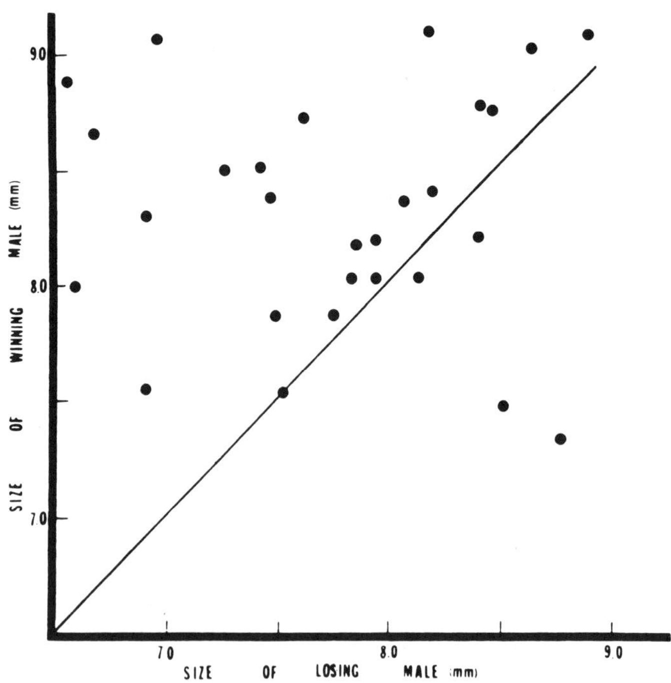

Fig. 1. Body sizes (in mm) of winning and losing members of pairs of adult male *Tetraopes* captured following antagonistic interactions in the field. The diagonal is provided to help illustrate the relative sizes of the participants in each fight.

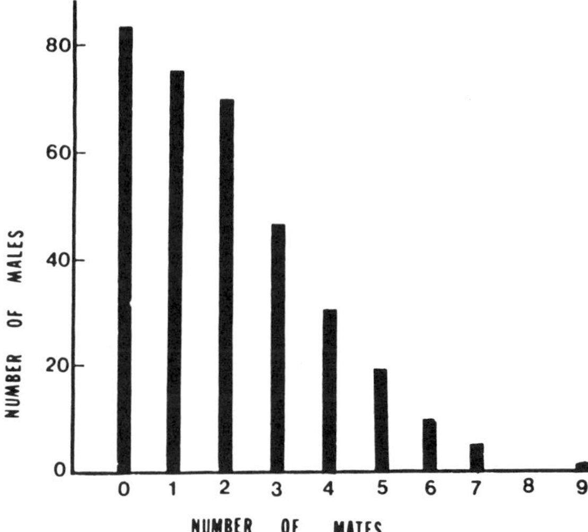

Fig. 2. The distribution of numbers of observed mates associated with individual male milkweed beetles.

There is sufficient differential survivorship during the reproductive period to generate considerable variation in mating success.

Males could vary in lifetime mating success not only due to differential survivorship, but also due to differences in the probability of mating at a given time that are related to fighting, courtship, etc. If we define a term "mating efficiency" (E) as the number of times an individual was observed mating divided by the total number of times that individual was observed, then variation in the courtship and defense of females should be reflected in variation in mating efficiency. This information also can be obtained from the data set. The mean of this value is 0.359 and the standard deviation is 0.158. It should be noted that the standard deviation was estimated by a technique that accounts for the high level of sampling variance associated with those individuals only recaptured a few times. This is described in detail in McCauley (1983).

The data suggest that total variation in male mating success is caused by variation in longevity and variation in mating efficiency. It would be valuable to look at the relative contribution of the two component variables to total variation. The lifetime mating success of an individual male must be related to the probability that he will mate during any brief period times the total length of time he is reproductively active. This is the product of the two variables presented above. In trying to estimate how variance in two component variables contributes to variance in their products one can use the propagation law of errors if the component variables are not correlated (Larsen and Marx 1981). That is:

$$V_T = L^2V_E + E^2V_L \tag{1}$$

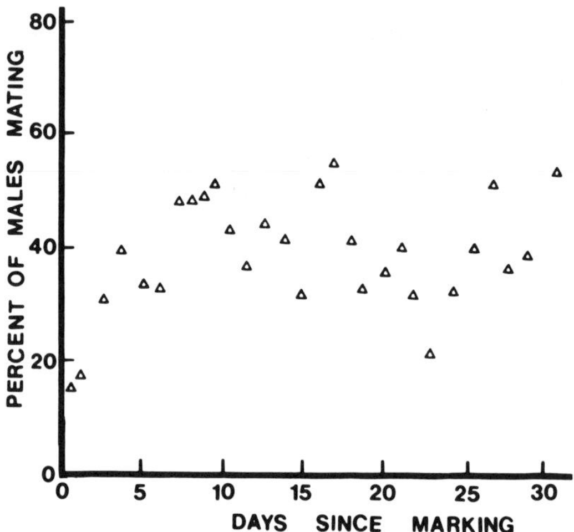

Fig. 3. Percent of adult male *Tetraopes tetraophthalmus* found in copula up to 29 days after marking in the field.

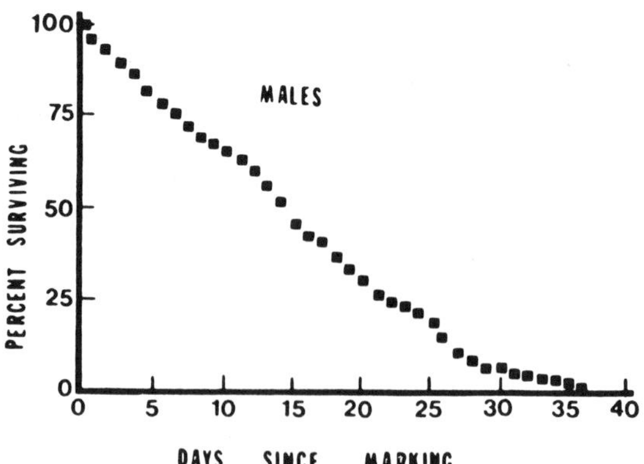

Fig. 4. Longevity during the adult stage of male milkweed beetles based on mark-recapture data.

where V_T, V_L and V_E are variance in total mates, longevity, and mating efficiency respectively and E and L are mean efficiency and longevity. This equation can be rearranged to yield:

$$\frac{V_T}{T^2} = \frac{V_E}{E^2} = \frac{V_L}{L^2} \qquad (2)$$

where T represents mean total mates. Since the index of opportunity for selection (I) is defined as the variance in fitness divided by the square of the mean (Crow 1958, Wade and Arnold 1980), equation (2) can be interpreted to say that the overall opportunity for selection due to variance in lifetime mating success (I_T) is equal to the sum of opportunities for selection due to variable survivorship (I_L) and variable mating efficiency (I_E).

The product moment correlation of L and E is 0.152, small enough to warrant use of this approximation. I_L is calculated from the data presented above to be 0.355. I_E is likewise calculated to be 0.194. The total opportunity for selection (I_T), then, is equal to 0.549. Nearly twice as much opportunity for selection is derived from differential survivorship (natural selection) as from variable probability of mating at any one time (sexual selection). This points out the importance of lifetime studies to understanding fully selection pressures generated by a given mating system.

Implications

These indices of selection are of value in comparisons of different populations of the same species and perhaps in comparisons of different species with mating systems of varying similarity. Unfortunately, few data sets based on lifetime studies of invertebrates are available at the present time. One important exception to this is a study of lifetime mating success in a natural population of the damselfly, *Enallagma hageni* (Walsh) by Fincke (1982). Based on her Figure 2 and personal communication, the total opportunity for selection in the population of male damselflies studied was about 1, or nearly twice as great as in *T. tetraophthalmus*. Significantly, male-biased operational sex ratios, sometimes as high as 9:1, were recorded. One might expect increased male-male competition for mates with male-biased sex ratios. It would be interesting to compare both the beetle and damselfly results to species in which breeding biology ranges from nearly "explosive" where all mating is completed in a few days to species in which reproductive life extends to months. One would predict that the survivorship component would become of increasing importance as reproductive lifespan increases.

It also would be important to compare these results taken for one population of milkweed beetles for one generation to other conspecific populations. That could give some indication of the temporal and spatial variation that might exist in selection components. One attribute, mentioned previously, by which populations could vary and which could affect the operation of selection would be the sex ratio. While the sex ratio in the beetle populations studied was approximately 1:1, one population monitored in 1982 for an unrelated study had a male-biased adult sex ratio of 2:1. This observation prompted a laboratory experiment in which individually marked beetles were placed into cages and their mating success monitored for three censuses a day for four days. In eight cages the sex ratio was set at 1:1 by the introduction of nine males and nine females. In seven other cages a 2:1 sex ratio was obtained by introducing 12 males and 6 females to each cage. Since few beetles died during the experiment, variation in mating success is equivalent to the mating efficiency (E) parameter scored in the field studies. I_E was scored separately for each cage. It averaged 0.25 in the 1:1 sex ratio cages and 0.50 when the sex ratio was male-biased. A Box-Cox transformation (Sokal and Rohlf 1981) revealed that a log transformation of the I values made them nearly normally distributed and allowed a t-test of means of treatment differences. The means of the treatments are indeed statistically different ($t = 2.41$, d.f. $= 13, <0.05$). Sex ratio variation has a measurable effect on I in the laboratory. This would certainly be worth investigating in the field should further between-population variation in sex ratio be discovered.

This study begins to show how both natural and sexual selection can be generated by variation in behaviors and life-history, characteristics associated with mating. It illustrates the need for lifetime studies to supplement information gained from logistically less demanding cross-sectional studies. It has not addressed the important question of the heritability of the various traits measured, so we cannot yet make any predictions as to the potential for evolutionary response to the observed selection.

Literature Cited

Arnold, S. J. 1983. Sexual selection: the interface of theory and empiricism. *In* P.P.G. Bateson, ed. Mate Choice. Cambridge Univ. Press.

Brown, L., and B. D. Siegfried. 1983. Effects of horn size on courtship activity in the forked fungus beetle, *Bolitotherus cornutus* (Coleoptera: Tenebrionidae). Ann. Entomol. Soc. Amer. 76: 253.

Cade, W. 1981. Alternate male strategies: genetic differences in crickets. Science 212: 563.

Crow, J. F. 1958. Some possibilities for measuring selection intensities in man. Hum. Biol. 30: 1.

Eberhard, W. G. 1982. Beetle horn dimorphism: making the best of a bad lot. Amer. Natur. 119: 420.

Fincke, O. M. 1982. Lifetime mating success in a natural population of the damselfly, *Enallagma hageni* (Walsh) (Odonata: Coenagrionidae). Behav. Ecol. Sociobiol. 10: 293.

Larsen, R. J., and M. L. Marx. 1981. An Introduction to Mathematical Statistics and Its Applications. Prentice Hall, Englewood Cliffs.

Lawrence, W. S. 1982. Sexual dimorphism in between- and within-patch movements of a monophagous insect: *Tetraopes* (Coleoptera: Cerambycidae). Oecologia 53: 245.

Mason, L. G. 1964. Stabilizing selection for mating fitness in natural populations of *Tetraopes*. Evolution. 18: 492.

McCauley, D. E. 1979. Geographic variation in body size and its relation to the mating structure of *Tetraopes* populations. Heredity 42: 143.

McCauley, D. E. 1982. The behavioral components of sexual selection in the milkweed beetle *Tetraopes tetraophthalmus*. Anim. Behav. 30: 23.

McCauley, D. E. 1983. An estimate of the relative opportunities for natural and sexual selection in a population of milkweed beetles. Evolution 37: 701.

McCauley, D. E., J. R. Ott, A. Stine, S. McGrath. 1981. Limited dispersal and its effect on population structure in the milkweed beetle *Tetraopes tetraophthalmus*. Oecologia 51: 145.

McCauley, D. E., and M. J. Wade. 1978. Female choice and the mating structure of a natural population of the soldier beetle, *Chauliognathus pennsylvanicus*. Evolution. 32: 771.

Sokal, R. R., and F. J. Rohlf. 1981. Biometry. W. H. Freeman, San Francisco.

Thornhill, R. 1980. Sexual selection within mating swarms of the lovebug, *Plecia neararctica*. Anim. Behav. 28: 405.

Thornhill, R. 1981. *Panorpa* (Mecoptera: Panorpidae) scorpionflies: systems for understanding resource-defense, polygyny and alternative male reproductive efforts. Ann. Rev. Ecol. Syst. 12: 355.

Wade, M. J., and S. J. Arnold. 1980. The intensity of sexual selection in relation to male sexual behavior, female choice, and sperm precedence. Anim. Behav. 28: 446.

Hybridization as a Causal Mechanism of Mixed Color Broods and Unusual Color Morphs of Female Offspring in the Eastern Tiger Swallowtail Butterflies, *Papilio glaucus*

J. Mark Scriber

Mark H. Evans

David B. Ritland

Department of Entomology
University of Wisconsin
Madison, Wisconsin 53706

Introduction

The eastern tiger swallowtail butterfly, *Papilio glaucus* L., is comprised of three putative subspecies (*P. g. canadensis* R & J, *P. g. glaucus* L. and *P. g. australis* Maynard) which have a total range of approximately one billion hectares across the eastern half of the U. S. and nearly all of Canada (Fig. 1). Based upon multivariate discriminant analyses of adult morphometric traits, diapause differences, differential foodplant detoxication abilities of larvae, and reproductive incompatibilities of the hybrids and backcrosses (i.e., fecundity, fertility and viability) we believe that *P. g. canadensis* may warrant species status (Scriber et al., unpublished, and Fig. 2). The zone of potential hybrid interaction between *P. g. canadensis* and *P. g. glaucus* is felt to be restricted to the zone between latitudes of 40 and 45° north (Scriber 1982, 1983, and Fig. 1). However, the actual degree of reproductive isolation that exists between tiger swallowtails in nature is unknown. One factor which some researchers suggest should have an influence upon mating preferences of tiger swallowtails is female color (Brower 1957, Brower 1959a and 1959b, Burns 1966, Makielski 1972, Pliske 1972, Levin 1973).

The purported subspecies *Papilio glaucus glaucus* has monomorphic yellow males and a sex-limited color dimorphism, with distinct yellow and distinct black adult female forms that generally produce daughters of the same color as themselves regardless of the origin of the male parent. The putative *P. g. canadensis* and three closely related species in the western half of the U.S.A. (*P. rutulus* Lucas, *P. multicaudatus* Kirby, and *P. eurymedon* Lucas) are completely monomorphic for yellow females. Mate recognition and courtship behavior within and between members of the western species was studied by Brower (1957, 1959b). However, nothing is known about the significance of the dark female morph in reproductive isolation between *P. g. canadensis* and *P. g. glaucus* across their entire 3000 km strip of sympatry/hybridization (Fig. 1).

In addition to potential non-preference for dark morph females by *P. g. canadensis* males, known differences in the habitat preferences, foodplants utilized, and diapause tenden-

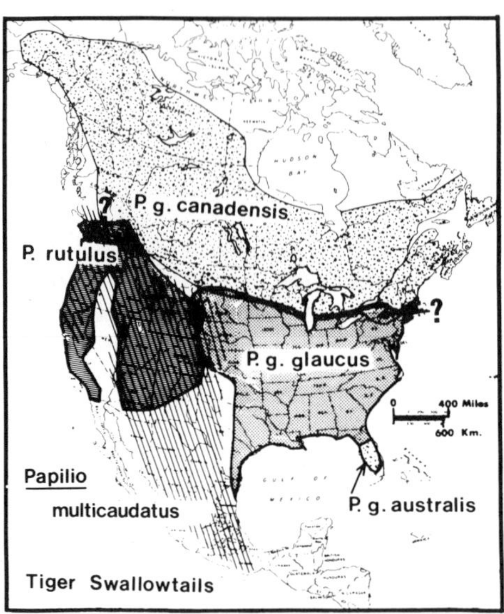

Fig. 1. The geographic distribution of the tiger swallowtail group. *Papilio rutulus* and *P. eurymedon* (not labeled) are basically sympatric while *P. multicaudatus* extends into Mexico. The three putative subspecies of *Papilio glaucus* are basically allopatric. (Data from Brower 1957, Ferris and Brown 1981, Scriber et al. 1982).

cies between *P. g. canadensis* and *P. g. glaucus* certainly could contribute to restricted gene flow and the apparently narrow zone of overlap/hybridization in Wisconsin (Remington 1968, Ebner 1970, Scriber 1982, 1983, Scriber et al. 1982, and Fig. 2). Laboratory hybridization studies have permitted us to investigate the genetic basis of these and other differences, including the control of dark morph expression in adult females. Based on these studies, we suggest that introgression from *P. g. canadensis* may largely explain a variety of unusual and some extremely rare color variants which have been reported occasionally in the literature (Edwards 1884, Skinner and Aaron 1888, Ehrman 1893, Howard 1899, Clark 1932, Clark and Clark 1951, Walsten 1977, Ehle 1981, Clarke and Clarke 1983, and D. Robacker, pers. comm.).

The rare occurrence (i.e., low frequency) of color variants and intermediate "dusted" yellow females (see Fig. 3) may be related to mating non-preference by males of both *P. g. canadensis* and *P. g. glaucus*, or it may be related to other ecological factors. Different predation rates upon the two morphs (suggested as early as 1884 by Edwards) has been cited as one factor determining relative proportions of the morphs throughout its range (Fig. 2). Dark morph females of *P. g. glaucus* are presumed mimics of the pipevine swallowtail, *Battus philenor* L. It has been suggested that such selection against the yellow morph may be most intense where the model is most abundant (Remington 1958, Brower and Brower 1962). Evidence of differential selection by birds has been provided in laboratory studies (J. V. Z. Brower 1958) and in the field (Sternburg et al. 1977, Jeffords et al. 1979, G. P. Waldbauer, pers. comm.). It is interesting that dark morph females of *P. g. glaucus* do not occur north of 45 °N latitude (Platt and Brower 1968, Shapiro 1974, Scriber et al. 1982), which also corresponds roughly to the northernmost limits of the model species. It is unknown whether the mosaic or intermediate forms are more intensely predated upon.

The possibility, in adult *P. g. glaucus* females, of color determination by an autosomal gene which is either dominant or recessive and sex-controlled has been ruled out by Clarke and Sheppard (1959). Since the female is the heterogametic sex in Lepidoptera (with XY sex chromosomes), it has been suggested that the black and yellow female forms of *P. g. glaucus* are probably controlled by a locus on the Y sex chromosome (Clarke and Sheppard 1959). Very rarely, yellow daughters may originate from black morph mothers and black morph daughters may originate from yellow morph mothers; a genetic mechanism explaining this result has been offered by Clarke and Sheppard (1959, 1962). The various alternative genetic

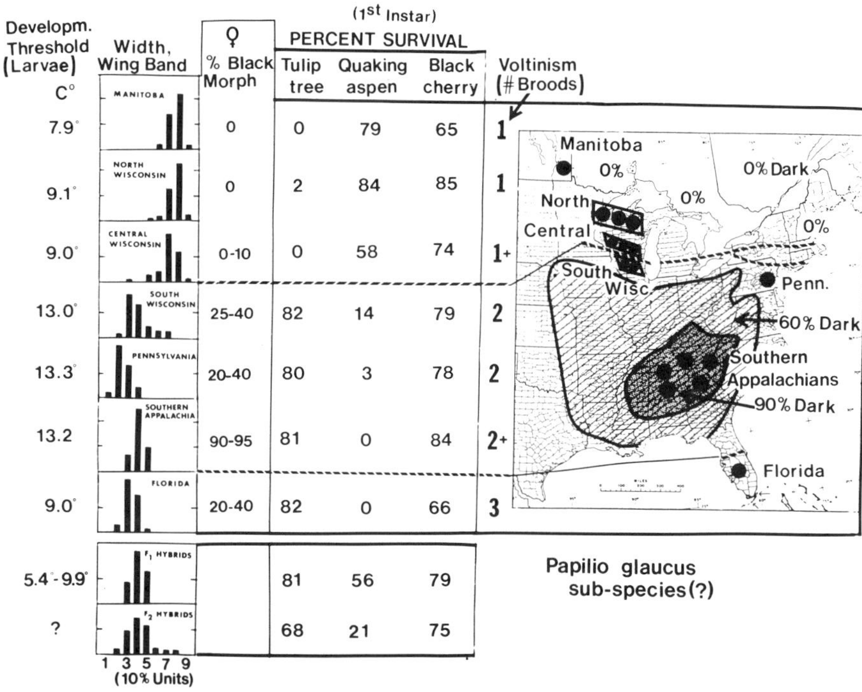

Fig. 2. Dark morph female frequencies of *P. g. glaucus* (areas with < 90% dark morphs are indicated with heavy shading; > 60% with cross hatching). No dark morphs are found north of central Wisconsin, central Michigan, or central New York. A variety of life history traits have been observed to differ rather drastically at the areas where the putative subspecies separate in central Wisconsin. Relative frequencies of the various black band width categories on the ventral hind wing anal cell illustrates one of many morphological differences between the taxa (H. Luebke, M. Collins, J. M. Scriber, unpublished) for 445 males. Differential foodplant utilization abilities are illustrated from Florida to Manitoba for nearly 200 females and nearly 8,000 larvae (J. M. Scriber and M. Evans, unpublished). The *P. g. canadensis* subspecies is believed to be everywhere univoltine. (J. M. Scriber and J. H. Hainze, unpublished).

models have not been adequately tested and the genetic basis of the sex-limited color dimorphism has remained an enigma for decades.

The study of inheritance using hybrids has been hampered by the tendency of hybrid

Fig. 3. Rare color morphs of *Papilio glaucus* adults; an (see text) "intermediate" dusted-yellow from (A) Dane Co., Wisconsin and (B) Lancaster Co., Pennsylvania. (C) An irregular mosaic with yellow streaking/blotching. This individual is a sib of the intermediate at the left (from Dane County, Wisconsin). (D) A low level polymorphism (?) in a male *P. g. canadensis* (see discussion in Scriber and Lintereur 1983).

females to enter "permanent" diapause. However the recent use of α-ecdysone to break diapause of female hybrids of a dark *P. glaucus* female crossed with a male of the western monomorphic *P. rutulus* produced F_1 hybrid females intermediate between the black and yellow forms in that the normal yellow pattern is present, but heavily suffused with black

scales (see figures in Clarke and Willig 1977). These and similar intermediate females from the backcross between male hybrids and dark *P. glaucus* are interpreted by Clarke and Willig (1977) as a demonstration that "*rutulus* carries an autosomal or perhaps X-linked gene (or genes) which modifies the effect of the Y-linked allelomorph controlling black." It has recently been reported that the "intermediate" *glaucus* x *rutulus* F_1 hybrid females have a Smith body (heteropyknotic sex chromatin) on the Y chromosome as is the case with all dark morph *P. g. glaucus* females thus far examined (see Clarke et al. 1976, Cross and Gill 1979). The presence or absence of the Smith body is still unknown in the closely related western species, which lack dark morph females (*P. rutulus*, *P. eurymedon*, and *P. multicaudatus*), as well as in the Canadian (northern) populations of *P. glaucus canadensis*. It is also not known, but considered doubtful, if crossing over between the X and Y chromosomes, suggested as a mechanism for producing daughters of the "unexpected" morph (Clarke and Sheppard 1959), ever occurs in female butterflies (Robinson 1971, Suomalainen et al. 1973, Turner and Sheppard 1975).

Cytoplasmic inheritance (e.g., an unusual case where cytoplasmic factors might be carried in the sperm) has not yet been totally ruled out as an explanation for the occurrence of a form unlike the female parent (Clarke et al. 1976). The rare record of a bilateral *P. glaucus* female with one side black and one side yellow, figured by Edwards (1884), has been interpreted as evidence against this hypothesis since cytoplasmic genetic factors would be provided to both halves of the insect (Clarke and Sheppard 1959, 1962). In contrast, loss of a Y chromosome or somatic mutation of the gene at the first cleavage division could produce such a bilateral mosaic adult if the inheritance is Y-linked. These authors further suggest that cytoplasmic factors in the sperm should be "all or nothing" in the case of segregation of a female form that was unlike the female parent (otherwise dosage-dependent intermediate forms would exist) and that, therefore, if cytoplasmic factors exist, they must be particulate with one carried in the sperm being sufficient to switch the balance. If only a few particles are involved and fail to pass into both cells at mitosis, the yellow/black bilateral mosaic female of Edwards (1884, plate V) could be explained by this mechanism, except that more frequent occurrence of bilateral mosaics or other aberrent dark morph mosaics (with irregular blotches/streaks of yellow) would be expected (Clarke and Sheppard 1959). These bilateral and irregular mosaics are extremely rare and their existence is the subject of considerable attention by lepidopterists. Walsten (1977) and Clarke and Clarke (1983) review and figure a variety of forms from the Herman Strecker collection including both bilaterally divided and anterio-posteriorly divided individuals (part dark and part yellow, see also Skinner and Aaron 1888) and more irregular mosaics (see also Ehrman 1893, Howard 1899, Ehle 1981).

Clarke and Sheppard (1962) later suggested that the few cases of females which are unlike their mothers could be the result of a chromosome abnormality (i.e., a "contamination" preventing straight Y-linked inheritance) which appears to be unique to one closely related batch of insects which they obtained from pupae supplied by E. Dluhy near Chicago, Illinois (Clarke and Sheppard 1962, see also Clarke et al. 1976). Since the half-yellow, half-black female mosaic is the strongest evidence favoring the Y-linked inheritance hypothesis (i.e., all or none) over the cytoplasmic (continuous variation in colors), it is ironic that an additional record of one yellow/black irregular mosaic female and also one dusted-yellow "intermediate" female of *P. glaucus glaucus* should turn up at the Milwaukee Public Museum in the collection of none other than E. Dluhy (from Chicago)! The dusty-yellow colored female might be considered an "intermediate," which once again (weakly) suggests the possibility of cytoplasmic factors that might not act in an "all or none" fashion. Interestingly, while it might favor the Y-linked hypothesis (Clarke and Sheppard 1962), the irregular mosaic and intermediate of Dluhy in Milwaukee (suggesting gradual and perhaps cytoplasmic mechanisms) might also be related to the "contaminated" (chromosomally abnormal?) Chicago stock used in the original study by Clarke and Sheppard (1959). We suggest that these three hypotheses (i.e., cytoplasmic, Y-linked, and "contamination") are not mutually exclusive.

We present data from field captures, laboratory hybridization studies, and mass-rearings of butterflies, which may contribute significantly to a better understanding of the genetic bases of dark morph inheritance in *Papilio glaucus*. From these studies we have seen how modifications of dark potential can be altered by environmental factors (rearing temperature; Ritland 1983) and by factors in the genome of the northern *P. g. canadensis*. We suggest that hybridization or introgression may explain the rare segregation of both female color morphs from either a yellow female or a dark female. In addition, rearing observations of offspring from field-captured females suggest that natural hybridization (gene introgression from *P. g. canadensis* to *P. g. glaucus* populations) may be partially responsible for the rare occurrence of irregular and bilateral mosaics and dusted intermediates among the typically distinct yellow or black morph females.

Results and Discussion

In the last two years we have captured, reared and bred several thousand tiger swallowtail butterflies (including all three purported subspecies; *P. g. canadensis*, *P. g. glaucus*, and *P. g. australis*; see Tyler 1975) in order to gain a better understanding of the genetic basis for known differences in diapause, color, size and foodplant detoxication abilities within and between subspecies and populations (Scriber 1982, 1983, Scriber et al. 1982). During several generations of mass-rearing we have (as expected) observed that virtually all individual females of the three purported subspecies of *Papilio glaucus* breed true and produce daughters only of the same color morph as themselves. Our only exception thus far is a case in which a normal-appearing dark female captured in Dane County, Wisconsin (June 1981) gave rise to a range of variations including normal dark daughters, a mosaic, and intermediate or dusted-yellow female types (Fig. 3A and C). (These specimens are currently in the research collection of J. M. Scriber, at the University of Wisconsin.) Our laboratory hand-pairings and mass-rearings have permitted us to examine experimentally some of the hypotheses that have been advanced previously to explain the genetic basis of these inheritance patterns with particular relevance to those exceptional cases described by Clarke and Sheppard (1959, 1962). Results in Table 1 are compatible with the hypothesis that the gene(s) controlling the color is simply Y-linked (on the female heterogametic sex chromosome). In other words, yellow females (XY) and dark females (XY^D) should breed true regardless of the mate's origin.

In order to assess the hypothesis that the rare intermediate form (with the usual yellow background pattern suffused heavily with black scales; i.e., "dusty-yellow") might be produced by a modifier gene (or genes) as suggested by Clarke and Willig (1977), and Clarke and Clarke (1983), we made a variety of hand-pairings using males of the northern subspecies, *Papilio glaucus canadensis*, and dark morph females from Wisconsin, Pennsylvania, Georgia and Florida. It should be noted that the results of these hybrid pairings do not closely fit any previous hypothesis (Table 2). The production of only yellow females suggests that a modifier from the *P. g. canadensis* male totally suppresses the expression of the dark wing background and dark body color, unlike the partial (intermediate color) suppression in the F_1 *glaucus* x *rutulus* hybrid (Clarke and Clarke 1983). The general rarity of female offspring in such race or species crosses is not unusual (see Haldane 1922).

At this point, it is uncertain whether the modifier is associated with an autosome or a sex chromosome; our data do not conclusively favor one possibility over the other. For the purposes of this discussion, we have hypothesized a sex-linked modifier associated with the X chromosome, with X^MM representing the X chromosome carrying the modifier. In this terminology, then, a *P. g. canadensis* male would exhibit genotype X^MX^M, and a *P. g. glaucus* male would be XX. Thus, X-linkage could account for the presence of the modifier in the F_1 male offspring X^MX.

We first tested for the postulated *P. g. candensis* modifier with a backcross of a dark morph female *P. g. glaucus* (XY^D) and a hybrid male (XX^M). The expected ratio of yellow (X^MY^D) to dark (XY^D) females would be 1:1. Three fertile backcrosses of this type were obtained and the results in Table 3 support the hypothesis ($x^2 = 1.84$, $.1<p<.2$).

In order to pursue this hypothesis, we obtained fertile crosses of the yellow B_1 females (X^MY^D) with normal *P. g. glaucus* males (XX). We can see that the potential of dark morph expression is once again realized by the daughters in each of five such crosses (Table 4). This suggests that there is not loss of a Y chromosome (i.e., XO which will be yellow) as suggested by Sir Cyril Clarke (pers. comm.). The explanation for the occurrence of some unexpected yellow daughters is uncertain, but the presence of these yellow females favors the idea of an autosomal modifier. Alternative explanations may involve segregation/modification of an autosomal modifier, crossing over in the hybrid, preferential loss of the heteropyknotic (Smith) body, or other unexplored possibilities such disruption of a normal developmental process (see Clarke and Clarke 1983).

Additional information was obtained from another sequence of fertile crosses involving the yellow morph F_1 hybrid females (X^MY^D) with their male sibs (X^MX) or other F_1 male hybrids between *P. g. glaucus* and *P. g. canadensis* also (X^MX). All resulting F_2 females were yellow in coloration (Table 5). The total lack of F_2 dark morphs from F_1 females with the dark potential (crosses 272, 279, 288) was not expected. Among the possible explanations are the following: association with the *P. g. canadensis* diapause tendencies (possibly the diapausing pupae are dark females which may emerge later in 1983); inadvertent selection against dark morph females in our rearing process; genetic incompatibilities in dark morph expression (i.e., the "target" of modifiers being different in *P. g. canadensis* and *P. g. glaucus*); or loss of the heteropyknotic (Smith) body (or the entire Y chromosome) by females (C. A. Clarke, pers. comm.).

We have observed unusual extensive alterations in female wing patterns (increased area of dark patterning) which have never before been reported (Scriber and Evans, unpublished). None of the male sibs exhibit these patterns and the explanation is unknown, although these individuals are known to have some *P. g. canadensis* genes from the male parent (a sib of the female in cross #15). Another odd color pattern has been reported several times over the last century only in males, and almost entirely near the 40-45° north latitude zone of sympatry/hybridization of *P. g. canadensis* and *P. g. glaucus* (see Scriber and Lintereur 1983 for review and Fig. 3D). We cannot rule out the possibility of introgression in these cases, and it emphasizes the need for additional studies to better understand the genetic and physiological mechanisms as well as the behavioral and ecological implications of these color polymorphisms.

Regardless of the role of predation in determining the frequency of rare variants and of yellow *versus* dark morphs in *P. g. glaucus*, it is evident that the proportion of yellow females is significantly greater at the range periphery of this subspecies (Fig. 2; Bruner 1877, Edwards 1884, Brower 1957, Remington 1958, Makielski 1972, Pliske 1972, Levin 1973, Shapiro 1974, Tyler 1975, Scriber et al. 1982, M. Scriber and R. Lederhouse, unpublished). Our hand-paired crosses of dark morph females with *P. g. canadensis* males resulted in only yellow females (or prolonged diapause or death of all other females). These results suggest that it is possible that

Table 1. Results of intrapopulation lab crosses of *Papilio glaucus*.[a] Madison, Wisconsin 1981, 1982.

Female color (origin)		Male's origin & his Mother's color	Number of Fertile Pairings	Males	Offspring Females (Yellow)	(Dark)
Y(PA)	x	Y(PA)	40	377	(537)	(0)
Y(WI)	x	Y(WI)	2	34	(21)	(0)
Y(PA)	x	?(PA)	34	88	(106)	(0)
Y(WI)	x	?(WI)	2	6	(10)	(0)
[b]Y(PA)	x	Y(WI)	1	15	(8)	(0)
Y(PA)	x	D(PA)	1	27	(14)	(0)
Y(WI)	x	D(WI)	1	11	(16)	(0)
D(PA)	x	Y(PA)	1	42	(0)	(35)
D(WI)	x	Y(PA)	2	1	(0)	(2)
D(PA)	x	D(PA)	1	25	(0)	(25)
D(WI)	x	D(WI)	2	29	(0)	(26)
D(PA)	x	?(PA)	12	3	(0)	(11)
D(WI)	x	?(WI)	2	7	(0)	(12)
D(AL)	x	D(AL)	3	13	(0)	(15)
[b]D(PA)	x	D(WI)	1	33	(1)	(32)

[a] The eclosed larvae/total ova represents the percent viability: for all Wisconsin x Wisconsin crosses (1296/1721 = 75.3%); for all Pennsylvania x Pennsylvania crosses (6544/9076 = 72.1%); for all Pennsylvania x Wisconsin crosses (388/548 = 70.8%).

[b] Interpopulation crosses: Pennsylvania female x Dane Co. Wisconsin male. The Wisconsin male parent at the bottom originated from the same dark female that produced the normal dark, intermediates, and irregular mosaic (see Fig. 2 and text for further details), hence it could contain *P. g. candensis* factors (e.g., autosomal modifiers of dark morph expression) or chromosomal abnormality.

some of the higher proportion of yellow morph females at the northern limits of the *P. g. glaucus* range could be partially due to introgression of modifiers from *P. g. canadensis* males which suppress and/or alter dark morph expression. On the other hand, "permanent" diapause of hybrid females suggests that incompatibilities exist that may minimize introgression (Scriber et al., unpublished).

We are uncertain of the actual degree of natural hybridization between *P. g. glaucus* and *P. g. candensis*, however preliminary morphological and physiological evidence (Scriber 1982, 1983, Scriber et al. 1982, Scriber et al., unpublished) suggests that some hybridization is possible. This zone of potential hybrid interaction in Wisconsin extends diagonally across the

Table 2. Results of primary hybrid crosses between *P. g. canadensis* males and seven dark *P. g. glaucus* females from Pennsylvania, Wisconsin, Georgia, Florida. Madison, Wisconsin, 1982.

Pairing Code #	Parentage[a] (glaucus x canadensis)	Eclosed larvae/ Total ova	%	Males	Females (Yellow)	Females (Dark)	Pupae remaining in diapause[b]
74	D(PA) x Pgc	32/34	94.1	2	1	0	0
73	D(WI) x Pgc	136/163	83.4	18	17	0	7
112	D(WI) x Pgc	10/21	47.6	4	0	0	5
4	D(GA) x Pgc	19/51	37.2	1	3	0	3
39	D(GA) x Pgc	117/152	76.9	18	4	0	26
75	D(GA) x Pgc	156/358	43.6	49	2	0	54
113[c]	D(FL) x Pgc	54/71	76.1	7	0	0	10
TOTAL				99	27	0	67

[a] The geographic origins of *P. g. canadensis* males in crosses 74, 73, 112, 4, 39, 75, and 113 are Marinette, Marinette, Bayfield, Adams, Price, Marinette and Bayfield Counties, WI, respectively.

[b] These 1982 F^1 hybrid pupae appear to be in a prolonged diapause as of June 16, 1983. They did not eclose within 30 days of their 1983 return to ambient temperature (21-25 °C). Of the adult totals, 15 males and 10 females emerged in 1983. All 1982 female emergences were from a single pairing (#73).

[c] Some researchers feel that Florida populations should be considered as a separate subspecies (*P. g. australis*). It is interesting in this regard that the F_1 hybrid egg viability (eclosed larvae/total ova) is lower between *P. g. australis* and *P. g. glaucus* (Pga x Pgg = 52/159 = 32.7%; Pgg x Pga = 58/172 = 33.7%) than is observed for *P. g. australis* and *P. g. canadensis* (pairing #113 above).

Table 3. Results of three fertile backcrosses of dark *Papilio glaucus* females (XY^D) with hybrid males of *P. g. candensis* x *P. g. glaucus* origin (XX^M). Madison, WI, 1981, 1982.

Pairing Code #	Parentage[a] (XY^D) x (XX^M)	Total pupae[b]	Males	Females (Yellow) (X^MY^D)	Females (Dark) (XY^D)
6	D(PA) x F_1 hybrid	18	8	1	3
14	D(PA) x F_1 hybrid	21	7	5	4
15	D(PA) x F_1 hybrid	88	40	11	20
TOTAL		127	55	17[c]	27[c]

[a] All females are from dark morph mothers from either Schuylkill Co. or Dauphin Co., Pennsylvania and the males are from a cross of *P. g. canadensis* female x *P. g. glaucus* male (see Scriber 1982). It should be noted here that 100% viability was observed in each of these pairings (i.e., eclosed larvae/total ova = 336/336).

[b] Some pupae died or are still in diapause.

[c] These data support the expected ratio of 1 yellow : 1 dark; x^2 = 1.84, $.10<p<.25$.

state and into the Chicago, Illinois area (Fig. 1). The odd-segregating broods of Clarke and Sheppard (1959), which were all from one group of Chicago pupae, may therefore have been the result of introgression from *P. g. canadensis*. It may also be reasonable that introgression of the putative *P. g. canadensis* modifier of the gene for dark female morphs may also be responsible in part for the eastern cases of mosaics, intermediates, and odd-segregation of dark/yellow morphs (i.e., throughout the northern Appalachian mountain regions; see Edwards 1884, Skinner and Aaron 1888, Ehrman 1893, Clark and Clark 1951, Ehle 1981, D. Robacker, pers. comm. and Fig. 3) as well as they might be responsible for our Wisconsin (and Chicago) cases.

In addition to the introgression from the biologically and morphologically distinct *P. g. canadensis* (Scriber et al., unpublished), other explanations for dusty-yellow "intermediates" such as environmental effects have not been ruled out. In fact, higher temperatures during the rearing of dark morph *P. g. glaucus* from Wisconsin (Dane County) and Pennsylvania (Dauphin and Schuylkill Counties), produce greater proportions of intermediate types of females (Ritland 1983). We are at this time uncertain if these intermediates are also possible from the central Appalachians or elsewhere in the *P. g. glaucus* range. The degree of differential canalization of phenotype in different geographic regions and other environmental effects deserve additional study (see e.g., Shapiro 1980a and 1980b, Bonner 1982, Caswell 1983).

The significance of color polymorphism and its coevolution with a suite of other life history traits (including morphological, physiological, and toxicological, as well as behavioral) have implications at the ecological and evolutionary levels of organization (Stearns 1982, Gilbert 1983). The specific behavioral consequences of color polymorphisms in the tiger swallowtails are virtually unknown. Habitat preferences, courtship behavior, and oviposition preferences of adult females would be of particular interest in regard to host race

Table 4. Results of fertile backcrosses of yellow backcross females?[a] with dark potential (genotype X^MY^D) crossed to various *P. g. glaucus* males (XX). Madison, WI, 1982.

Pairing Code #	Parentage (X^MY^D) x (XX)	Total pupae[b]	Males	Females (Yellow) (X^MY^D)	Females (Dark) (XY^D)
148	(B_1) x ('D(PA) x D(PA)')	4	2	0	2
154	(B_1) x ('D(PA) x D(PA)')	8	1	1	6
157	(B_1) x ('Y(PA) x Y(PA)')	20	2	2	5
225	(B_1) x ('Y(PA) x Y(PA)')	11	4	0	1
155	(B_1) x ('Y(PA) x D(PA)')	14	3	0	1
TOTAL		57	12	3	15

[a] The five mothers are yellow with dark potential (genotype X^MMY^D) from the 'D(PA) x F_1 hybrid' crosses 6, 14, and 15 in Table 3. The viability (eclosed larvae/total ova) for all B_1 x Pgg crosses is 302/443 = 68.2%.

[b] Some pupae died or are still diapause.

formation (see Futuyma 1983). However, we suggest that it will also be critical to understand these behaviors in the context of all other adaptations (i.e., morphological, physiological, etc.) and with particular concern for the effects of introgression and, perhaps, hybrid dysgenesis (genetic incompatibilities) where individuals of these taxa might interbreed naturally.

Acknowledgments

This research was supported by the National Science Foundation (DEB-7921749 and BSR-8306060), the College of Agricultural and Life Sciences at the University of Wisconsin at Madison (Hatch 5134), and the Graduate School Committee on Research. We wish to thank the following people for their assistance in obtaining living specimens: Robert Dowell, Les Ferge, Bruce Giebink, Walt Gould, Susan Heg, William Houtz, Phil Kingsley, Greg Lintereur, Heidi Luebke, Syafrida Manuwoto, James Maudsley, Brian Mohr, Ric Peigler, Sherry Roberts, Dave Robacker, Jane Schrimpf, Bill Warfield and Tom Uhlman, We are grateful to Lincoln Brower, Cyril Clarke, Michael Collins, Heidi Luebke, Jane Schrimpf, Art Shapiro and Steve Sims for discussion. Sue Borkin and Allen Young were generous in their offering of the Milwaukee Public Museum research collection for analysis. We are grateful to Jane Schrimpf for her help in the 1982 F_2 experiments.

Table 5. Results of three fertile F_2 crosses between yellow F_1 hybrid females[a] (with dark potential; genotype X^MMY^D) and F_1 hybrid males. Madison, WI, 1982.[c]

Pairing Code #	Parentage (X^MY^D) x (X^MX)	Total pupae[b]	Males (X^MX) & (XX)	Females (Yellow) (X^MY^D)	Females (Dark) (XY^D)
272	(D(WI) x Pgc)^2	71	29	28	0
288	(D(WI) x Pgc)^2	38	19	12	0
279	(D(WI) x Pgc) x (Y(PA) x Pgc)	26	11	6	0
TOTAL		135	59	46	0

[a] The 3 yellow mothers, with a hypothesized modifier (X^MY^D), are from cross 73 of Table 2. The egg viability, eclosed larvae/total ova, for these 3 crosses is 378/504 = 75.0%.

[b] Some pupae died, and others appear to be in a prolonged diapause (failed to eclose within 30 days of return to ambient temperatures of 21-25 °C in 1983). Of the adults listed, 5 of the males and 7 of the females eclosed in 1983 (as of June 16th).

[c] While most F_1 pupae in diapause (Table 2) died, we did observe the eclosion of twenty additional yellow females and no dark females subsequent to June 1983. Results of 54 additional fertile pairings of virgin dark morph females of *P. g glaucus* x male *P. g. canadensis* in 1984 support this conclusion that only yellow (or rare dusted "sooty" intermediate) females and males are produced (J. M. Scriber and M. H. Evans, unpublished).

Subsequent to June 1983, we have obtained dark morph females as well as yellow females from F_2 crosses additional to those of Table 5. Whle these results are more in agreement with our initial expectations for genetic segregation of color morphs, the explanation for all yellow F_2 female offspring in Table 5 is still undetermined. We now strongly suspect that crossovers involving the female gene for melanism do occur in this species (J. M. Scriber, M. H. Evans, and C. Clarke, unpublished). Also, a number of additional "intermediates," irregular (asymmetrical) yellow-blotched patches on dark females, bilateral gynandromorphs, and other color aberrant female forms have been obtained from our laboratory rearings and from field collections in 1984 and 1985 (Scriber and Evans, unpublished).

Literature Cited

Bonner, J. T. 1982. Evolution and Development. Springer-Verlag, Berlin.

Brower, J. V. Z. 1958. Experimental studies of mimicry in some North American butterflies. II. *Battus philenor* and *Papilio troilus*, *P. polyxenes* and *P. glaucus*. Evolution 12: 123.

Brower, L. P. 1957. Speciation in the *Papilio glaucus* Group. Ph.D. Dissertation. Yale University. New Haven, Connecticut.

Brower, L. P. 1958. Larval foodplant specificity in butterflies of the *P. glaucus* group. Lepid. News 12: 103.

Brower, L. P. 1959a. Speciation in butterflies of the *Papilio glaucus* group. I. Morphological relationships and hybridization. Evolution 13: 40.

Brower, L. P. 1959b. Speciation in butterflies of the *Papilio glaucus* group. II. Ecological relationships and interspecific sexual behavior. Evolution 13: 212.

Brower, L. P., and J. V. Z. Brower. 1962. The relative abundance of model and mimic butterflies in natural populations of the *Battus philenor* mimicry complex. Ecology 43: 154-8.

Bruner, J. 1877. Black variety of *P. turnus*. Can. Entomol. 9: 20.

Burns, J. M. 1966. Preferential mating versus mimicry: Disruptive selection and sex-limited dimorphism in *Papilio glaucus*. Science 153: 551.

Caswell, H. 1983. Phenotypic plasticity in life-history traits: Demographic effects and evolutionary consequences. Amer. Zool. 23: 35.

Clark, A. H. 1932. The butterflies of the District of Columbia and vicinity. Smithsonian Inst. USNM Bulletin 157: 172.

Clark, A. H., and L. F. Clark. 1951. The butterflies of Virginia. Smithsonian Inst. Misc. Coll. 116 (no. 7): 124.

Clarke, C. A., and F. M. M. Clarke. 1983. Abnormalities of wing pattern in the Eastern Tiger Swallowtail butterfly, *Papilio glaucus*. Syst. Entomol. 8: 25.

Clarke, C. A., and P. M. Sheppard. 1959. The genetics of some mimetic forms of *Papilio dardanus* Brown and *Papilio glaucus* Linn. J. Genetics 56: 236.

Clarke, C. A., and P. M. Sheppard. 1962. The genetics of the mimetic butterfly, *Papilio glaucus*. Ecology 43: 159.

Clarke, C. A., P. M. Sheppard, and U. Mittwoch. 1976. Heterochromatin polymorphism and colour pattern in the tiger swallowtail butterfly, *Papilio glaucus* L. Nature 263: 585.

Clarke, C. A., and A. Willig. 1977. The use of α-ecdysone to break permanent diapause of female hybrids between *Papilio glaucus* L. female and *Papilio rutulus* Lucas male. J. Res. Lepid 16: 245.

Cross, W., and A. Gill. 1979. A new technique for the prospective survey of sex chromatin using larvae of Lepidoptera. J. Lepid. Soc. 33: 50.

Ebner, J. A. 1970. Butterflies of Wisconsin, Milwaukee Public Mus. Pop. Sci. Handbook No. 12.

Edwards, W. H. 1884. The Butterflies of North America (Vol. II). Houghton Mifflin Co.

Ehle, G. 1981. Letter to the editor. News of the Lepidopterists' Society (No. 5, Sept/Oct). p. 63.

Ehrman, G. A. 1893. A few remarkable variations in Lepidoptera. Can. Entomol. 26: 292.

Ferris, C. D., and F. M. Brown. 1981. Butterflies of the Rocky Mountain States. University of Oklahoma Press.

Futuyma, D. J. 1983. Evolutionary interactions among the herbivorous insects and plants. *In* D. J. Futuyma and M. Slatkin, eds. Coevolution. Sinauer Associates. Sunderland, Massachusetts.

Gilbert, L. E. 1983. Coevolution and mimicry. *In* D. J. Futuyma and M. Slatkin, eds. Coevolution. Sinauer Associates, Sunderland, Massachusetts.

Haldane, J. B. S. 1922. Sex ratio and unisexual sterility in hybrid animals. J. Genetics 12: 101.

Howard, L. O. 1899. An abnormal tiger swallowtail. Insect Life 7: 44.

Jeffords, M. R., J. R. Sternberg, and G. P. Waldbauer. 1979. Batesian mimicry: field demonstration of the survival value of pipevine swallowtail and monarch color patterns. Evolution 33: 275.

Levin, M. P. 1973. Preferential mating and the maintenance of the sex-limited dimorphism in *Papilio glaucus*: Evidence from laboratory matings. Evolution 27: 257.

Makielski, S. K. 1972. Polymorphism in *Papilio glaucus* L. (Papilionidae): Maintenance of the female ancestral form. J. Lepid. Soc. 26: 109.

Platt, A. P., and L. P. Brower. 1968. Mimetic versus disruptive coloration in intergrading populations of *Limenitis arthemis* and *astyanax* butterflies. Evolution 22: 699.

Pliske, T. E. 1972. Sexual selection and dimorphism in female tiger swallowtails, *Papilio glaucus* L. (Lepidoptera: Papilionidae): A reappraisal. Ann. Entomol. Soc. Amer. 65: 1267.

Remington, C. L. 1958. Genetics of populations of Lepidoptera. Proc. 10th Intern. Congr. Entomol. 2: 787.

Remington, C. L. 1968. Suture-zones of hybrid interactions between recently joined biotas. Evol. Biol. 2: 321.

Ritland, D. B. 1983. The Effects of Temperature on Larval Growth and Adult Phenotype in *Papilio glaucus glaucus* L., *P. glaucus canadensis* R & J, and their Hybrids. M.S. Thesis. University of Wisconsin. Madison, Wisconsin.

Robinson, R. 1971. Lepidoptera genetics. Pergamon, Oxford.

Scriber, J. M. 1982. Foodplants and speciation in the *Papilio glaucus* group. *In* Proc. 5th Intern. Symp. Insect-Plant Relationships. Pudoc, Wageningen, The Netherlands.

Scriber, J. M. 1983. The evolution of feeding specialization, physiological efficiency, and host races.*In* R. F. Denno and M. S. McClure, eds. Variable Plants and Herbivores in Natural and Managed Systems. Academic Press, New York.

Scriber, J. M., and R. C. Lederhouse. 1983. Temperature as a factor in the development and feeding ecology of tiger swallowtail caterpillars, *Papilio glaucus* (Lepidoptera). Oikos 40: 95.

Scriber, J. M., and G. L. Lintereur. 1983. A melanic aberration of *Papilio glaucus canadensis* from northern Wisconsin. J. Res. Lepid. (21:199).

Scriber, J. M., G. L. Lintereur, and M. H. Evans. 1982. Foodplant utilization and a new oviposition record for *Papilio glaucus canadensis* R & J (Papilionidae: Lepidoptera) in northern Wisconsin and Michigan. Great Lakes Entomol. 15: 39.

Shapiro, A. M. 1974. Butterflies and skippers of New York State. Search 4: 1. Cornell Expt. Stat. Publ.

Skinner, M. D., and E. M. Aaron. 1888. A list of the butterflies of Philadelphia, Pa. Can. Entomol. 20: 126.

Stearns, S. C. 1982. The role of development in the evolution of life histories. *In* J. T. Bonner, ed. Evolution and Development. Springer-Verlag, Berlin.

Sternburg, J. G., G. P. Waldbauer, and M. R. Jeffords. 1977. Batesian mimicry: Selective advantage of color pattern. Science 195: 681.

Suomalainen, E., L. M. Cook, and J. R. G. Turner. 1973. Achiasmatic oogenesis in the Heliconiine butterflies. Hereditas 74: 302.

Turner, J. R. G., and P. M. Sheppard. 1975. Absence of crossing-over in female butterflies (*Heliconius*). Heredity 34: 265.

Tyler, H. 1975. The swallowtail butterflies of North America. Naturegraph. Healdsburg, California.

Walsten, D. M. 1977. Tigers without their stripes. Chicago Field Museum Natural History Bulletin (May 1977).

Genetic Variation in the Maternal Defensive Behavior of the Lace Bug *Gargaphia solani*

Douglas W. Tallamy

Department of Entomology & Applied Ecology
University of Delaware
Newark, Delaware 19711

Hugh Dingle

Department of Entomology
University of California
Davis, California 95616

Introduction

The evolution of parental care is an extraordinary breakthrough in the adaptation of organisms to their environment. The step from simple deposition of offspring to parent-offspring interactions requires an array of intricate behavioral patterns and forms the central core of more complex social systems. Because it is so fundamental, we need to understand the genetic basis of parental behavior in order to gain insight into its sensitivity to selection and thus the potential rate of its evolution. Furthermore, before the interrelationships between parental behavior, life history and natural selection can be better understood, it is necessary to assess the natural variation in parental responses and the effects of such variation on other life history traits. For these reasons we undertook an empirical analysis of the variation in parental behavior, including its ontogenetic, environmental and genetic components.

Parental behavior in insects is an adaptation to adverse environmental conditions that reduce juvenile survivorship. Without parental care, the young of social or subsocial insects usually fall victim to predators, extreme environmental conditions, or competition (Bro Larsen 1952, Eberhard 1975, Wood 1976, Tallamy and Denno 1981a). In view of its association with fitness, little variation is expected in parental quality within a given population (Lewontin 1965, Maynard Smith 1979). If it is adaptive to defend offspring aggressively from predators, for example, selection should eliminate passive parents, leaving little variation in the behavior. Yet this is not the case in the subsocial lace bug *Gargaphia solani* Heidemann (Hemiptera: Tingidae). Within a given population, there is an entire spectrum of variation in maternal behavior, ranging from mothers that defend their young aggressively to mothers that abandon them. Much of this variation has been attributed to age-specific changes in maternal defensive behavior that are a function of the increasing reproductive value of offspring and the decreasing reproductive value of parent females (Tallamy 1982). Young mothers with high reproductive value take fewer risks while defending offspring and defend less aggressively than older mothers whose potential contribution to future generations is limited. In this study we investigate variation in lace bug maternal behavior in terms of its genetic basis, maintenance in natural populations, and adaptive significance.

Maternal Behavior

Gargaphia solani is a prolific, iteroparous lace bug with large egg-clutches and short generations, completing as many as eight generations/yr in Virginia (Bailey 1951). This species is common on horsenettle (*Solanum carolinense* L.; Solanaceae) in the middle Atlantic states and ranges as far west as Iowa (Drake and Ruhoff 1965). Maternal behavior in *G. solani* has apparently evolved as a means of protecting eggs and nymphs from invertebrate predators (Fink 1915, Tallamy and Denno 1981a, 1981b). Females cement their eggs in discrete masses to the undersurface of horsenettle leaves over a period of 2 to 3 days. As the mass is completed, females undergo a physiological suppression of egg production and begin to exhibit maternal behavior. Mothers remain with and aggressively defend their eggs and gregarious nymphs throughout all five nymphal stadia. As the young mature and disperse, females produce a second egg mass and resume their maternal duties. Two to three clutches are reared before senescence.

Maternal care in *G. solani* consists of at least five distinct behavioral components that serve to detect and repel predators. The speed with which mothers react aggressively to an approaching predator often determines the number of nymphs that will be lost before the predator can be driven away. When nymphs are attacked by a predator, mothers launch an aggressive display that can include (1) a wing fanning response, (2) ramming the predator head on, and (3) chasing the predator some distance from the nymphal aggregation. To protect against persistent predators, females must respond several times, often in rapid succession. The ability to do so further increases their success in predator defense.

Genetic Variability

To estimate the extent to which variation in the maternal behavior of *G. solani* is determined genetically, we created 19 isofemale lines from mated females collected at several locations in September 1980 in College Park, Prince Georges Co., Maryland. Since each line consisted of the descendants of a single female, consistent behavioral differences between lines under controlled conditions were considered genetic in nature (Parsons 1980). Each line was inbred four generations prior to testing. In a controlled greenhouse environment (25 °C, 15L/9D), ten females per line were mated and reared in net sleeves on potted horsenettle plants. Each female was permitted to guard her first clutch until maturity.

Beginning on the day of first oviposition, components of maternal defensive behavior were quantified at 3-day intervals in the following way: natural encounters with predators were simulated by tethering a dead adult coccinellid beetle, a common lace bug predator, on a thin wire and "guiding" it for 30 sec in a standardized pattern toward, around and away from female lace bugs. A single encounter with the predator consisted of five such passes around the female, providing the opportunity for up to five separate responses. For each encounter we measured the latency of the first aggressive response from the female (maximum of five), the number of times the female rammed the beetle, and the maximum distance the female chased the beetle from her young during the encounter. Since *G. solani* females remain with each clutch approximately 19 days, defensive behavior was quantified up to six times per female. Variation between lines in each of the behavioral components quantified is graphically summarized in Figure 1. The results are conclusive; there were significant differences among lines for all five components of the defensive behavior (ANOVA: $F_{18,164}$: $p < .001$). In particular, females from lines 6 and 10 responded to predators faster and more often, fanned and rammed predators more aggressively, and chased predators farther from their offspring than

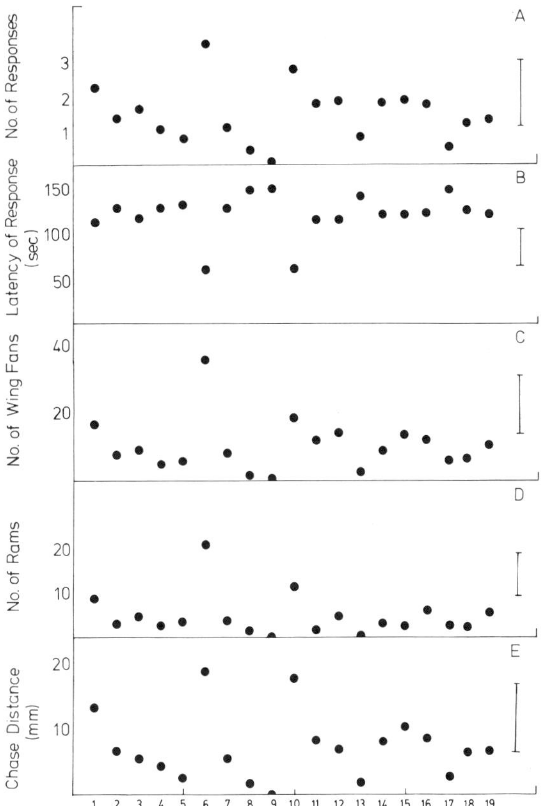

Fig. 1. Mean response of five components of maternal defense in *Gargaphia solani* to simulated predator attack in 19 isofemale lines. Interval about means = Standard error of the mean. Statistical interval on the right = Scheffe's multiple comparison test. Means differing by this interval or more are significantly different ($p < .05$).

females from other lines. In contrast, females from lines 9 and 13 showed little or no defensive behavior and either passively ignored predators or retreated entirely from the encounter. The remaining lines exhibited more or less intermediate levels of defense.

Ecological Considerations

The results appear paradoxical for, in spite of its relation to fitness, there is considerable genetic variation in the defensive behavior of *G. solani* females. One explanation of this paradox can be found in an analysis of the costs and benefits of lace bug maternal care. In *G. solani*, maternal care has evolved in response to intense predation on the juvenile stages of development (Tallamy and Denno 1981a). If predators are excluded from this system, nearly 80% of the nymphal lace bugs survive to maturity with or without maternal protection (Fig. 2). When predators are not excluded, however, maternal defense improves nymphal survivorship sevenfold. Clearly, maternal care is an effective solution to the predation problem facing *G. solani*, but it is also an ecologically expensive solution (Tallamy and Denno 1982). The protection of a clutch of young through all five nymphal stadia requires a maternal time invest-

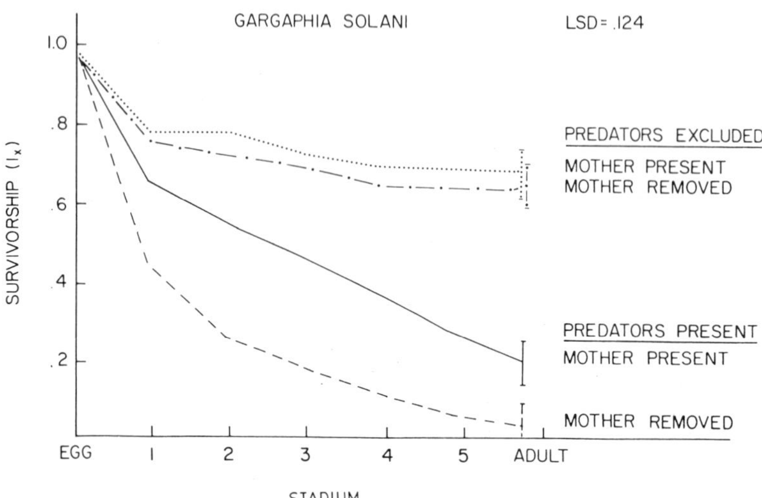

Fig. 2. Effects of maternal care on the juvenile survivorship of *Gargaphia solani* in the presence and absence of predators. LSD = Least Significant Difference multiple comparison test. Means differing by this interval or more are significantly different ($p < .05$) (From Tallamy and Denno 1981b).

ment of 19 days, nearly one-half of an adult female's life span. Since egg production is physiologically suppressed throughout this period, maternal care reduces fecundity significantly. In fact, females that consistently abandon their eggs lay more than twice as many eggs as females that guard and, more importantly, lay most of their eggs much earlier in life. Thus, if females that do not guard manage to avoid or escape predation, they enjoy a considerable advantage in terms of progeny surviving to maturity over females that protect their young.

The evidence suggests that maternal behavior in *G. solani* is balanced between two powerful selection pressures. When predators are abundant, there is pressure to protect young, for without protection nymphal survivorship is negligible. When predation is less intense, however, the reduction in fecundity that is associated with maternal care selects against such behavior. It follows, then, that if predation is sufficiently variable in space and/or time, considerable genetic variation in maternal "quality" can be maintained in lace bug populations.

Populations of invertebrate lace bug predators are subject to significant seasonal fluctuations that are largely unpredictable in nature. Spring densities of predators are a function of overwintering success and may vary widely from year to year. Further fluctuations result as predator populations track changes in the density of numerous prey species (Hassell 1978).

Predator density also may vary spatially, creating "safe sites" in which lace bug maternal behavior is not advantageous. Horsenettle is a biennial, stoloniferous *Solanum* that characteristically grows in patches among early successional vegetation. Ramets (individual members of a clone) can be found in at least two structurally different microhabitats. Many stand isolated from surrounding vegetation; that is, they do not physically touch other plants, particularly in areas that have been grazed recently or otherwise disturbed. Other ramets are enmeshed within dense patches of vegetation. Although *G. solani* densities are high, generalist predators such as Nabidae, Reduviidae, Coccinellidae, Chrysopidae and jumping spiders in the family Salticidae may be more numerous in overgrown patches because of an abundance of invertebrate prey items and refugia.

We investigated the possibility that predator density varies significantly between horsenettle microhabitats and on a seasonal basis at two study sites: Newark, New Castle Co., Delaware, and College Park, Prince Georges Co., Maryland. To determine if predators are more numerous in overgrown horsenettle patches, we recorded the number of known lace bug predators on 400 overgrown and 400 isolated horsenettle ramets in June, July, and August of 1982 at the College Park study site (Fig. 3). The results depict no consistent pattern in predator abundance on isolated and overgrown ramets. Early in the season there were significantly more predators on isolated ramets, but by midseason predators were found in significantly higher numbers on overgrown plants (ANOVA: $F_{12,394}$;$p < .01$). In August, predator density did not differ between the two microhabitats. The data are too erratic to evaluate the exact effect of microhabitat on predator density. They do suggest, however, that predator density varies significantly and possibly in an unpredictable way between isolated and overgrown horsenettle ramets.

The density of lace bug predators also varied significantly through time at the Delaware study site in 1981 (Fig. 4). Predators were significantly more numerous in July and August than in June, suggesting that progeny of females that did not guard their young aggressively had a better chance of escaping predation early in the season. These and the data presented above support the contention that predation in horsenettle patches is variable in time and space. To extrapolate the effects of such variation on the reproductive success of nonaggressive females, we compared the survivorship of juvenile lace bugs with and without maternal protection in June, July, and August, and on isolated and overgrown horsenettle ramets at the Delaware site in 1982 (Fig. 5). Seasonal differences in nymphal survivorship were not significant (ANOVA: $F_{2,112}$;$p < .093$), but differences between the survival of nymphs on isolated and overgrown ramets were highly significant (ANOVA: $F_{3,112}$;$p < .0001$). With and without maternal protection, more young survived to maturity on isolated ramets than on ramets growing within dense vegetation.

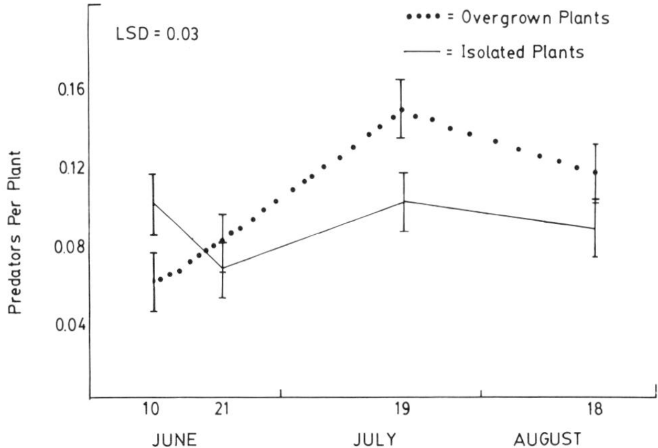

Fig. 3. Seasonal variation in lace bug predator density on overgrown and isolated horsenettle ramets. LSD = Least Significant Difference multiple comparison test. Means differing by this interval or more are significantly different ($p < .05$).

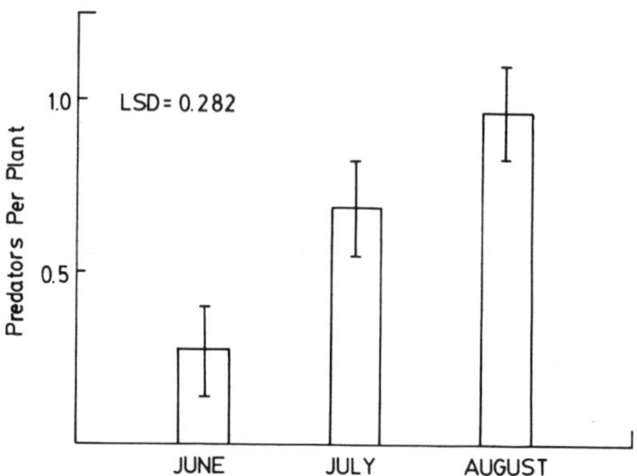

Fig. 4. Seasonal variation in the density of lace bug predators inhabiting patches of horsenettle. LSD = Least Significant Difference multiple comparison test. Means differing by this interval or more are significantly different ($p < .05$).

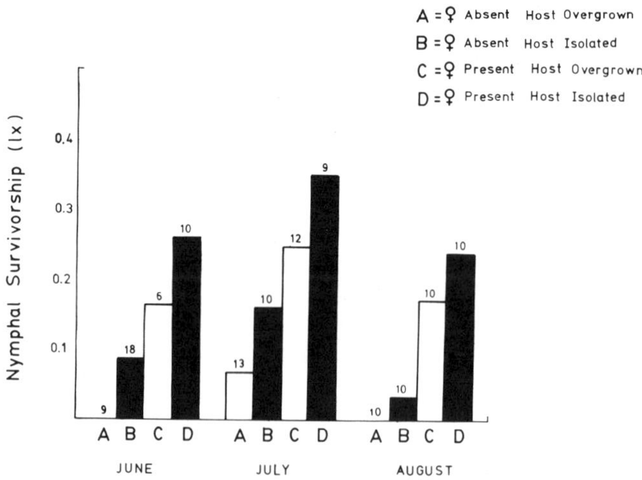

Fig. 5. Seasonal survivorship of *Gargaphia solani* nymphs through 5th instar with and without maternal protection on isolated and overgrown horsenettle ramets. Least Significant Difference multiple comparison intervals for month and microhabitat are 0.053 and 0.141 respectively.

Discussion

Our analysis of maternal defensive behavior in isofemale lines of *G. solani* reveals more genetic variation than is generally predicted for fitness traits under uniform, directional selection. When relevant environmental conditions are constant, selection is thought to reduce genetic variation with time (Lewontin 1965, Maynard Smith 1979), though Lande (1976) presents a theoretical argument suggesting that recurrent mutation alone can adequately preserve variability in populations. If, however, there are repeated fluctuations in one or more of the environmental variables that affect a trait, selection will preserve genetic variability in the trait, assuring its adaptive expression in any given state of a changing environment (Istock 1981a). Extensive variation in behavioral traits linked strongly to fitness is not an unrecognized phenomenon. Considerable environmental and genetic variation has been reported in the migration and diapause responses of several insect species (Dingle et al. 1977, Istock 1978, 1981b, Lynch and Hoy 1978, Tauber and Tauber 1978, Showers 1981). In each case, this variation has been attributed to variation in the environmental conditions to which these species must adapt.

The results presented here are far from definitive, but field surveys and manipulations support the contention that genetic variation in the maternal defensive behavior of *G. solani* is maintained within populations by spatial and temporal variation in the intensity of predation. Diurnal lace bug predators can be more numerous as the season progresses and in overgrown areas of horsenettle patches. We do not, however, wish to stress the spatial and temporal patterns observed in predator populations; the data are as yet too few for such predictions. We do emphasize the degree to which predation and thus lace bug survival vary within local populations, for these fluctuations provide a means by which the high ecological costs (reduced fecundity) of maternal care in *G. solani* can be minimized. Passive maternal behavior or desertion of young results in the maturation of fewer progeny per clutch, even when predation on juveniles is less severe. There is no advantage to such behavior when predation is intense, but it is entirely possible that, because of the inverse relationship between maternal behavior and fecundity, reductions in maternal care are favored when predation pressure slackens. If predation is relaxed temporally or spatially, "bad mothers" may be more than able to compensate for predation losses through increased fecundity. We conclude that repeated fluctuations in selection intensity have interacted with the ecological constraints of maternal care in *G. solani* to create a genetically based continuum of maternal defensive behaviors, each adaptive at a given level of predation.

Acknowledgments

We gratefully acknowledge the comments and criticisms of George Eickwort, Joseph Hegmann, Conrad Istock and James Leslie, as well as the statistical advice of Leonard Freed. This project was supported by a Neurobehavioral Science Postdoctoral Fellowship and is published with the approval of the Director of the Delaware Agricultural Experiment Station as Misc. Paper No. 1017, Contribution No. 538 of the Department of Entomology and Applied Ecology, University of Delaware, Newark, Delaware.

Literature Cited

Bailey, N. S. 1951. The Tingoidea of New England and their biology. Ann. Entomol. Soc. Amer. 31: 1.

Bro Larsen, E. 1952. On subsocial beetles from the saltmarsh, their care of progeny and adaptation to salt and tide. Trans. Int. Congr. Entomol. 11th, 1951. 1: 502.

Dingle, H., C. K. Brown, and J. P. Hegmann. 1977. The nature of genetic variance influencing photoperiodic diapause in a migrant insect, *Oncopeltus fasciatus*. Amer. Natur. 111: 1047.

Drake, C. J., and F. A. Ruhoff. 1965. Lacebugs of the World: A Catalog (Hemiptera: Tingidae) Smithson. Bull. 243.

Eberhard, W. G. 1975. The ecology and behavior of a subsocial pentatomid bug and two scelionid wasps: strategy and counter-strategy in a host and its parasitoids. Smithson. Contrib. Zool. 205: 1.

Fink, D. E. 1915. The eggplant lace-bug. U. S. Dept. of Agric. Bull. 239: 1.

Hassell, M. P. 1978. The Dynamics of Arthropod Predator-Prey Systems. Princeton University Press, Princeton.

Istock, C. A. 1978. Fitness variation in a natural population. *In* Dingle, ed. Evolution of Insect Migration and Diapause. Springer-Verlag, New York.

Istock, C. A. 1981a. Natural selection and life history variation: Theory plus lessons from a mosquito. *In* Denno, Dingle, eds. Insect Life History Patterns: Geographic and Habitat Variation. Springer-Verlag, New York.

Istock, C. A. 1981b. The extent and consequences of heritable variation for fitness characters. *In* King, Dawson, eds. Population Biology Retrospect and Prospect. Oregon State University Colloquium: Biology. Columbia University Press, New York.

Lande, R. 1976. The maintenance of genetic variability by mutation in a polygenic character with linked loci. Genet. Res. 26: 221.

Lande, R. 1976. The maintenance of genetic variability by mutation in a polygenic character with linked loci. Genet. Res. 26: 221.

Lewontin, R. D. 1965. Selection for colonizing ability. *In* Baker, Stebbins, eds. The Genetics of Colonizing Species. Academic Press, New York.

Lynch, C. B., and M. A. Hoy. 1978. Diapause in the gypsy moth: environmental specific mode of inheritance. Genet. Res. 32: 129.

Maynard Smith, J. 1979. The effects of normalizing and disruptive selection on genes for recombination. Genet. Res. 33: 121.

Parsons, P. A. 1980. Isofemale strains and evolutionary strategies in natural populations. *In* Hecht, Steere, Wallace, eds. Evolutionary Biology. Vol. 13. Plenum Press, New York

Showers, W. B. 1981. Geographic variation of the diapause response in the European corn borer. *In* Denno, Dingle, eds. Insect Life History Patterns: Geographic and Habitat Variation. Springer-Verlag, New York.

Tallamy, D. W. 1982. Age specific maternal defense in *Gargaphia solani* (Hemiptera: Tingidae), Behav. Ecol. Sociobiol. 11: 7.

Tallamy, D. W., and R. F. Denno. 1981a. Maternal care in *Gargaphia solani* (Hemiptera: Tingidae). Anim. Behav. 29: 771.

Tallamy, D. W., and R. D. Denno. 1981b. Alternative life history patterns in risky environments: an example from lace bugs (Hemiptera: Tingidae). *In* Denno, Dingle, eds. Insect Life History Patterns: Geographic and Habitat Variation. Springer-Verlag, New York.

Tallamy, D. W., and R. F. Denno. 1982. Life history trade-offs in *Gargaphia solani* (Hemiptera: Tingidae): the cost of reproduction. Ecol. 63: 616.

Tauber, M. J., and C. A. Tauber 1978. Evolution of phenological strategies in insects: a comparative approach with ecophysiological and genetic considerations. *In* Dingle, ed., Evolution of Insect Migration and Diapause. Springer-Verlag, New York.

Wood, T. K. 1976. Alarm behavior of brooding female *Umbonia crassicornis* (Homoptera: Membracidae). Ann. Entomol. Soc. Amer. 69: 340.

Behavioral Analysis of Male-Induced Interstrain Differences in Realized Fecundity in *Callosobruchus maculatus*

Steven S. Wasserman

Department of Biology
University of Virginia
Charlottesville, Virginia 22901

Introduction

The role of sexual selection in the evolution of insect mating systems has recently come under close scrutiny. Intrasexual selection, due to competition for mates among members of a single sex, has been demonstrated in a variety of insects (cf. Blum and Blum 1979). Several morphological, physiological and behavioral traits have been shown to have arisen by this process. Wilson (1975) cites the weaponry of hercules beetles used in combat between males courting a single female. Several strategies have evolved to enhance the likelihood that a male's sperm will fertilize his mate's eggs: a passive phase of courtship (Parker 1970), secretions of the male that lower receptivity of recently-mated females (Craig 1967) and sperm precedence/displacement systems (Walker 1980), among others.

Realized fecundity, the number of eggs laid, is generally considered a trait of females: the manifestation of an individual's propensity to garner energy for oogenesis, to mate and to find suitable oviposition substrates. The influence of a male on the realized fecundity of his mate is rarely considered. I examine here not the contribution of the female but of her mate, to the number of eggs she lays.

This study has a very different approach from many of the reports in this volume. Rather than observing variation in nature and attempting to show its underlying genetic framework, I start with two populations showing differences in realized fecundity, and ask whether these differences are due to behavioral causes.

Two independent lines of research have led to this study. First, Mark (1981) demonstrated a peculiar sex-related mode of inheritance in the bruchid beetle *Callosobruchus maculatus* (F.); two day oviposition of hybrids between strains differing in oviposition rate was more similar to the paternal than the maternal strain. Because his females were mated with males of their own treatment, he could not distinguish which sex was responsible for the behavioral differences. Second, in crosses among three geographic strains of *C. maculatus*, I noted an effect of strain of origin of males on their mates' daily and overall realized fecundities. Females mated with one strain laid up to 10 (and in a replicate experiment, 20) eggs more than their counterparts mated with a different strain.

A male effect on oviposition could have several explanations. First, males could interfere behaviorally with their mates, perhaps by interrupting oviposition behavior. This has been shown to be not a difference between strains but, rather, a density effect on realized fecundity (e.g., Pearl 1932). However, Menusan (1935) denies a density effect in bruchids. Under this hypothesis, then, males of strains yielding high numbers of eggs should be less active than low-fecundity males. Second, there might exist positive non-mating behavioral interactions between mates; for example, females disturbed while resting might move to a host and oviposit. Under this hypothesis, more active males would cause higher fecundity in their mates. Third, if sperm are limited (unlikely) or have low viability, the length and number of copulations would affect fecundity. Hence, males of strains yielding more fecund mates should mate more often or for longer periods per copulation. Fourth, it is possible that non-behavioral interactions could effect these patterns (cf. Thornhill 1976, Walker 1980). Hence, strains could differ in the contribution of accessory gland secretions incorporated into yolk by their mates (Friedel and Gillott 1977) or in the speed at which females mature their egg complements (Pratt and Davey 1972, Davey 1980). Finally, and least likely, the differences in realized fecundity might be attributed to a male effect on egg size. It should be noted, however, that in many insects the chorion is produced before mating occurs.

In this preliminary study, I examine the three behavioral hypotheses (numbers 1, 2 and 3 above). I ask whether males of two strains differ behaviorally in a way that would explain differences in oviposition by their mates. In this report, I use as shorthand the term "fecundity" in place of "realized fecundity." I have no record of the potential fecundity of these beetles, as I did not dissect females after their deaths.

Materials and Methods

Callosobruchus maculatus: life history

The southern cowpea weevil attacks grain legumes in both fields and warehouses. Eggs are glued singly to the surface of the seed coat; larvae hatch inward, excavating a feeding and pupation cavity within the endosperm. Pupation and eclosion occur in this chamber. Adults chew through the testa to emerge.

Mating, occurring soon after emergence, can be divided into three distinct phases: palpation (antennal palpation for about a minute after contact with the female: intromission occurs just after the initial antennal palpation); stability (both partners motionless for several minutes); and kicking (female kicks the male's aedeagus, he retracts his legs and sits passively until dislodged). Occasionally, the male turns away from the female during the kicking stage, and the two walk in opposite directions.

Females can oviposit within 10 mins of mating. Oviposition behavior encompasses four phases: the female slowly patrols a bean, turning often to examine its surface for previously laid eggs and to assess its quality as a substrate; after the final turn, she extrudes her internal genitalia and rubs the seed coat; she stands motionless for about 1 min; the egg is deposited. The entire cycle takes about two or 3 min. Females live 7 to 12 days, during which they lay 60 to 150 eggs. Virgin females lay less than five viable eggs.

The life cycle is completed in about three weeks at 29 °C and 73% R.H., the conditions used for this study. Beetles were reared in constant light, on crowder beans (*Vigna unguiculata* (L.) Walp. var. Mississippi silver) or blackeye peas (*V. unguiculata* var. blackeye).

Interstrain crosses

Three strains of *C. maculatus* were reared on a single host for one generation and paired in all possible combinations. The *A* strain has been reared in the laboratory since the 1940s

(Wasserman and Futuyma 1981); *W* was collected recently in Wisconsin (see Qi and Burkholder, 1982); *F* was provided from a field infestation in Florida by Dr. David W. Hagstrum.

Individual newly emerged (aged 0 to 24 h) virgin males and females were placed in 60x15 mm petri dishes with 10 beans. Beans were replaced daily with fresh ones; daily oviposition was recorded. Data collection was discontinued for a pair when the female died or laid no eggs on two successive days. Females not laying more than a total of 10 eggs were considered uninseminated and discarded.

The first interstrain cross used blackeye peas (BE) as hosts and oviposition substrates; for the second, crowder beans (CR) were used.

Behavioral tests

All of the treatments used a similar regimen. Each strain was reared on CR for one generation before these tests. A pair of newly emerged beetles was placed in a 60x15 mm petri dish with 10 CR on one side. I observed at least the first mating. After the observation period, another 30 CR were added to the dish. Fecundity was measured after 24 h and after the death of the female.

Treatment PAIR beetles were observed during the first h of their confinement. I recorded the number of matings, time to first mating, time elapsed from mounting to kicking, total duration of each copulation, number of male-female encounters, length of courtship chases, number of eggs laid in 1 h, and number of ovipositions interrupted by the male. The positions (on dish/bean) and activities (resting/patrolling) of females were recorded before and after several of their encounters with males. Two sub-treatments were run in PAIR: *A* males x *A* females, and *F* males x *A* females.

Treatment ONE (i.e., one copulation) males were allowed to mate once with their females and discarded. This treatment examines the relative availability or longevity of sperm. Only *A* male x *A* female matings were run.

Treatment MANY (i.e., many copulations) males mated once the first day, and were removed from the dish, but retained in a one dram vial. Daily, until the female died, males and females were reunited in a petri dish for 20-30 min and allowed to mate no more than once. Mating success and duration were recorded daily. MANY females differ from PAIR females by not encountering males other than during a short mating period. Hence, differences in fecundities of these females should be due to non-mating behavioral interactions between mates. Only *A* male x *A* female matings were run.

Results

Interstrain crosses

In the BE experiment, there were differences in fecundity attributable to the strain of both males and females (two-way anova; Table 1a). Relatively large numbers of eggs are laid by females during the first four days of oviposition; hence, daily oviposition records were analyzed separately. A similar pattern was seen in day 1 and 3 fecundities but not for days 2 or 4, where only differences among female strains were detected. Although females of any of the strains did not lay statistically different numbers of eggs when mated to *F* or *A* males (sum of squares-STP test), *F*-mated females laid more eggs for all strains (Table 1a). These trends were consistent over the first five days of oviposition; females mated to *F* males laid more eggs on 3 of 5 (*A* females) or 4 of 5 (*W* and *F* females) days than did their A male-mated counterparts.

In the CR experiment, overall fecundity differed due to both males and females (interaction significant; two-way anova; Table 1b). This pattern is repeated for days 3 and 4, but for

Table 1. Realized fecundities of females in two interstrain crosses: standard error, sample size.

Strain of female	Strain of male *A*	*F*	*W*
a. Blackeye peas as host			
A	64.7 4.6 13	73.8 2.9 15	74.1 2.5 12
F	94.6 3.7 14	101.6 4.4 13	104.6 4.5 9
W	75.2 2.9 13	80.4 3.1 14	78.1 3.5 13
b. Crowder beans as host			
A	83.1 3.1 17	86.2 2.2 22	88.3 3.3 15
F	100.4 4.9 11	122.8 4.6 17	112.2 3.7 16
W	102.0 5.1 15	108.4 2.6 17	99.1 3.5 11

days 1 and 2 of oviposition only the strain of female affected the number of eggs laid. Females of each strain, mated to males of *A* and *F*, did not differ in overall fecundity (Dunn-Sidak test, $p > .05$, Sokal and Rohlf, 1981).

Behavioral tests

I first examined the effect of non-mating behavioral interactions on fecundity. If such interactions were important, then PAIR females should lay fewer eggs than females of treatment MANY. On the first day, there was no difference in oviposition of these treatments (Mann-Whitney two-tailed test, $.10 < p < .20$). Overall, fecundities for these differed greatly (two-tailed $p = .001$); there is a 30 egg difference in the expected direction. These patterns support an overall negative effect of male-female encounters on fecundity, the first hypothesis above. Moreover, there is no evidence that females move to beans when encountered on the dish by their mates (2x2 contingency G-test with Williams' correction, $G = .127$; $p = .722$), though they are more likely to patrol after a disturbance ($G = 4.96$; $p = .026$). Only one female was observed to move directly from resting on the dish to laying an egg after an encounter with her mate. Hence, these data do not support hypothesis two.

In light of these trends, I abandon the search for positive non-mating behavioral interactions, in exchange for greater ease in detecting their negative counterparts. Hence, I apply one-tailed hypotheses to my non-mating data: *A* males are expected to encounter their mates more often (i.e., more active patrolling), and chase them longer per encounter (and overall) than *F* males. This was observed: for number of encounters, total length of courtship chases, and chase time per encounter, *A* males show greater activity (Mann-Whitney test, $p < .05$ for each). The number of disrupted oviposition events was greater for *A* males than *F* males, though not significantly so (Table 2).

Patterns of copulation also differed between strains of males (Table 2). Because of *a priori* hypothesis three above, one-tailed tests are used here as well. There was no difference between strains in the time to first copulation. *F* males mated for a longer period of time in their first copulation ($p < .05$), and, on the average, more times in l h ($p < .001$) than did *A* males. The interstrain difference in length of copulation appears to be due not to the stable phase ($.10 < p < .20$), but to the length of time a male remains coupled during the kicking phase ($p < .025$). Females mated to *F* or *A* strains showed no difference in average number of copulations by Mann-Whitney test (Table 2), but a higher proportion of *F*-mated females mated more than once (t-test on equality of percentages; $t_s = 1.606$, one-tailed $p = .054$).

Fecundity patterns differed marginally, at best, between *A*- and *F*-mated females. Though not statistically significant with sample sizes of 15, *F*-mated females showed higher 24-h and overall egg deposition, as predicted from the interstrain cross data. During the first h, similar number of eggs were laid (Mann-Whitney test, l-tailed $p > .10$). After 24 h, there was a three-egg difference between these matings ($.05 < p < .10$). Overall fecundity differed by an average of over five eggs ($p > .10$).

There was evidence that more than a single copulation was required to fertilize the entire egg complement of a female. Treatment MANY females laid more eggs (119.0 ± 8.1) than did females mated once (treatment ONE, 95.3 ± 5.3 eggs) (Mann-Whitney one-tailed test, $p < .0025$).

Discussion

The two strains of males, which cause different fecundities in their mates, also differ in mating behavior; differences in overall mating period are probably due to difference in the length of the kicking phase. From two lines of evidence, I believe the interstrain differences in the kicking period (and hence the duration of the entire copulation) does not cause a detectable differences in fecundity. First, if during the kicking phase substantial amounts of sperm were transferred to the female, then for ONE females there should be a positive correlation between fecundity and duration of kicking. Although copulation period and fecundity appear positively correlated (Spearman's $r_S = .482$, $p < .05$), this is due to two outliers. When these points are removed from the analysis, the covariance disappears ($r_S = .143$). There is no covariance between fecundity and duration of kicking even with the outliers included ($r_S = -.045$). Second, females mate several times during their lives: MANY included one pair that mated for nine consecutive days, though the probability of mating on a given day dropped below 50% on day six; in other studies similar to PAIR, matings were observed as late as day six. Hence, it is unlikely that interstrain differences in mating behavior are responsible for observed fecundity patterns.

Table 2. Summary statistics for variables measured in behavioral tests (treatment PAIR): mean, (standard error). Sample sizes are 15 for *A* males, 14 for *F* males, unless noted by superscript. Mann-Whitney test was performed on PAIR except for durations of phases of copulation, where *A*-males are lumped with data of ONE and MANY. Probabilities shown are one-tailed (see text).

Variable	*A* males	*F* male	p value
Time to first copulation (sec)	87.7 (15.3)	65.3 (18.5)	p > .10
Number of copulations	1.27 (0.12)	1.50 (0.20)	p > .10
Duration of first copulation (sec)	420.8 (21.0)	550.1 (33.0)	p < .005
Duration of non-kicking phase (sec)	343.7 29.2)	396.1 (33.0)	.10 < p < .20
Duration of kicking phase (sec)	74.7 (14.9)	153.9 (28.2)	p < .025
No. of male-female encounters	21.5 (2.5)	15.1 (1.8)	p < .05
Total length of courtship chases (sec)	250.2 (36.8)	145.9 (28.5)	p < .05
Chase time per encounter (sec)	12.1 (1.5)	9.0 (1.7)	p < .05
Number of oviposition events disrupted	0.53 (0.26)	0.36 (0.20)	p > .10
Number of eggs laid, 1 hour	0.67 (0.30)	0.62** (0.33)	p > .10
Number of eggs laid, 24 hours	33.1* (2.1)	36.6** (1.5)	.05 < p < .10
Overall fecundity	89.2*** (2.5)	94.5*** (4.1)	p > .10

*N = 14; **N = 13; ***N = 12

Strain *A* and *F* males also differ in the proportion of females mating with them more than once during the first h. In *F*-male crosses of PAIR, the overall fecundity of females was related to their first h copulation frequency (Mann-Whitney one-tailed test, $p < .005$). (The same analysis is not significant in the *A* crosses of PAIR). This suggests either that males copulating more than once passed more sperm during the critical early reproductive life of females, or that this high mating frequency is maintained throughout the male's lifespan, or that day one mating frequency is positively correlated with some other parameter promoting high egg production.

Non-copulatory encounters between mates can also account for a portion of interstrain differences in fecundity. *A* males patrolled their dishes more (i.e., made more contacts with their females), and were more persistent at chasing after the initial contact. However, this persistence seldom paid off in a second copulation during the first h. Within strains, males mating more than once in 1 h encountered females *less* often than did males mating once only, for both strains (Mann-Whitney one-tailed $p < .05$); this probably reflects differences in activity after the initiation of the second copulation. *A* males averaged about 24 min to the second copulation, *F* males some 15 min. To test whether more active males copulate more often, I compared, within each strain, the number of encounters in the first 24 or 15 min for once- and multiply-mated males. No differences were found (Mann-Whitney test, $p > .10$ for each strain). Interstrain differences in numbers of encounters, when controlled for the effect of second mating on activity of males, remain as reported above (Mann-Whitney test, $p < .025$). Therefore, differences in male activity within strains do not affect mating success; rather, qualitative differences in courtship or in previous insemination of the female govern mating frequency. Between strains, both differences in male activity and male quality may explain patterns of mating success.

It is tempting to speculate as to why these interstrain genetic differences in male behavior exist. However, since these experiments are based on two populations, a conclusive answer is impossible; genetic differences might be due to differential adaptation to the laboratory milieu, population bottlenecks, or other ecological differences experienced by the strains. Still it might be asked why a seemingly counterproductive trait like highly active males might arise. One answer is that high activity is being selected against, but without a response in the *A* strain. In contrast, it is likely that high-activity female chasers are at a selective advantage for other reasons. For instance, in *C. maculatus*, copulations can be disrupted during the first min (S. P. Martindale and S. S. Wasserman, unpublished); active males may be adapted to discover and supplant other males *in copula*. Second, behavioral observations indicate that virgin females overwhelmingly accept the courtship of the first male encountered. Also, since multiple matings can raise realized fecundity by about 25 percent, a previously mated female should eventually become receptive to courting males. In either case, highly active males may leave greater than average numbers of offspring.

In conclusion, interstrain differences in realized fecundity are due to both male and female contributions. Certain strains of males were consistently more fecund than others. Behavioral tests on one such pair of strains indicated two likely sources of male-induced interstrain variation in realized fecundity: differential frequency of copulation and differential numbers of non-copulatory encounters between mates. Differences in the mating behaviors of these strains are probably not responsible for patterns of realized fecundity. The analyses reported here do not rule out the possibility of physiological mechanisms influencing these fecundity patterns.

Literature Cited

Blum, M. S., and N. A. Blum, eds. 1979. Sexual selection and reproductive competition in insects. Academic Press, New York.

Craig, G. B., Jr. 1967. Mosquitoes: female monogamy induced by male accessory gland substance. Science 156: 1499.

Davey, K. G. 1980. The physiology of reproduction in *Rhodnius* and other insects: some questions. *In* M. Locke and D.S. Smith, ed. Insect Biology in the Future. Academic Press, New York.

Friedel, T., and C. Gillott. 1977. Contribution of male-produced proteins to vitellogenesis in *Melanoplus sanguinipes*. J. Insect Physiol. 23: 145.

Mark, G. A. 1981. Anomalous sex-related inheritance of oviposition rate in *Callosobruchus maculatus* Fab. (Coleoptera: Bruchidae). Behav. Genet. 11: 145.

Menusan, H. 1935. Effects of constant light, temperature, and humidity on the rate and total amount of oviposition of the bean weevil, *Bruchus obtectus* Say. J. Econ. Entomol. 28: 448.

Parker, G. A. 1970. Sperm competition and its evolutionary consequences in the insects. Biol. Rev. 45: 525.

Pearl, R. 1932. The influence of density of population upon egg production in *Drosophila melanogaster*. J. Exp. Zool. 63: 57.

Pratt, G. E., and D. G. Davey. 1972. The corpus allatum and oogenesis in *Rhodnius prolixus*. III. The effect of mating. J. Exp. Biol. 56: 223.

Qi, Y. -T., and W. E. Burkholder. 1982. Sex pheromone biology and behavior of the cowpea weevil, *Callosobruchus maculatus* (Coleoptera: Bruchidae). J. Chem. Ecol. 8: 527.

Sokal, R. R., and F. J. Rohlf. 1981. Biometry. 2nd ed. W.H. Freeman, San Francisco.

Thornhill, R. 1976. Selection and parental investment in insects. Amer. Natur. 110: 152.

Walker, W. F. 1976. Sperm utilization strategies in nonsocial insects. Amer. Natur. 115: 780.

Wasserman, S. S., and D. J. Futuyma. 1981. Evolution of host plant utilization in laboratory populations of the southern cowpea weevil, *Callosobruchus maculatus* Fabricius (Coleoptera: Bruchidae). Evolution 35: 605.

Wilson, E. O. 1975. Sociobiology: the New Synthesis. Belknap Press, Cambridge, Massachusetts.

A Genetic Analysis of Habitat Selection in the Cactophilic Species, *Drosophila mojavensis*

Katherine L. Lofdahl

Committee on Evolutionary Biology
University of Chicago
Chicago, Illinois 60637 USA

Introduction

Past studies of habitat selection in insects have addressed various relationships among variation, heredity and natural selection; three basic factors for the process of evolution. Most research on habitat choice in insects concentrates on the selection of host plants as feeding and oviposition sites because these resources are demonstrably related to fitness (Whitham 1980). Indeed, the concept of host plant races even implies a degree of coevolution of an insect with such resources (Jaenike 198l). Demonstration that two populations or species currently using different host plant species also have different loci or alleles at a single locus that exerts control over habitat choice is a frequent test for past coevolution with the respective host plants (Huettel and Bush 1972). These studies are helpful in comparing the outcome of evolution with theoretical models of population subdivision and their associated speciation patterns (Maynard Smith 1966, Bush 1969). To predict future evolutionary possibilities, however, an investigation of genetic polymorphism for habitat preference behavior within a single population is valuable. Establishing that individual genetic differences in host plant preferences exist allows the prediction of a response to any form of natural selection: directional, stabilizing or disruptive. Quantification of the amount of genetic variation in behavior, which is the trait's heritability when measured as a precentage of the phenotypic variance, reveals the rate of evolution of behavior when information on the intensity of natural selection is available also (Falconer 1981). The aims of this study are: (1) to learn whether a monophagous insect population has genetic variation for acceptance of an unfamiliar host plant species; and (2) to quantify the amount of such genetic variation in behavior that is available to produce a response to natural selection in future generations.

A search for genetic variation can usefully begin with a species that shows geographic variation in the relevant behavioral phenotype. An ecologically well known insect that exhibits geographic variation in host plant utilization is the cactus breeding species, *Drosophila mojavensis* Patterson and Crow. Most populations use only one or two host cactus species, but there is geographic replacement in the species of cacti used (Fellows and Heed 1972). Thus, *D. mojavensis* uses five to six cactus species as breeding sites when all populations are considered together. Within the center of its distribution, in the Sonoran Desert of North America, the Arizona and Sonora (Mexico) populations use organpipe cactus, *Stenocereus thurberi* (Engelm.) Buxbaum, with a possible seasonal shift to saguaro cactus, *Carnegiea gigantea* (Engelm.) Br. & R. In Baja California, *D. mojavensis* breeds almost exclusively in agria cactus, *Stenocereus gummosus* (Engelm). Britt. & Rose, despite the presence of organ-pipe cactus there. A peripheral population in the Mojave Desert uses barrel cactus, *Ferocactus acanthodes* (Lemairz) Br. & R. A second isolated population, located on Santa Catalina Island, California, breeds in the only available cactus (genus *Opuntia*). The present research asks whether this monophagous *Opuntia*-breeding population possesses selectable genetic variation for use of another cactus, agria, which is highly preferred by *D. mojavensis* wherever the insect and agria occur together.

Experimental studies of habitat selection require a definition of this behavior in terms relevant to the ecology of the species investigated. In *D. mojavensis*, the behavior is designated correctly as habitat rather than as host plant selection. *Drosophila mojavensis* breeds only in cactus necroses in which the actual resource is the microbial community of bacteria and cactophilic yeasts that serve as food for both adults and larvae. The fresh tissue of the host cactus is not itself the direct source of most nutrients for *D. mojavensis*. These bacteria and yeasts also provide chemical cues that can attract *D. mojavensis* to the cactus necroses (Fogleman 1982). Each cactus species may produce a characteristic array of such stimuli (Vacek 1979). *Drosophila mojavensis* can therefore potentially discriminate among host cacti on this basis. Both because oviposition behavior is more easily quantified for an individual than is chemotaxis and because it is more closely tied to research on fitnesses of progeny reared on different cactus species (Mangan 1978), egg-laying behavior is the focus of the present assay for genetic variation in habitat selection.

The experiment tested the oviposition preference of an *Opuntia*-breeding population of *D. mojavensis* on a simulated agria cactus necrosis, representing a cactus habitat that the population never experiences in nature (Heed 1982). The method used allows the estimation of genetic variance for oviposition acceptance measured in two complementary ways: (1) in the initial acceptance of agria cactus as an oviposition substrate and; (2) as the number of eggs laid on agria given that initial acceptance has occurred. The results demonstrate that significant genetic variance exists for habitat selection measured in either way. The amount of genetic variation that is available for response to natural selection is, in both cases, equal in magnitude to 10-20% of the phenotypic variance in the behavior. The biological meaning of these estimates concerns the genetic consequences of coevolution of the Santa Catalina Island population with its natural *Opuntia* host. Regardless of the nature of any genetic changes that may have occurred in this *D. mojavensis* population after it colonized *Opuntia*, genetic variation for acceptance of a cactus species used by other *D. mojavensis* populations is still present. The monophagous population studied here is therefore not genetically restricted to oviposition on its usual host cactus.

Methods and Materials

Establishment of the Laboratory Stock

The population sample of *D. mojavensis* used was University of Arizona stock no. A826, which was collected by William B. Heed in November, 1981. This stock was about six generations removed from nature when heritabilities were estimated. Originally, this stock was taken as two separate collections from Santa Catalina Island, California. Each sample was reared from one of the following substrates: *Opuntia "demissa"* cladodes (pads) (producing 28 eclosed adult *D. mojavensis*) or *O. "demissa"* fruits (969 adults eclosing from 20-30 fruits). The sole host cactus on Santa Catalina Island is *Opuntia "demissa"* Griffiths, which is thought to be a hybrid swarm. The introduced mission cactus, *Opuntia ficus-indica* (L.) Miller, has apparently been heavily introgressed by a smaller native *Opuntia* species to form *O. "demissa"* (Thorne 1967). The population of *D. mojavensis* on Santa Catalina Island is inferred to be a recent colonist because its host cactus, *O. ficus-indica*, is recently introduced, and *D. mojavensis* is not known to breed in the smaller *Opuntia* species (Heed 1982).

These two samples were maintained on standard banana *Drosophila* food inoculated with live baker's yeast (*Saccharomyces cerevisiae*). In early January, 1982, these two stocks were mixed, using about 100 flies from each stock, to form a new stock designated A826 M. The females of this pooled stock were allowed to lay eggs. When these eggs produced adults, males and virgin females were collected to be used as parents in the heritability experiment. Transient linkage disequilibrium, arising from mixing the two collections, should not seriously bias the estimates of genetic variance made here (Lewontin 1974).

Behavioral Assay Procedure

The fresh agria cactus used for the oviposition medium was collected in Baja California near San Ignacio in January, 1982. The sample comprised two to three arms of this cactus. All tissue was removed from the woody core, coarsely chopped and then autoclaved. This was then inoculated with *Erwinia carnegieana* (strain no. 1-12, Department of Plant Pathology, University of Arizona), which is the saguaro cactus necrosis bacterium. This bacterium also occurs in agria rots in nature (Vacek 1979). The cactus tissue was then placed in a 30 °C incubator for two weeks. At this point, the cactus was a bright yellow color that characterizes a natural agria rot when it has high concentrations of chemicals suitable for attracting *Drosophila* (Vacek 1979). By behavioral assay, this medium also elicited oviposition from most females. The cactus medium was then put through a fine (2 x 2 mm mesh) sieve to standardize the texture of the medium. The same batch of agria was used to test all females. The medium was kept refrigerated at 10 °C except when aliquots were taken for the day's experiment. To make the eggs more easily visible, 5 to 10 drops of green food coloring (McCormick & Co.) were added to each liter of cactus homogenate. This dye has no effect on oviposition behavior in *D. mojavensis* (W. B. Heed, pers. comm.). Scoring is improved, since the eggs absorb the dye and become blue-green, contrasting with the yellow-green medium.

The test apparatus was a pair of Microtest III plates (Falcon Plastics). One plate of each pair had each of the 96 wells filled level with the surface with the cactus medium. Each well in the plate will hold approximately 0.6 ml of liquid. This plate with the medium was refrigerated until 6 to 8 h prior to the test period when it was placed at room temperature. The second Microtest plate of each pair was left empty as a chamber for the flies.

The test regime began when single females were aspirated without anaesthesia into empty individual vials to which an aged male was added. These females were then starved for 6 h without water to increase their response to cactus stimuli. It is important to note that *D. mojavensis* is not induced to oviposit in the absence of cactus stimuli merely by starvation, nor are the females rendered nondiscriminating in oviposition behavior by being held as virgins for several weeks (Lofdahl, unpublished). During these 6 h, each pair of flies was lightly anaesthetized with CO_2 and transferred to a well in the empty Microtest plate. This plate was then covered with microscope slides that were secured with rubber bands. At the end of the starvation period, the entire Microtest plate with the flies was put under CO_2 anaesthesia for several minutes, the slides removed, and the Microtest plate with the cactus medium inverted over it. The paired Microtest plates were held together with rubber bands. Within one min, all pairs had awakened. After 15 min, the test apparatus was inverted, placing the cactus medium on the bottom. The apparatus was then placed in the dark at room temperature (21-23 °C) for 18 h. This method excludes visual cues, so the test is for genetic variation on chemical and tactile cues. At the end of the test period, the paired Microtest plates were put in the CO_2 chamber, the flies tapped into the empty Microtest plate, and then all flies were discarded. The oviposition medium plate was covered with plastic film and refrigerated until the eggs were counted. Those cases in which the female was trapped in the medium were not scored.

Genetic Design and Statistical Analysis

The experimental design was a sire-dam breeding analysis as is outlined by Falconer (1981) for the study of polygenic traits. This method relies on the establishment of either full- or half-sibling families; the degree of resemblance in the scores for the trait among and within families is then used to arrive at an estimate of genetic variance in relation to the total (phenotypic) variance in the trait. An analysis of variance is used to do this. The present study employed a nested ANOVA because full-and half-sibling families were both created in the breeding design. Because the parents used are considered a sample of a natural population, interest is in knowing the amount of genetic variance present in the population rather than in estimating the genotypes of the parents. Since population parameters (the heritabilities) are estimated, the ANOVA is a Model II or variance components model. In the nested ANOVA, the among-sires component of variance refers to the degree of resemblance of members of half-sibling families. Likewise, the among-dams component of variance measures the similarity of members of full-sibling families in relation to the total variance in the behavior.

Each of 38 males was randomly mated to 42 different females. Allowing for failed matings and nonovipositing females, this gave approximately 1100 full-sibling families. The female parents were each placed in an individual 1-dram vial of banana food that had been inoculated with live baker's yeast. All members of a full-sibling family shared the same larval environment. The progeny for testing in the heritability experiment were collected from the rearing vials as virgin females. Females from each full-sibling family were kept together in a vial of banana food until tested. All males collected were pooled into larger groups and aged for 10 days to allow them to become sexually mature. The females tested were of various ages, but all were approximately 2-4 weeks past eclosion. For the first test series, one female was chosen at random from each full-sibling family and allowed to oviposit. The second replicate took one female from each of the remaining families. This process was continued until all the females were tested. Approximately 300-400 progeny were tested per day.

Data from the sire-dam analysis were examined in two ways: as a metric trait and as a threshold trait. Since the number of eggs laid was recorded for each female, the heritability can be calculated for these continuously distributed data. The percentage of females ovipositing, however, depends on how long the cactus medium has been fermented as this determines the concentrations of volatile oviposition cues such as ethanol. In the present experiment, only 67% of all females tested laid eggs. The NESTED program for the analysis of variance of hierarchic designs from the Statistical Analysis System (Helwig and Council 1979) was therefore used to estimate the variance components excluding cases where the female failed to lay any eggs. Formulae for calculating the heritability from a sire-dam breeding analysis, given by Falconer (1981), were used. The among-sires component of variance was preferred to the among-dams component as a means of estimating heritability, for, unlike estimates from the dam component, the estimate is not biased with dominance effects and the influences of a common environment on the progeny. The heritability estimated from the sire component thus gives an accurate value for the amount of genetic variance in the number of eggs laid on agria that, in combination with knowledge of the intensity of selection, determines the rate of response to natural or artificial selection (Falconer 1981).

The data also gave the percentage of each sire's progeny ovipositing on agria. The trait can therefore be treated as a threshold character with the female's initial acceptance of the oviposition medium occurring if she is beyond a physiological threshold for egg laying on agria. The individual females are then scored as 1 (one or more eggs laid on agria) or as 0 (no eggs laid). This gives a 2 x N Chi-square table where N = number of sires and there are two response categories (oviposition *vs.* no oviposition). An estimate of the heritability of this threshold trait, initial acceptance of agria for oviposition, can be made from the sample Chi-square value that is used to test for heterogeneity among sire percentages (Robertson and Lerner 1949).

Results

Genetic Variation for Initial Acceptance of Agria Cactus

The formula for estimating the amount of selectable genetic variance for the threshold trait, initial acceptance of agria for oviposition, is given by Robertson and Lerner (1949). The percentage of each sire's female progeny ovipositing on agria is given in Table 1. A Chi-square test for homogeneity of these percentages reveals that different genotypes (sires) have different genetic predispositions to oviposit on agria ($p < 0.001$). Although males do not express oviposition behavior, as in most sex-limited traits, they do transmit genes for initial acceptance of agria to their daughters. A heritability estimate for the initial acceptance of agria is obtained from the sample Chi-square value. Robertson and Lerner (1949) suggested that the arc-sin transformation of these percentages can be dispensed with when the amount of genetic variance is small, as is the case here. The untransformed percentages were therefore used. The Chi-square test is robust to small cell sizes in a 2 x N table (Lewontin and Felsenstein 1965), so no sire's data were excluded on this basis. The Chi-square values are virtually identical whether the cells with fewer than five females ovipositing are exluded or not. Using Robertson and Lerner's (1949) formula, this gives a heritability estimate of $h^2 = 0.11$. Although the same authors give a formula for the standard error of heritability estimates for threshold traits, this was not applied here because it relies on equal cell sizes for the estimate. The main conclusion is that significant genetic variation, equal in magnitude to 11% of the phenotypic variance, exists for the initial acceptance of agria cactus as an oviposition substrate.

The problem with estimating the heritability of a threshold trait on the observed phenotypic (binomial) scale, as given above, is that the magnitude of the heritability depends on the incidence of the trait (in this case, the mean percentage of females ovipositing). The incidence in the present study was 67%. To avoid this difficulty, Falconer (1981) gives a transformation to an estimate of heritability on an underlying physiological scale where gene action is additive and the tendencies for initial acceptance of agria are normally distributed. When this is applied to the estimate of h^2 on the binomial scale reached above, this gives h^2 *(underlying scale)* = 0.18. Measured on either the phenotypic or the underlying physiological scale, the *Opuntia*-breeding population of *D. mojavensis* has genetic variation for initial acceptance of agria cactus for oviposition. Natural selection can act on this genetic variation to increase or decrease the percentage of females in the population using agria for egg laying.

Genetic Variation for Number of Eggs Laid on Agria Cactus

Before performing the analysis of variance, the data were square-root transformed. This is appropriate because count data are often Poisson distributed, and this transformation helps remove the dependence of the variance on the mean (Sokal and Rohlf 1981). This data transformation also significantly reduced the degree of kurtosis in comparison with the raw or logarithm-transformed scores, which is important in drawing valid inferences from variance components models (Scheffe 1959).

The F-test for significance of the sire component of variance, i.e., for genetic variance in the number of eggs laid on agria, can be performed on the nested ANOVA (Table 2) in several forms. The first is the simple test assuming equal cell sizes, which gives a highly significant

Table 1. Oviposition of *Drosophila mojavensis* on agria cactus (*Stenocereus gummosus*): analysis of oviposition substrate acceptance as a threshold trait.

Male number	Total number of females tested	Percentage females ovipositing
1	91	0.82
2	52	0.65
3	113	0.71
4	94	0.52
5	38	0.58
6	39	0.62
7	77	0.77
8	32	0.66
9	29	0.62
10	75	0.64
11	66	0.59
12	24	0.67
13	52	0.81
14	72	0.65
15	24	0.79
16	43	0.65
17	33	0.76
18	49	0.88
19	33	0.52
20	31	0.55
21	44	0.77
22	64	0.66
23	23	0.70
24	37	0.57
25	65	0.68
26	57	0.72
27	61	0.77
28	45	0.76
29	39	0.64
30	20	0.65
31	2	1.00
32	9	0.67
33	67	0.76
34	54	0.74
35	94	0.46
36	40	0.68
37	23	0.61
38	29	0.72

Chi-square test for homogeneity of binomial samples:

$x^2_{37} = 84.5$ $\qquad$ $x^2_{37}\ (.001) = 69.4$

Table 2. Nested analysis of variance using square-root transformed oviposition scores of *Drosophila mojavensis* on agria cactus (*Stenocereus gummosus*).

Source of variation	D.F.	Mean squares	Variance component	Percent of total variance
Sires	36	14.89	0.23	3.4
Dames (within sires)	539	6.88	0.25	3.6
Error	665	6.36	6.36	93.0
Total	1240	6.83	6.84	100.0

F-test for significance of sire component of variance using Satterthwait's approximation:

$F_{(36,237)} = 2.09 \quad p < 0.005$

F-test for significance of dam component of variance:

$F_{(539,665)} = 1.08$ n.s.

result ($p < 0.005$). In the case of highly unequal cell sizes, as in the present experiment, an F-test employing Satterthwaite's approximation is preferred. Although one of the conditions suggested for the application of this approximation is violated, the result is merely to give a test that is overly conservative (Sokal and Rohlf 1981). The fact that the *F*-test with Satterthwaite's approximation still gives a very high significance level ($p < 0.005$) is strong evidence for the presence of additive (selectable) genetic variance in the number of eggs laid by *D. mojavensis* on agria cactus.

The analysis of variance on the sire component is quite significant while the dam component of variance is just below the $p =$ significance level. The heritability as estimated from the sire component is $h^2 \pm$ S.E. $= 0.136 \pm 0.062$ on the square-root scale (Falconer 1981, Turner and Young 1969). If an estimate of heritability were to be made on the raw data, the value would be almost identical ($h^2 \pm$ S.E. $= 0.127 \pm 0.062$). Thus, genetic variance, equal in amount of 14% of the phenotypic variance, is present and available for response to either natural or artificial selection on the number of eggs laid on agria cactus.

Discussion

Before relating the present estimates of genetic variance to the evolutionary potential of the *D. mojavensis* population studied, it must be demonstrated that these significant results are not a laboratory artifact. The acceptance of agria cactus as an oviposition substrate is consistent with studies of chemotaxis in wild *D. mojavensis* populations. The species is most attracted to agria baits in the field, even in areas where this cactus is wholly absent (Fellows and Heed 1972). Agria preference itself is therefore not laboratory-generated, but rather characterizes a number of *D. mojavensis* populations (Mangan 1978).

Because the amount of genetic variation, i.e., the heritability, partially determines the rate of response to natural selection, it is valuable to know if the magnitudes of the heritabilities are comparable in the laboratory and in nature. Certainly, the population sample assayed in the oviposition tests was only a few generations removed from nature. Yet this is not sufficient to infer a lack of change in the genetic parameters of the population. Natural selection in the laboratory can, for example, increase the tendency of *Drosophila melanogaster* females to lay eggs on a sucrose medium over a period of only a few generations (Mazing 1946). To the extent that continued selection exhausts the genetic variance (Bulmer 1971), the percentage of genetic variance in relation to the phenotypic variance can decrease during laboratory maintenance of a stock. Natural selection based on differential fitnesses in the laboratory environment can alter both the oviposition phenotype and in some cases also the amount of genetic variation present for response to selection in future generations. The amount of genetic variation for oviposition on agria in nature may be greater or less than that estimated here, but the presence of genes for habitat acceptance is not put in doubt. The fact that all the genes influencing agria acceptance are not fixed at frequencies of 100%, but that individual genetic differences exist for this behavior, is also a laboratory result that may well hold true in nature. Natural selection acting in the Santa Catalina Island population could therefore increase the preference of *D. mojavensis* for agria if this cactus were introduced and fitnesses of offspring were higher on agria than on *Opuntia* cactus.

A second relevant issue in evaluating these results on oviposition behavior against empirical studies on other species and models of habitat selection, is whether the genetic differences measured are in fecundity or in degree of agria acceptance. Fecundity is here defined as the number of mature eggs present in a virgin female. Indirect evidence suggests that the present experiment has employed a valid measure of agria acceptance. Thorax length, an index of body size, is known to have a high phenotypic correlation with ovariole number (Mangan 1978), which is itself highly correlated with the number of mature eggs present in virgin females (David 1970). Therefore, thorax length is a good index of fecundity in *Drosophila*. Since the number of eggs laid on agria gave a nonsignificant regression on thorax length ($p > 0.83$) for the first 150 females tested in the present experiment, acceptance of agria necroses is being measured. The number of eggs laid on agria is at least partially independent of the number of eggs stored by virgin females. The heritability of number of eggs laid on agria can be regarded as a true measure of genetic variation in cactus acceptance.

Often it is useful to ask how a behavior in a temporal sequence, such as that involved in courtship or prey-catching, evolves in relation to others in the sequence. Then one needs to know the amount of genetic variation present for that element in the series given that all previous elements have occurred. For example, *D. mojavensis* females may not lay eggs readily unless they have mated (Markow 1982). To determine the degree of expression of oviposition acceptance in response to host plant traits, it is reasonable to discard those females not laying eggs. This is merely measuring heritability in a more defined environment; an environment in which causes not relevant to preference, such as whether the female has mated, are removed. The opposite tactic is to measure genetic variation for the initial acceptance of agria as an oviposition substrate to estimate the penetrance of oviposition on agria. Knowing that heritability values for both the penetrance and expressivity of oviposition on agria are statistically significant, demonstrates additive genetic (i.e., selectable) variance for the initial decision and for the degree of acceptance in the agria stimulus environment. This dual knowledge gives a clear impression of the ways in which natural selection can change genetically based habitat selection.

Establishing the existence of genes for oviposition on agria reveals nothing about the ontogeny of this behavior. Perhaps variation in oviposition acceptance may be due wholly or in part to genetic variation among individuals in tendency to undergo habituation or associative olfactory conditioning to chemical cues in the larval or adult environments (Jaenike 1982). Alternatively, these preferences may be due to congenital genetic variation in oviposition responses. The design of the present experiment cannot determine whether the genetic variances for the initial acceptance and for the number of eggs laid refer to genes for learning, for congenital preferences or both. The need to consider genetic variation in learning abilities in studies of insects is demonstrated by recent successful artificial selection for classical conditioning in the black blow fly, *Phormia regina* Meigen (McGuire and Hirsch 1977). Fuyama (1976, 1978) documented genetic variation for congenital olfactory preferences in *Drosophila*. Further research is needed to estimate the relative influences of genes for learning and for congenital preferences on the oviposition behavior of *D. mojavensis*. The knowledge of how cactus preferences are acquired in ontogeny is critical because learned preferences and congenital predispositions can have very different implications for the extent of host plant restriction.

The evidence for genetic variation in host acceptance in *D. mojavensis* has a fundamental bearing on allopatric host colonization in this species. Since the Santa Catalina Island population represents a recent colonization, it is accurate to claim that *D. mojavensis* has extended its degree of polyphagy. This is similar to the situation in *Drosophila grimshawi* Oldenberg to the extent that both colonizing populations added a new host plant or plants to the species' list of habitats (Carson and Ohta 1981). The genetic prerequisites for *Opuntia* colonization by *D. mojavensis* are unknown. Perhaps they involve physiological adaptations to the new host or genetic predispositions to feed and oviposit on *Opuntia*. Yet such hypothesized genetic predispositions do not require the loss of genetic variation for use of an earlier host cactus species. Addition of a new host cactus need not involve the interchange of genes for agria use and genes for *Opuntia* use. The same genes creating a predisposition to use agria also may produce a tendency to successfully colonize *Opuntia* due to pleiotropic genetic effects or to the existence of common oviposition cues in the two cactus species. The present analysis of the genetics of allopatric host colonization allows a prediction: if the agria- and *Opuntia*-breeding populations of *D. mojavensis* are brought in contact, absolute niche separation in breeding sites is not likely to be present, at least in the early stages. Colonization of a new host plant species has not promoted genetic differentiation in cactus use of the sort that can lead to immediate, total population subdivision during secondary contact.

Acknowledgments

I am extremely grateful to Dr. William B. Heed who kindly provided me with *D. mojavensis* stocks and laboratory facilities during my research visit at the University of Arizona. Drs. W. B. Heed and James C. Fogleman also provided many helpful discussions on the evolutionary ecology of *Drosophila*. I also thank Tom Orum of the Department of Plant Pathology, University of Arizona, for the culture of *Erwinia carnegieana*. The manuscript benefited greatly from comments by S. Arnold, M. Huettel, R. Lande, M. Rauscher, M. Rose and W. T. Starmer. This research was supported by a Nierman Research Award from the University of Chicago.

Literature Cited

Bulmer, M. G. 1971. The effect of selection on genetic variability. Amer. Natur. 105: 201.

Bush, G. L. 1969. Sympatric host race formation and speciation in frugivorous flies of the genus *Rhagoletis* (Diptera, Tephritidae). Evolution 23: 237.

Carson, H. L., and A. T. Ohta. 1981. Origin of the genetic basis of colonizing ability. *In* G. G. Scudder and J. L. Reveal, eds. Evolution Today. Hunt Institute Botany Document, Pittsburgh.

David, J. 1970. Le nombre d'ovarioles chez *Drosophila melanogaster*: relation avec la fecondite et valeur adaptive. Arch. Zool. Exp. Gen. 111: 357.

Falconer, D. S. 1981. Introduction to Quantitative Genetics. Longman, London.

Fellows, D. P., and W. B. Heed. 1972. Factors affecting host plant selection in desert-adapted cactiphilic *Drosophila*. Ecology 53: 850.

Fogleman, J. C. 1982. The role of volatiles in the ecology of cactophilic *Drosophila*. *In* J. S. F. Barker and W. T. Starmer, eds. Ecological Genetics and Evolution. Academic Press, New York.

Fuyama, Y. 1976. Behavior genetics of olfactory responses in *Drosophila*. I. Olfactometry and strain differences in *Drosophila melanogaster*. Behav. Genet. 6: 407.

Fuyama, Y. 1978. Behavior genetics of olfactory responses in *Drosophila*. II. An odorant-specific variant in a natural population of *Drosophila melanogaster*. Behav. Genet. 8: 399.

Heed, W. B. 1982. The origin of *Drosophila* in the Sonoran Desert. *In* J. S. F. Barker and W. T. Starmer, eds. Ecological Genetics and Evolution. Academic Press Inc., New York.

Helwig, J. T., and K. A. Council. 1979. SAS User's Guide. SAS Institute, Inc., Cary, North Carolina.

Huettel, M. D., and G. L. Bush. 1972. The genetics of host selection and its bearing on sympatric speciation in *Procecidochares* (Diptera: Tephritidae). Entomol. Exp. Appl. 15: 465.

Jaenike, J. 1981. Criteria for ascertaining the existence of host races. Amer. Natur. 117: 830.

Jaenike, J. 1982. Environmental modification of oviposition behavior in *Drosophila*. Amer. Natur. 119: 784.

Lewontin, R. C. 1974. The Genetic Basis of Evolutionary Change. Columbia University Press, New York.

Lewontin, R. C., and J. Felsenstein. 1965. The robustness of homogeneity tests in 2 x n tables. Biometrics 21: 19.

Mangan, R. L. 1978. Competitive interactions among host plant specific *Drosophila* species. Ph.D. Dissertation, University of Arizona, Tucson.

Markow, T. A. 1982. Mating systems of cactophilic *Drosophila*. *In* J. S. F. Barker and W. T. Starmer, eds. Ecological Genetics and Evolution. Academic Press, New York.

Maynard Smith, J. 1966. Sympatric speciation. Amer. Natur. 100: 637.

Mazing, R. A. 1946. Inheritance of the ability to select media for egg-laying in *Drosophila melanogaster*. Dokl. AN SSSR 51: 543.

McGuire, T. R., and J. Hirsch. 1977. Behavior-genetic analysis of *Phormia regina*: conditioning, reliable individual differences, and selection. Proc. Natl. Acad. Sci. USA 74: 5193.

Robertson, A., and I. M. Lerner. 1949. The heritability of all or none traits: viability of poultry. Genetics 34: 394.

Scheffe, H. 1959. The Analysis of Variance. John Wiley & Sons, New York.

Sokal, R. R., and F. J. Rohlf. 1981. Biometry. W. H. Freeman & Co., San Francisco.

Thorne, R. F. 1967. A flora of Santa Catalina Island, California. Aliso 6: 1.

Vacek, D. C. 1979. The microbial ecology of the host plants of *Drosophila mojavensis*. Ph.D. Dissertation, University of Arizona, Tucson.

Whitham, T. G. 1980. The theory of habitat selection: examined and extended using *Pemphigus* aphids. Amer. Natur. 115: 449.

Genetic Differences in Oviposition Preference Between Two Populations of *Heliothis virescens*

John C. Schneider

Richard T. Roush

Department of Entomology
Mississippi State University
Mississippi State, Mississippi 39762

Introduction

Heliothis virescens (F.) (Lepidoptera: Noctuidae), the tobacco budworm, is a multivoltine, polyphagous, highly mobile and economically important insect indigenous to the Americas. A majority of larvae of a given *H. virescens* population feed on different host plant species in different generations and different geographic locations, apparently depending largely on the local relative abundance of host plants (Table 1). In Mississippi, for example, adults eclose in May from overwintered pupae to oviposit primarily on *Geranium dissectum* L. and *G. carolinianum* L. (wild geraniums), *Vicia villosa* Roth. (winter vetch), and *Trifolium resupinatum* L. (Persian clover), as well as at least six other apparently less important species (Stadelbacher 1981, Snow and Brazzel 1965). The second through fourth generation larvae are primarily on *Gossypium hirsutum* L. (cotton) even though four other hosts are acceptable (Snow and Brazzel 1965). Cotton probably does not contribute as significantly as wild hosts to the fifth and final (overwintering) generation due to agricultural defoliation and tillage (Roach and Hopkins 1979; however, see also Laster and Furr 1971). In contrast, in North Carolina, *Nicotiana tabacum* L. (tobacco) is the major host in each generation although cotton is attacked also. Tobacco is an acceptable host in Mississippi, but very little is grown there (Snow and Brazzel 1965).

There are at least two geographical areas, however, where cotton supports an apparently disproportionately low fraction of the *H. virescens* population. *Heliothis virescens* is not commonly found on cotton on St. Croix, U. S. Virgin Islands, where the primary hosts are - *Cajanus cajan* (L.) (pigeon pea) and *Bastardia viscosa* (Kth.) (Snow et al. 1974, Proshold pers. com.). In addition, although *H. virescens* can be collected on cultivated geraniums in the San Joaquin Valley (Twine and Reynolds 1980), it is rarely, if ever, collected on cotton (T. F. Leigh, R. L. Coviello, L. T. Wilson, W. W. Allen, pers. comms.). In contrast, *H. virescens* is a pest of cotton in the Imperial Valley of southern California (Sharma et al. 1977).

Fox and Morrow (1981) reviewed the literature on intraspecific variation in the diets of insects. They discuss at least five mechanisms for local variation in ratios of abundances of a given insect species on available host plants. These include spatial variation in the following: competitors, repellent nonhosts, host quality, predators, and preference. Only the last

Table 1. Host plants by larval generation of *Heliothis virescens* in Mississippi and North Carolina[a].

Month	May		June		July		Aug.		Sept.	
Generation	I		II		III		IV		V	
Hosts	MS	NC	MS	NC	MS	NC	MS	NC	MS	NC
Geranium spp.	*[b]	-								
Vicia vilosa	*	-								
Trifolium resupinatum	*	-								
Nicotiana tabacum	-	*	-	*	-	*	-	*	-	*
Linaria canadensis	-	*								
Gossypium hirsutum			*	+	*	+	*	+	+	+
Rhexia mariana					-	*				
Desmodium rigidum							+	?	*	?
Haplopappus divaricatus									*	?
Polygonum sp.									*	?

[a] Based on Neunzig (1963, 1969), Stadelbacher (1981), and Snow and Brazzel (1965).
[b] Symbols: (*) - primary, (+) - secondary, (-) - tertiary, (?) - uncertain hosts.

mechanism, preference, could be the result of genetic differentiation among populations of the insect, expressed as geographically variable host utilization. The other mechanisms constitute ecological forces of selection that might eventually result in differentiation of preferences given favorably low gene flow rates, etc. (Endler 1977). We decided to test the hypothesis that geographic differences in importance of cotton as a host of *H. virescens* are due to genetic differences in oviposition preferences. As an adjunct to the oviposition study, we also tested for differences in ability of larvae to feed and grow on cotton.

Differences in Host Plant Utilization

Oviposition Preferences

Cultures of *H. virescens* were obtained from Mississippi and the Virgin Islands in 1981. (We were unable to obtain any cultures from northern California.) The Mississippi strain (MS) was collected as larvae from cotton at four sites (near Starkville, Greenville, Belzoni, and Greenwood), each of which was at least 50 km from any other site. About 80 adults were reared successfully from the field collected larvae. The Virgin Island strain (VI) was collected as larvae from *Bastardia* and pigeon pea on St. Thomas, U. S. Virgin Islands, and Vieques (a nearby island that belongs to Puerto Rico). Five and 15 adults, respectively, eclosed from the two collections, and were combined into one colony. (Collections of *H. virescens* from St. Croix were not desirable because male-sterile *Heliothis* hybrids have been released there

[Proshold et al. 1983]). The MS and VI colonies were reared in the laboratory for seven and four generations, respectively, before any experiments were begun. The census population size was at least 500 per generation. All life stages were maintained under similar environmental conditions (27 ± 3 °C, 40-80% RH, 14L:10D). Larvae were reared on a wheat germ diet (Bio-Serv, Inc., Frenchtown, N. J.). Adults were held in 4-liter glass jars and fed a vitamin fortified 5% sucrose-water solution. Eggs were collected from paper towels and muslin hung inside the jars.

The relative oviposition preferences of the two populations for cotton and two species of Geraniaceae were tested on 11 nights beginning April 30 and ending July 12, 1982. The cotton (Stoneville 213) was planted on one half of each of several 1.3 × 2 m flats in a greenhouse at Mississippi State University (MSU). The Geraniaceae were placed on the unplanted half of each flat. We used bouquets in jars of water of field collected *G. dissectum*, while it was available (first three trials), and potted ornamental geraniums (*Pelargonium graveolens* L'Her.) for the last eight trials. The tests were run only when both cotton and the geraniums were flowering, but the total leaf mass of the former was considerably greater than that of the latter (very roughly 20:1).

Female moths were prepared for oviposition tests by holding them for 3-5 days with equal numbers of males in 4-liter glass jars. Approximately 70-100% of 3-day-old females are mated under these conditions (Guerra et al. 1972, Proshold and LaChance 1974, Roush, unpublished). At dusk on test evenings, two greenhouse flats were covered with 1.3 × 1.3 × 2 m polyethylene plastic cages each of which had numerous small ventilation holes. A group of four to six female moths (mean ± S. D. = 5.2 ± 1.2) from each population was then released into one of the two cages. Moth populations and geraniums were both randomly assigned to flats. Only one pair of groups was tested on any given night. On the morning following each introduction, we discarded the moths and counted and removed all of the eggs oviposited on each of the host plant species.

Results of the oviposition choice tests indicate clearly that the MS and VI populations differed in their relative oviposition preferences (Table 2). In every trial, the MS moths oviposited a greater proportion of their eggs on cotton than did the VI moths ($p < .0005$, one tailed sign test). There are two ways of looking at these results. One is to ignore the relative abundances of foliage of the two species in the cages and to conclude that MS moths vastly preferred cotton over geranium while VI moths were approximately equally attracted to each. The other point of view is to note that the proportion of cotton foliage in the cages was very roughly equal to the proportion of eggs oviposited on cotton by the MS moths. Thus, one would conclude that the MS moths were approximately equally attracted to geranium and cotton while the VI moths showed a considerable preference for geranium over cotton. The latter interpretation seems more reasonable. The test cage situation was rather artificial, so these results should not be extrapolated to infer absolute degrees of preference in the field. For example, cage tests minimize the effects of any long range, host plant attractants.

The results of the tests are variable. We believe that much of the variation among nights can be attributed to differences in host plant quality, particularly in the quality of the geraniums. In trial 3, for example, where the greatest fractions of eggs were oviposited on cotton by VI moths, the field collected *Geranium dissectum* was senescing. Other experimental artifacts do not appear to have influenced the results significantly: eggs were not oviposited on the cages; and there was no correlation between total number of eggs oviposited by VI females and fraction of eggs on cotton.

During the last four oviposition preference trials, we tested the hypothesis that the differences between the two populations were genetically determined. Two additional test cages were prepared, and female progeny of both reciprocal F_1 crosses were tested simultaneously

Table 2. Relative oviposition preferences of *Heliothis virescens* from Mississippi (MS) and Virgin Islands (VI) and their reciprocal F_1 hybrids for cotton and geraniums; mean ± S.E.M. number eggs/trial: MS- 338 ± 72; VI- 302 ± 46.

Fraction Oviposition on Cotton				
	Population		F_1 hybrids	
Trials	MS	VI		
1	1.00	0.15		
2	0.91	0.34		
3	1.00	0.98		
4	0.99	0.59		
5	0.90	0.75		
6	0.96	0.92	MS ♀ × VI ♂	VI ♀ × MS ♂
7	0.83	0.65		
8	0.98	0.53	.87	0.67
9	0.99	0.11	0.87	0.91
10	0.80	0.68	0.61	0.72
11	0.99	0.74	0.59	0.57
8-11	0.94 ± 0.05[a]	0.52 ± 0.14	0.74 ± 0.08	0.72 ± 0.07
	0.75 ± 0.05[b]		0.73 ± 0.05[c]	
1-11	0.94 ± 0.02	0.59 ± 0.08		

[a]Mean ± S.E.M.

[b]Parental midpoint

[c]Pooled F_1

with the parental strains. The oviposition preferences of the F_1 females indicate that the difference between the two populations was determined genetically. Both reciprocal F_1 crosses were very close to the midpoints of the two parental strains during the last four trials (Table 2). This is consistent with additive inheritance. The F_1's were significantly different from the Mississippi parents (probability that the MS strain oviposited a greater proportion on cotton than both F_1's is ⅓ in each trial; probability that this happens in all four trials = $(1/3)^4$ = .012) but not the VI parents.

In addition to the studies described above, we tested individual VI females for oviposition preference to determine whether this population consisted of individuals that preferred either cotton or geranium, or whether each individual oviposited on both hosts. Of 14 females tested, seven did not oviposit, three oviposited less than eight eggs each, and four oviposited 58 to 123 eggs each (female *H. virescens* can oviposit up to 300 eggs per night [Proshold et al. 1982]). The fractions of eggs oviposited on cotton by the latter four females were: 0.37, 0.63, 0.83, and 1.00. These data indicate that oviposition preference is not absolute. Individual VI females oviposited on both geranium and cotton in our cage tests.

Larval Performance

After discovering differences in oviposition preference, we decided to determine if there were correlated differences in larval performance. At weekly intervals for 5 weeks beginning July 16, 1982, 30 (first 2 weeks) or 15 (last 3 weeks) neonate larvae each from the MS and VI colonies were randomly assigned to the terminals of Stoneville 213 cotton field grown at Mississippi State University. The terminals were enclosed within organdy mesh bags to exclude predators and to confine the larvae for 1 week. At the end of each trial the larvae were weighed.

The MS larvae attained a significantly greater weight ($p < .05$, 1-tail sign test) than did the VI larvae (Table 3). The genetic basis for this difference was investigated during the last 3 weeks by testing 15 hybrid larvae from each reciprocal cross in an identical fashion. There appeared to be no maternal effect, and the pooled hybrid performance was identical to that of the MS parental colony (Table 3).

As a partial control for differences in growth rates of the MS and VI colonies independent of host plant, their 5-day weights when reared on wheat germ diet in the laboratory were determined for sets of 50 larvae from each colony on each of four occasions between August 29 and October 2, 1982. An analysis of variance adjusting for disproportionate subclass numbers by the method of weighted squares of means (Steel and Torrie 1960) was performed to test for an interaction between diet (wheat germ *vs.* cotton) and population (MS *vs.* VI). The VI larvae did not grow as fast as MS larvae whether compared on wheat germ diet or the cotton plant. However, the decrement in weight gain was significantly greater on cotton than that on wheat germ diet (Table 4).

This evidence for genetic differences in larval growth rates parallels our data for oviposition preference and suggests that there is a correlation between adult preference and larval performance. Unfortunately, we realize that we did not control for the possibility of a diet (host) by environment interaction since the wheat germ and cotton diets were tested under different environments. However, oviposition preference and larval performance are not always correlated. An example of this is given by Wasserman and Futuyma (1981) who observed changes in oviposition preference of a bruchid beetle in a laboratory selection program but were unable to increase larval adaptation.

Discussion

Intraspecific, geographic differentiation of populations is a widespread phenomenon (Mayr 1970, Endler 1977). The rates of formation of, and the selective forces responsible for, this differentiation are usually obscure. Agriculturally important pests can provide a rare opportunity to obtain such information for differentiation of host plant preferences, because some historical records of crop availability and pest abundance often exist, as in the case of *H. virescens*.

Table 3. Seven day larval weights of *Heliothis virescens* from Mississippi (MS) and Virgin Islands (VI) and their reciprocal F_1 hybrids reared on cotton in the field.

	Larval Weight (mg) (Mean ± S.E.M.;*N*)			
	Population		F_1 hybrids	
Weeks	MS	VI	MS♀ X VI♂	VI♀ X MS♂
1	25.5 ± 4.7;14	7.2 ± 1.9;13		
2	13.2 ± 3.9;13	3.4 ± 1.8; 8		
3	9.8 ± 4.2; 7	7.9 ± 2.7; 8	3.1 ± 0.4; 7	17.7 ± 2.6;10
4	11.4 ± 3.4; 9	8.2 ± 3.2;10	7.1 ± 1.0; 8	7.8 ± 2.1; 7
5	15.2 ± 3.1; 9	3.8 ± 2.2; 5	16.9 ± 4.9;13	16.2 ± 4.9; 6
3-5	12.3 ±2.0;25	7.1 ± 1.7;23	10.6 ± 2.5;28	14.3 ± 2.0;23
			12.3 ± 1.7; 51[a]	
1-5	16.1 ± 2.0;52	6.5 ± 1.1;44		

[a]Pooled F_1

Table 4. Analysis of variance of five day larval weights of *Heliothis virescens* as a function of diet and population.

	Wt. (mg) (Mean ± S.E.M.;*N*)	
	Diet (D)	
Population (P)	Wheat germ	Cotton
VI	19.0 ± 0.6; 194	6.5 ± 1.1; 44
MS	23.6 ± 0.7; 192	16.1 ± 2.0; 52

Source	d.f.	SS	MS	F	P
Trtmt	3	11266.0	3755.3	38.96	0.001
DXP	1	485.5	485.5	5.04	0.025
Error	478	46066.2	96.4		

Cotton has been cultivated extensively in the southeastern US for about 170 years (Handy 1896). It has been present in the Virgin Islands for at least 200 years, but was very abundant as a cultivated crop for less than 20 years, and has not been grown commercially for over 50 years. The first botanical collection of living *Gossypium* was established on St. Croix, V.I. in the 1780's (Fryxell 1979). Sea Island cotton, *G. barbadense* L., was introduced to St. Croix for cultivation in 1908 (Briggs 1933) and now grows wild there (Ricks 1932, Proshold, pers com.). Very little cotton was cultivated on neighboring islands (including St. Thomas and Vieques from which the culture tested was obtained). The pink bollworm, *Pectinophora gossypiella* (Saunders), caused serious damage to cotton on St. Croix beginning in 1921 (Smith 1922). The pink bollworm and low cotton prices eliminated cotton production on the Virgin Islands by 1927 (Ricks 1932).

The first unequivocal observation of *H. virescens* on cotton was in the U. S. Virgin Islands on St. Croix in 1922 (Wilson 1923, Hambleton 1944). The first unequivocal reference for *H. virescens* on cotton in the southeastern U. S. was in Louisiana in 1934 (Folsom 1936). However, there is an inconclusive reference to *Aspila virescens* (F.) (= Heliothis virescens) on cotton in Georgia in 1879 (Riley 1885). The facts that larvae of *H. virescens* are commonly mistaken for those of *H. zea* (Boddie) (Riley 1885, Folsom 1936) and that *H. virescens* was found on St. Croix after the increased attention cotton pests received with the advent of the pink bollworm in 1922 support the hypothesis that *H. virescens* occurred on cotton, at low densities, for some indeterminate period of time before the 1920's and 30's. Unfortunately, it is not known whether the status of *H. virescens* as a pest on cotton in the last 50 years is due to an increase in preference for cotton or to exogenous changes such as availability of alternate host plants or differences in cultural practices.

We conclude that the greater oviposition preference by *H. virescens* for cotton in Mississippi relative to the population on the Virgin Islands probably developed because of the greater availability of cotton in the former over a period of about 150 years.

In summary, *H. virescens* from Mississippi has a greater preference for cotton than *H. virescens* from the U. S. Virgin Islands. Based on reciprocal F_1 crosses between the two populations, this difference was shown to be inherited genetically. We also obtained evidence suggesting genetic differences in larval performance between the two populations that are related adaptively to adult oviposition preference. Early instar larvae from Mississippi and reciprocal F_1 crosses grow faster on cotton in the field during the midsummer in Mississippi than do Virgin Island larvae. The genetic differences between the two populations appear to have arisen in the last 150 years in response to a difference in the abundance of cotton in the two locations.

Future research on the host plant preferences of this species should include investigation of possible differences among other populations, such as those from northern California, and more complete studies of the genetic control of these differences.

Acknowledgments

We thank F. Proshold for providing the Virgin Islands culture, J. Willers for technical assistance, L. Branch for suggesting the diet control for larval growth and S. McDaniel for identification of the *Pelargonium* sp.

Literature Cited

Briggs, G. 1933. Report of the Virgin Islands Agric. Expt. Sta., 1932.

Endler, J. 1977. Geographic Variation, Speciation, and Clines. Princeton University Press, Princeton, NJ.

Folsom, J. W. 1936. Notes on little-known cotton insects. J. Econ. Entomol. 29: 282.

Fox, L. R., and P. A. Morrow. 1981. Specialization: species property or local phenomenon? Science 211: 887.

Fryxell, P. A. 1979. The Natural History of the Cotton Tribe. Texas A & M Univ. Press, College Station.

Guerra, A. A., D. A. Wolfenbarger, R. D. Garcia. 1972. Factors affecting reproduction of the tobacco budworm in the laboratory. J. Econ. Entomol. 65: 1341.

Handy, R. B. 1896. History and general statistics of the cotton plant. *In* A. C. True, ed. The Cotton Plant: Its History, Botany, Chemistry, Culture, Enemies, and Uses. USDA Bull. No. 33, Washington.

Hambleton, E. J. 1944. *Heliothis virescens* as a pest of cotton, with notes on host plants in Peru. J. Econ. Entomol. 37: 660.

Laster, M. L., and R. E. Furr. 1971. Relationship of regrowth cotton to overwintering populations of the bollworm complex. J. Econ. Entomol. 64: 974.

Mayr, E. 1970. Populations, Species and Evolution. Harvard University Press, Cambridge, Mass.

Neunzig, H. H. 1963. Wild host plants of the corn earworm and the tobacco budworm in eastern North Carolina. J. Econ. Entomol. 56: 135.

Neunzig, H. H. 1969. The biology of the tobacco budworm and the corn earworm in North Carolina. North Carolina Sta. Tech. Bull. 196.

Proshold, F. I., and L. E. LaChance. 1974. Analysis of sterility in hybrids from interspecific crosses between *Heliothis virescens* and *H. subflexa*. Ann. Entomol. Soc. Amer. 67: 445.

Proshold, F. I., C. P. Karpenko, and C. K. Graham. 1982. Egg production and oviposition in the tobacco budworm: effect of age at mating. Ann. Entomol. Soc. Amer. 75: 51.

Proshold, F. I., J. R. Raulston, D. F. Martin, and M. L. Laster. 1983. Release of backcross insects on St. Croix to suppress the tobacco budworm (Lepidoptera: Noctuidae): behavior and interaction with native insects. J. Econ. Entomol. 76: 626.

Ricks, J. R. 1932. Report of the Virgin Islands Agric. Expt. Sta. 1931.

Riley, C. V. 1895. Report of the United States Entomological Commission. U.S.D.A., Washington.

Roach, S. H., and A. R. Hopkins. 1979. *Heliothis* spp.: behavior of prepupae and emergence of adults from different soils at different moisture levels. Environ. Entomol. 8: 388.

Sharma, R. K., N. C. Toscano, H. T. Reynolds, K. Kido, R. M. Hannibal, and W. M. Quillman. 1977. Tobacco budworm invades Imperial Valley cotton. Calif. Agric. 31: 16.

Smith, L. 1922. Report of the Virgin Islands Agric. Expt. Sta., 1921.

Snow, J. W., and J. R. Brazzel. 1965. Seasonal host activity of the bollworm and tobacco budworm during 1963 in northeast Mississippi. Mississippi State Univ. Agric. Expt. Sta. Bull. 712.

Snow, J. W., W. W. Cantelo, A. H. Baumhover, J. L. Goodenough, H. M. Graham, and J. R. Raulston. 1974. The tobacco budworm on St. Croix. U. S. Virgin Islands: host plants, population survey and estimates. Fla. Entomol. 57: 297.

Stadelbacher, E. A. 1981. Role of early-season wild and naturalized host plants in the buildup of the F_1 generation of *Heliothis zea* and *H. virescens* in the Delta of Mississippi. Environ. Entomol. 10: 766.

Steele, R. G. D., and J. H. Torrie. 1960. Principles and Procedures of Statistics. McGraw Hill, New York.

Twine, P. H., and H. T. Reynolds. 1980. Relative susceptibility and resistance of the tobacco budworm to methyl parathion and synthetic pyrethroids in southern California. J. Econ. Entomol. 73: 239.

Wasserman, S. S., and D. J. Futuyma. 1981. Evolution of host plant utilization in laboratory populations of the southern cowpea weevil, *Callosobruchus maculatus* Fabricius (Coleoptera: Bruchidae). Evolution 35: 605.

Wilson, C. E. 1923. Report of the Virgin Island Agric. Expt. Sta., 1922.

Evolution of Host Plant Utilization in *Colias* Butterflies

Bruce E. Tabashnik*

Department of Biological Sciences
Stanford University
Stanford, California 94305

Introduction

Evolutionary biologists have recently been criticized for employing the "adaptationist program" — i.e., seeking explanations as to why things are exactly as they should be in the best of all possible worlds (Gould and Lewontin 1979). Gould and Lewontin (1979) criticize the approach of dissecting organisms into traits, each of which is then analyzed for optimality. They cite the human chin to illustrate their point: the chin itself has no adaptive function, but arises as a by-product of the structural features of the jaw; analysis of optimal chin design would be misguided.

Unlike the chin in humans, host plant utilization by herbivorous insects has a direct bearing on fitness. Because insect herbivores do not grow equally well on all plants, their survivorship, development and reproductive success depend on their ability to select appropriate plants to eat. Thus, an insect's host choice behavior might be expected to be correlated with its physiological abilities so that the most preferred plants are also the most nutritionally suitable. Numerous recent studies, however, show a mismatch between host preference and nutritional suitability (e.g., Wiklund 1975, Chew 1977, Singer 1983, Smiley 1978, Ohsaki 1979, Stanton 1980, Courtney 1981, Holdren and Ehrlich 1982, Messina 1982).

There are several potential explanations for the imprecise correspondence between host choice behavior and host plant nutritional suitability. First, host suitability may be affected by ecological factors such as parasitism and predation (Smiley 1978, Price et al. 1980, Atsatt 1981) or host phenology (Singer 1971, Holdren and Ehrlich 1982). Consequently, the most nutritionally suitable host may be unsuitable in the field due to ecological factors that are not evident in laboratory feeding tests.

Second, host finding and acceptance behavior may be influenced by foraging constraints not directly related to host suitability. For example, suboptimal hosts may be accepted because habitats containing suitable hosts are avoided (Singer 1971, Ohsaki 1979, Rausher 1979, Chew 1981). If time available for searching is a limiting factor, suboptimal hosts that are abundant may be used more frequently than more suitable hosts that are scarce (Courtney 1982, Jaenike 1978). Host choice behavior also may be affected by energy requirements; e.g.,

*Present address: Department of Entomology, University of Hawaii, Honolulu, HI 96822. Journal series no. 2955 of the Hawaii Institute of Tropical Argiculture and Human Resources.

oviposition choice in butterflies might be influenced by the proximity of larval host plants to nectar sources.

Third, the evolution of host utilization may be constrained by lack of genetic variation for either host choice behavior or digestive ability (Wasserman and Futuyma 1981). For example, ovipositing females may not be capable of discriminating between plants that differ in suitability for larval success, but are similar in most other respects (Dolinger et al. 1973).

Fourth, poor correspondence between host preference and host suitability may occur when the herbivore does not have sufficient time to evolve appropriate responses to changes in host quality or availability. For instance, Chew (1977) suggests that crucifer-feeding *Pieris* butterflies lay eggs on a crucifer that is toxic to larvae because this plant was introduced recently and there has not been enough time for discrimination against it to evolve.

Current behavior and physiology reflect selection under past environments. Thus, it may be important to interpret current host use patterns in the context of the evolutionary history of a plant-herbivore relationship. Unfortunately, information about historical aspects of plant-herbivore relationships in nature is usually lacking. In many cases involving introduced plants, however, it is possible to estimate when an herbivorous insect first used a novel plant (Bush 1969, Phillips and Barnes 1975, Hsiao 1978). Such situations provide a unique opportunity to study evolution of host utilization, because experimental results can be interpreted in both the current and historical contexts.

To better understand how host utilization evolves, I studied two types of populations of the butterfly, *Colias philodice eriphyle* (Edwards) (Lepidoptera: Pieridae), occurring in Colorado, U.S.A. Pest populations of *C. p. eriphyle* use the recently introduced legume crop, alfalfa (*Medicago sativa* L.), as their sole larval host plant (Tabashnik 1980, 1983a). Nonpest populations are found in regions where alfalfa does not occur; they use several noncrop legumes, including native species, as hosts (Watt et al. 1979, Stanton 1982, Tabashnik 1983a,). Pest and nonpest populations appear to be reproductively isolated due to geographic separation and limited dispersal (Tabashnik 1980).

Pest populations of *C. p. eriphyle* have several generations per year (Tabashnik 1980) and have been using alfalfa for about 90 years (Berube 1972). Thus, we might expect improved utilization of alfalfa by pest *C. p. eriphyle* to have evolved through changes in larval growth ability and host choice behavior. In contrast, nonpest *C. p. eriphyle* have not used alfalfa, but have presumably undergone continuing selection for improved utilization of their noncrop host plants. Evolutionary divergence in host-use between these populations is expected as each adapts to its own host(s). To determine if host-choice behavior has evolved in parallel with host suitability, I compared the larval growth and oviposition preference of pest and nonpest populations.

Larval Growth

If divergence has occurred, then larvae from each population should grow more successfully on their own host(s): on alfalfa, pest larvae should grow better than nonpest larvae. On nonpest host plants, the opposite is expected; nonpest larvae should grow better than pest larvae. To test this hypothesis, I reared larvae from both populations on both alfalfa and on nonpest host plants. The two nonpest host plants used were vetch (*Vicia americana* Muhl.), a primary native host, and *Lathyrus leucanthus* Rydb., a secondary native host (Stanton 1982). Eggs were obtained from field-caught females in the laboratory, and each female's brood was split so that her progeny were reared on both alfalfa and native hosts. Three types of larval growth parameters were measured: (1) "fitness" parameters, (2) consumption rates and conversion efficiencies underlying growth performance, and (3) response to variation in leaf nitrogen and water content.

Survivorship, development rate and pupal weight are measures of growth performance that are expected to be highly correlated with fitness in *Colias*. The differences between populations in these parameters support the hypothesis that each population has become adapted to its own host(s) (Fig. 1). Pest larvae fed alfalfa had higher survivorship ($p < 0.05$) and shorter development times ($p < 0.001$) than nonpest larvae fed alfalfa. There also were significant host plant x population interactions in both survivorship ($p < 0.05$) and development time ($p < 0.01$) indicating that each population grew best on its "own" host(s). No significant differences between populations in pupal weight were found when each host was considered separately, but nonpest pupae were significantly heavier than pest pupae on the two native hosts considered together ($p < 0.05$).

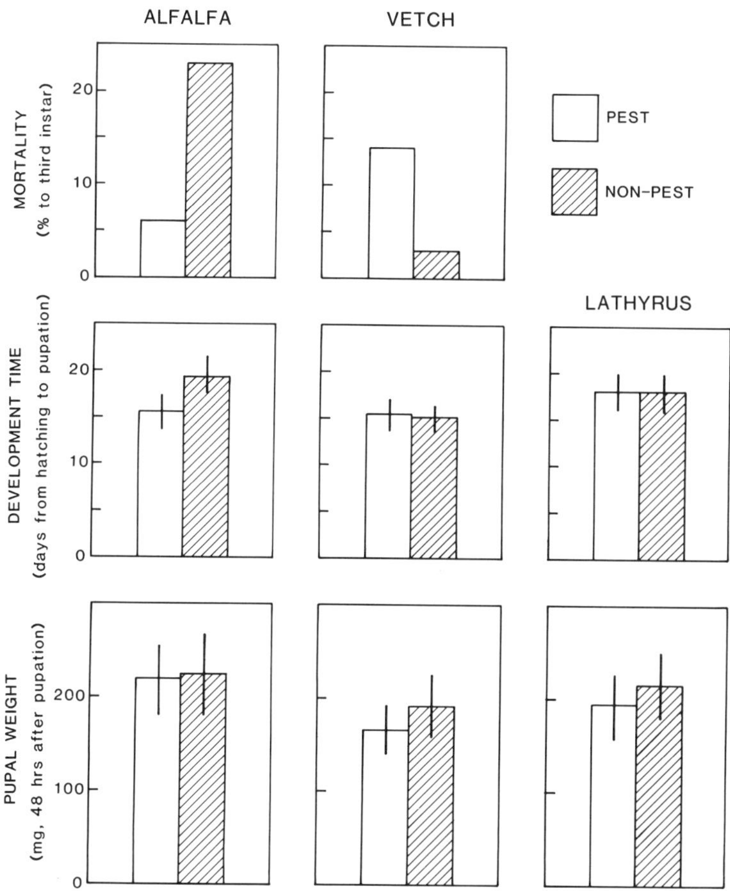

Fig. 1. Mortality, development time and pupal weight of pest and nonpest *C. p. eriphyle* larvae fed alfalfa and native hosts ($n = 244$ larvae total, 6 to 49 per treatment). Bars show means ± SD (mortality on *Lathyrus* not measured) (adapted from Tabashnik 1983a).

Growth is determined by the product of two factors: consumption x conversion efficiency (the amount of food consumed x the fraction of consumed food that is converted to larval biomass). Thus, the divergence between populations in the fitness-related parameters described above could be due to differences in consumption rate, conversion efficiency, or both. To determine which factors were most important, consumption rates and conversion efficiencies of fifth-instar larvae were estimated using gravimetric techniques (Waldbauer 1968). There were no differences between populations in consumption rates on either host plant (Fig. 2). Populations did differ, however, in approximate digestibility; pest larvae were able to digest alfalfa more efficiently than nonpest larvae ($p < 0.05$, Fig. 2). These results suggest that the differences between populations in larval growth performance are due to physiological adaptations improving the ability of pest larvae to digest alfalfa.

Many recent studies have shown that leaf nitrogen and water content can have a major impact on the growth performance of herbivores (see reviews by Mattson 1980, Scriber and Slansky 1981). Thus, I suspected that differences between pest and nonpest larvae might be due to differential responses to leaf nitrogen and water content. To examine larval responses to variation in these two key plant parameters, the nitrogen and water content of field-collected leaves fed to individual larvae were estimated (Tabashnik 1982). Correlation analysis was used to test for associations between larval growth indices and leaf nitrogen and water content.

Both populations responded alike to variation in the leaf nitrogen and water content of the native host, vetch, but differed in their responses to alfalfa (Table l). Growth rate and conversion efficiency were positively correlated with alfalfa leaf nitrogen content for nonpest larvae, but not for pest larvae. Conversely, growth rate, consumption rate and conversion efficiency were positively correlated with alfalfa leaf water content for pest larvae, but not for nonpest larvae. Thus, nonpest larvae were more sensitive to variation in alfalfa leaf nitrogen content than pest larvae, whereas pest larvae were more sensitive to variation in alfalfa leaf water content. Nonpest larvae may be less sensitive to variation in leaf water content because their hosts are subject to dehydration in the field (Stanton 1982, Hayes 1981), whereas alfalfa eaten by pest larvae is usually irrigated. The responses of pest *C. p. eriphyle* larvae to alfalfa leaf nitrogen content are similar to those of *Colias eurytheme* Boisduval, an alfalfa pest species (Tabashnik 1982), and may represent adaptation to alfalfa and/or to a pest population structure (Tabashnik 198l). Unlike the differences between populations in fitness parameters and conversion efficiency described above, the dissimilar responses of pest and nonpest larvae to variable leaf quality cannot readily be judged advantageous or disadvantageous. Nevertheless, the differences between populations in response to variation in alfalfa leaf water and nitrogen content are further evidence of physiological divergence between pest and nonpest *C. p. eriphyle*.

The evidence suggests that the observed phenotypic differences between pest and nonpest larvae are genetically based. Environmental conditions for rearing were identical for all larvae. Thus, nongenetic factors were the same for pest and nonpest larvae, with the possible exception of differences between populations in egg size or quality due to maternal influence. A comparison of egg weights, however, showed no difference between populations (Tabashnik 1983a). Moreover, maternal effects would be expected to influence growth on all hosts in a similar way, and would not be likely to cause host plant x population interactions, as were observed in survivorship and development time. Results from preliminary interpopulation hybridization experiments (Berube 1972) also support the hypothesis that genetic differences between populations enable pest larvae to develop faster on alfalfa than nonpest larvae.

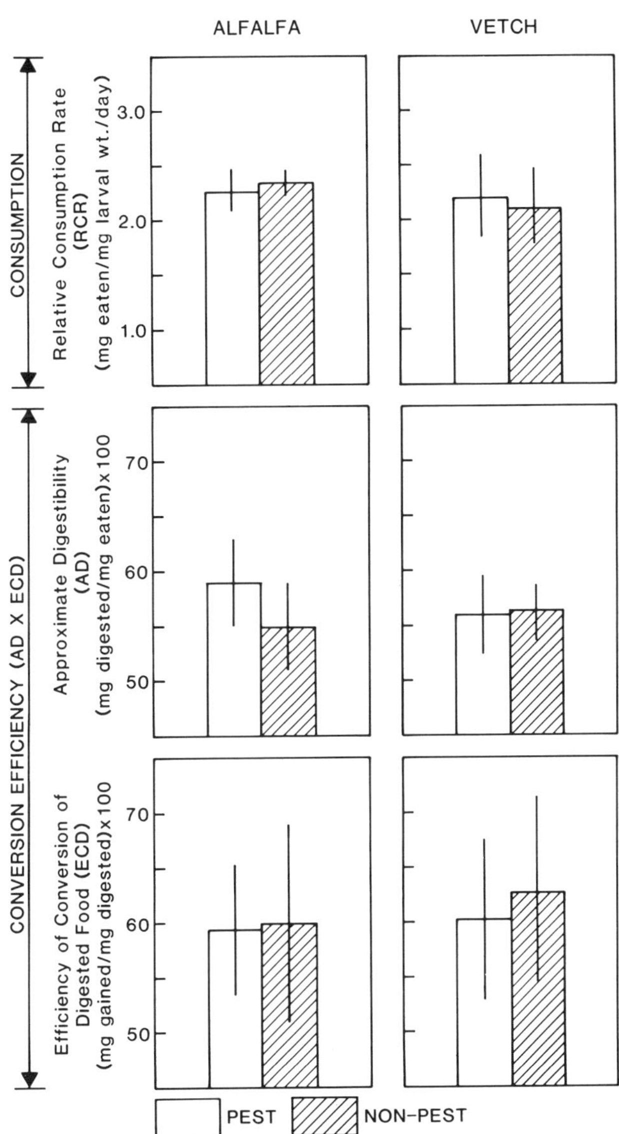

Fig. 2. Consumption and conversion of host plants by fifth-instar *C. p. eriphyle* larvae (n = 12 to 20 larvae per treatment). Bars show means ±SD (adapted from Tabashnik 1983a).

Table 1. Responses of *C. p. eriphyle* larvae to variation in leaf nitrogen and water content, $n = 12$ to 20 larvae per treatment (adapted from Tabashnik 1982).

Host plant	Population	RGR	RCR	ECI
		Correlations with leaf nitrogen content		
Alfalfa	Pest	+0.09	+0.07	+0.06
	Nonpest	+0.81**	+0.19	+0.85***
Vetch	Pest	−0.21	−0.75**	+0.83***
	Nonpest	−0.18	−0.86***	+0.79***
		Correlations with leaf water content		
Alfalfa	Pest	+0.62**	+0.56*	+0.46*
	Nonpest	−0.12	−0.50	−0.01
Vetch	Pest	+0.30	+0.35	−0.08
	Nonpest	+0.46	+0.27	+0.03

RGR = Relative growth rate = mg gained/mean larval wt/day.
RCR = Relative consumption rate = mg eaten/mean larval wt/day.
ECI = Efficiency of conversion of ingestion food (conversion efficiency) = mg gained/mg eaten.

*$p < 0.05$, **$p < 0.005$, ***$p < 0.001$.

Oviposition

If host-choice behavior has evolved in concert with host suitability, then pest females should have a stronger preference for alfalfa (*vs.* vetch) than nonpest females. To test this idea, oviposition preferences of field-caught pest and nonpest females were compared in the laboratory using two behavioral assays: (1) a chemical preference test, and (2) a whole plant preference test.

In the chemical oviposition preference assay, females in cylindrical glass chambers (15 cm high x 12 cm diam) were offered a choice between green construction paper "targets" (4 cm x 2 cm) soaked with extracts of either alfalfa or vetch (25 g fresh leaf material: 37.5 ml acetone: 37.5 ml water). This chemical preference test was chosen because plant chemicals seem to have played a major role in butterfly-plant coevolution (Ehrlich and Raven 1964) and because previous work showed that *Colias* females use chemical cues to discriminate among potential host plants (Stanton 1979). Stanton (1979) found that the preferences of *Colias meadii* Edw. females in the chemical assay were closely correlated with their preferences in the field. In my experiments with *C. p. eriphyle,* females laid significantly fewer eggs on control targets (soaked with acetone : water solvent) than on targets soaked with solvent and plant extracts (279 eggs on controls/3,308 total eggs, G test, $p < 0.001$). The chemical preference assay showed no difference between populations in their preference for alfalfa *vs.* vetch; pest females did not lay a significantly higher fraction of their eggs on alfalfa (62.6% ± 19.8% SD) than did nonpest females (60.1% ± 19.8% SD) ($p > 0.5$, $n = 58$ females, total eggs = 10,116).

In the whole plant assay, females in screen cages (28 cm x 22 cm x 22 cm) were offered a choice between live, intact alfalfa and vetch plants. This assay was chosen because *Colias* females may use nonchemical cues as well as chemical cues to distinguish among hosts (Stanton 1982). Thus, pest and nonpest females might differ in their responses to whole plants, even though they responded alike to plant extracts. In this assay, females laid less than 1% of their eggs on sunflower (*Helianthus*) plants that were included as controls ($p < 0.001$). Nonpest females laid a higher fraction of eggs on alfalfa (96.7% ± 2.4% SD) than pest females (80% ± 22.5% SD) ($n = 9$ females; 1,548 eggs). Although this difference is opposite to the direction expected, it is not significant (one-way ANOVA of arcsine-transformed data, $p < 0.1$, Sokal and Rohlf 1969).

The results suggest that populations have not diverged as expected in oviposition preference. It is possible, however, that pest females prefer alfalfa more strongly than nonpest females, even though this difference was not detected in my laboratory assays. For example, pest females laid a slightly higher fraction (62.5%) of their eggs on alfalfa than nonpest females (60.1%) in the chemical preference assay. If this observed difference represents the true difference between means, a greater than 10-fold increase in sample size (to > 600 females) would be needed to be 80% certain of detecting this difference at the 5% significance level (Sokal and Rohlf 1969). In the whole plant assay, however, pest females laid a lower fraction of their eggs on alfalfa than nonpest females, suggesting that the observed difference in the chemical assay was due to sampling error rather than a true difference between populations. Although no significant differences were detected between pest and nonpest females in either the chemical or whole plant preference assays, larvae differed as expected in survival, development rate, pupal weight and conversion efficiency. Therefore, it seems that pest and nonpest *C. p. eriphyle* have diverged more rapidly in larval growth ability than in oviposition preference.

Genetic Correlation

The results suggest that the ability of pest larvae to grow on alfalfa has improved, but there has been no concommitant change in oviposition behavior causing a stronger preference for alfalfa in pest females. This implies that larval growth and oviposition preference in *C. p. eriphyle* are genetically independent. To test the hypothesis of genetic independence more directly, I examined the correlation between oviposition preference and larval growth ability for five pest females and their progeny (Fig. 3).

Oviposition preferences of five pest females were measured by the chemical assay. Their broods were split, then reared on alfalfa and vetch. Two indices of larval growth were calculated for each family: (a) the growth rate ratio of early instar larvae = ln (mean wt. at 9 days after hatching for larvae fed alfalfa) ÷ (mean wt. at 9 days after hatching for larvae fed vetch). (b) The relative growth rate (RGR) ratio of fifth-instar larvae = (mean RGR on alfalfa) ÷ (mean RGR on vetch); where RGR = mg gained/mean larval wt/day. If oviposition preference is genetically correlated with larval growth ability, then the offspring of females that show the strongest preference for alfalfa should have the highest ratios of growth on alfalfa relative to vetch.

The lack of a positive phenotypic correlation between larval growth on alfalfa (relative to vetch) and oviposition preference for alfalfa *vs.* vetch (Fig. 3) is consistent with the hypothesis that larval growth and oviposition preference are controlled by different genes (Falconer 1981). Although the sample sizes were small, the negative trends suggest that a significant positive correlation is unlikely.

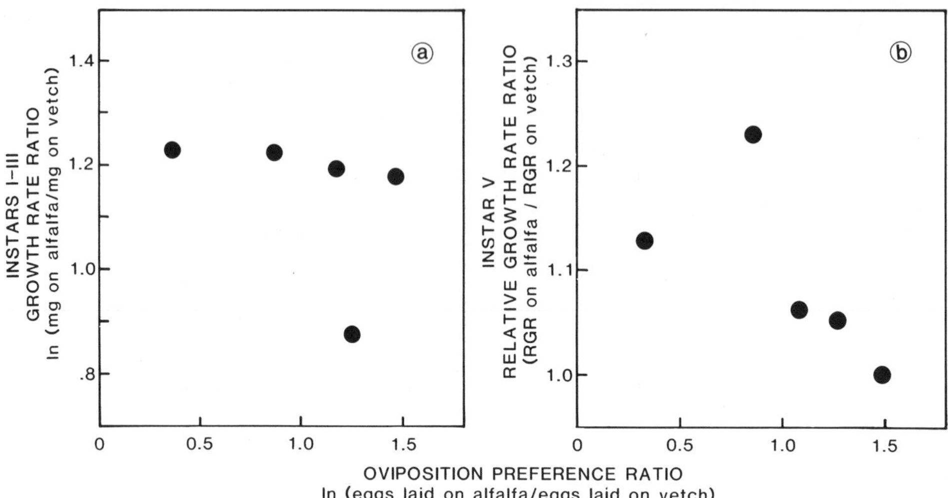

Fig. 3. Relationship between oviposition preference and larval growth in pest *C. p. eriphyle.* Oviposition preferences from five females (1,302 eggs) in chemical assay. (a) Growth rate ratio of early instar larvae (n = 79 larvae), r = −0.48, $p > 0.1$. (b) Relative growth rate ratio of fifth-instar larvae (n = 32 larvae), r = −0.66, $p > 0.1$.

Genetic Variation

Why has one component of host utilization evolved as expected while the other trait has not? Wasserman and Futuyma (1981) proposed that lack of suitable genetic variation for either the behavioral or physiological components of host utilization could lead to suboptimal host association. Hence, one possible explanation for the observed absence of evolutionary divergence between populations in choice behavior is that the genetic variation in oviposition preference is not sufficient to allow differentiation.

The results from the oviposition choice tests, however, suggest that there is considerable variation in preference within *C. p. eriphyle* populations (Fig. 4). Although the variation measured was phenotypic, several lines of evidence support the idea that oviposition preference in *Colias* is genetically based. First, pest and nonpest females did not differ in oviposition preference even though they ate different host plants as larvae and were exposed to different host plants as adults. These results suggest that neither larval nor adult experience have a major influence on oviposition preference. Second, there were consistent differences in preference among individual *C. p. eriphyle* females. In the chemical preference assays, females were tested for 2-6 trials, each trial lasting 1-2 days. For females that laid 15 eggs or more in two or more trials, the variation among females was significantly greater than the variation within females (among trials) ($F_{34,53} = 1.68$, $p < 0.05$).

In addition, the pattern of variation in oviposition preference in *C. p. eriphyle* is strikingly similar to the genetically based within-population variation seen in *Colias eurytheme*, a closely related legume-feeding species (Tabashnik 1981). In both species, the variation in preference was essentially continuous and not significantly different from a normal distribution. Experiments with *C. eurytheme* showed that, as in *C. p. eriphyle*, preferences were not influenced by larval or adult experience. There also were consistent differences in preference among *C. eurytheme* females; i.e., variation among females was significantly greater than variation within females (among trials) ($F_{34,136} = 2.18$, $p < 0.005$). Further, variation in

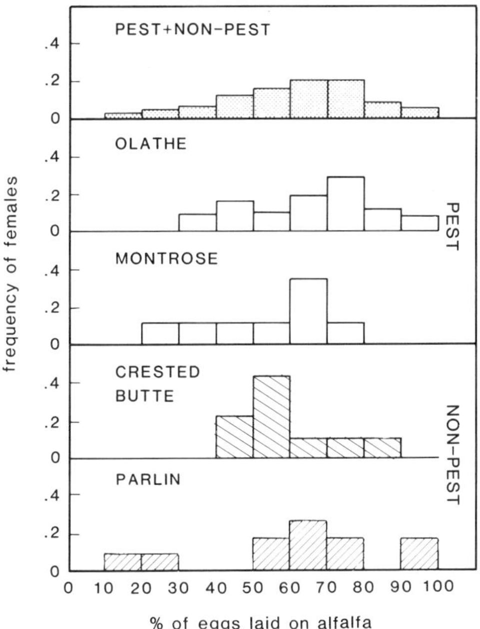

Fig. 4. Oviposition preferences of *C. p. eriphyle* in choice tests between alfalfa and vetch extracts ($n = 58$ females from four populations in Colorado). Females per population (from top to bottom) = 30, 8, 9, 11.

preference among females of *C. eurytheme* was significantly greater than variation within families ($F_{8,23} = 2.94$, $p < 0.025$), supporting the hypothesis that the preferences are genetically based (Tabashnik et al. 1981).

Historical Context

Examination of the historical relationship between *C. p. eriphyle* and its larval host plants suggests another explanation for why populations may have diverged more rapidly in larval growth ability than in oviposition preference. It may be that alfalfa pest populations were founded by females that were separated from native hosts and laid eggs on alfalfa in no-choice situations. Currently, pest females rarely leave alfalfa fields (Tabashnik 1980) and, thus, usually do not encounter native host plants. Therefore, pest females may have been selected for increased acceptance of alfalfa in no-choice situations, but they have rarely encountered alfalfa *vs.* native host-choice situations. Likewise, nonpest females do not encounter alfalfa and are not faced with a choice between alfalfa and native hosts.

Increased acceptance of a host plant in no-choice situations might be correlated with preference for that host in choice situations (e.g., Wasserman and Futuyma 1981), particularly for females that consistently reject one host during long periods when another host is accepted (see Singer 1982). Experiments with *Colias* females, however, support the idea that their behavior in no-choice situations with alfalfa or vetch alone is not correlated with their preference in choice tests between alfalfa and vetch (Tabashnik 1983b). Therefore, it may be that divergence in oviposition preference has not occurred because neither population has evolved in the context of a choice between alfalfa and native hosts.

In terms of larval growth, pest larvae have fed only on alfalfa, and thus have been selected, presumably, for better growth on alfalfa while being released from selection for ability to use other hosts. Meanwhile, nonpest larvae have been undergoing selection for improved utilization of hosts other than alfalfa. Therefore, to the extent that improved growth on alfalfa is not correlated with improved growth on nonpest host plants, divergence between populations in larval growth abilities is expected.

In conclusion, it appears that host-choice behavior and the ability of larvae to use host plants have not evolved in parallel in the case of pest and nonpest *C. p. eriphyle*. The results suggest that physiological adaptations have enabled pest larvae to digest alfalfa more efficiently, but pest females have not evolved increased preference for laying eggs on alfalfa. The high within-population variability in oviposition preference suggests that evolution of host-choice behavior is not constrained by insufficient genetic variation. The results are consistent with the hypothesis that oviposition preference and larval growth in *C. p. eriphyle* are genetically independent. It appears also that oviposition preference and larval growth have evolved at different rates because they have been under different selective regimes. Females have not encountered alfalfa *vs.* nonpest host-choice situations, but pest larvae have undergone selection for improved growth on alfalfa while nonpest larvae have been selected for better growth on hosts other than alfalfa.

I have emphasized the lack of precise correspondence between host-choice behavior and host suitability. Nonetheless, when one considers the vast array of potential hosts facing herbivorous insects, their successful exploitation of plants seems remarkable. Examples of extremely poor correspondence between host-choice behavior and host suitability are presumably eliminated by natural selection.

We need to remain aware, however, that the evolutionary processes that shape host utilization are subject to constraints. Evolution is finite in time and insect populations have finite numbers. Thus, the optimal combination of genes — derived by mathematical solutions or by intuition — may never occur in nature. Further, it is important to evaluate the correlation between host-choice behavior and host suitability in the broader context of the whole organism's fitness and in the historical context of the relationships between the insect and its host plants. With this in mind, we should not be surprised to find that a narrowly viewed correlation between host-choice behavior and host suitability is less than perfect.

Acknowledgments

I am grateful to Ward Watt for his support and advice. Todd Bierbaum, Dan Howard, Bill Mattson and Jim Miller gave thoughtful reviews. Special thanks to Rowena Krakauer for her invaluable assistance. This work was supported by an NSF graduate fellowship, an NIH traineeship, USDA HAW 947H and NSF grant DEB 75-23458 to Ward Watt.

Literature Cited

Atsatt, P. R. 1981. Ant-dependent food plant selection by the mistletoe butterfly, *Orgyris amaryllis* (Lycaenidae). Oecologia 48: 60.

Berube, D. F. 1972. Behavioral and physiological adaptations in the evolution of foodplant specificity in a species complex of *Colias* butterflies. Ph.D. Dissertation. Yale University, New Haven, Connecticut.

Bush, G. L. 1969. Sympatric host race formation and speciation in frugivorous flies of the genus *Rhagoletis* (Diptera, Tephritidae). Evolution 23: 237.

Chew, F. S. 1977. Coevolution of pierid butterflies and their cruciferous foodplants. II. The distribution of eggs on potential foodplants. Evolution 31: 568.

Chew, F. S. 1981. Coexistence and local extinction in two pierid butterflies. Amer. Natur. 118: 655.

Courtney, S. P. 1981. Coevolution of pierid butterflies and their cruciferous foodplants. III. *Anthocaris cardamines* (L.) survival, development and oviposition on different hosts. Oecologia 51: 91.

Courtney, S. P. 1982. Coevolution of pierid butterflies and their cruciferous foodplants. V. Habitat selection, community structure and speciation. Oecologia 54: 101.

Dolinger, P. M., P. R. Ehrlich, W. L. Fitch, and D. E. Breedlove. 1973. Alkaloid and predation patterns in Colorado lupine populations. Oecologia 13: 191.

Ehrlich, P. R., and P. H. Raven. 1964. Butterflies and plants: a study in coevolution. Evolution 18: 586.

Falconer, D. S. 1981. Introduction to Quantitative Genetics, 2nd ed. Longman, London.

Gould, S. J., and R. C. Lewontin. 1979. The spandrels of San Marco and the Panglossian paradigm: a critique of the adaptationist programme. Proc. Royal Soc. Lond. B, 205: 581.

Hayes, J. L. 1981. The population ecology of a natural population of the pierid butterfly *Colias alexandra*. 49: 188.

Holdren, C. E., and P. R. Ehrlich. 1982. Ecological determinants of food plant choice in the checkerspot butterfly *Euphydryas editha* in Colorado. Oecologia 52: 417.

Hsiao, T. H. 1978. Host plant adaptations among geographic populations of the Colorado potato beetle. Entomol. Exp. Appl. 24: 237.

Jaenike, J. 1978. On optimal oviposition behavior in phytophagous insects. Theor. Popul. Biol. 14: 350.

Mattson, W. J., Jr. 1980. Herbivory in relation to plant nitrogen content. Ann. Rev. Ecol. Syst. 11: 119.

Messina, F. J. 1982. Food plant choices of two goldenrod beetles: relation to plant quality. Oecologia 55: 342.

Ohsaki, N. 1979. Comparative population studies of three *Pieris* butterflies, *P. rapae, P. melete* and *P. napi*, living in the same area. I. Ecological requirements for habitat resources in the adults. Res. Popul. Ecol. 20: 278.

Phillips, P. A., and M. M. Barnes. 1975. Host race formation among sympatric apple, walnut and plum populations of the codling moth, *Laspeyresia pomonella*. Ann. Entomol. Soc. Amer. 68: 1053.

Price, P. W., C. E. Bouton, P. Gross, B. A. McPherson, J. N. Thompson, and A. E. Weis. 1980. Interactions among three tropic levels: influence of plants on interactions between insect herbivores and natural enemies. Ann. Rev. Ecol. Syst. 11: 41.

Rausher, M. D. 1979. Larval habitat suitability and oviposition preference in three related butterflies. Ecology 60: 503.

Scriber, J. M., and F. Slansky, Jr. 1981. The nutritional ecology of immature insects. Ann. Rev. Entomol. 26: 183.

Singer, M. C. 1971. Evolution of foodplant preference in the butterfly *Euphydryas editha*. Evolution 25: 383.

Singer, M. C. 1982. Quantification of host specificity by manipulation of oviposition behavior in the butterfly *Euphydryas editha*. Oecologia 52: 224.

Singer, M. C. 1983. Determinants of multiple host use by a phytophagous insect population. Evolution. 37:389

Smiley, J. 1978. Plant chemistry and host specificity: new evidence from *Heliconius* and *Passiflora*. Science 201: 745.

Sokal, R. R., F. J. Rohlf. 1969. Biometry. W. H. Freeman and Co., San Francisco.

Stanton, M. L. 1979. The role of chemotactile stimulation in the oviposition preference of *Colias* butterflies. Oecologia 39: 79.

Stanton, M. L. 1980. The dynamics of search: food plant selection by *Colias* butterflies. Ph.D. Dissertation, Harvard University, Cambridge, Massachusetts.

Stanton, M. L. 1982. Searching in a patchy environment: food plant selection by *Colias p. eriphyle* butterflies. Ecology 63: 839.

Tabashnik, B. E. 1980. Population structure of pierid butterflies. III. Pest populations of *Colias philodice eriphyle*. Oecologia 47: 175.

Tabashnik, B. E. 1981. Evolution into a pest niche: *Colias* butterflies and alfalfa. Ph.D. Dissertation. Stanford University, Stanford, California.

Tabashnik, B. E. 1982. Responses of pest and non-pest *Colias* butterfly larvae to intraspecific variation in leaf nitrogen and water content. Oecologia 55: 389.

Tabashnik, B. E. 1983a. Host range evolution: the shift from native legume hosts to alfalfa by the butterfly, *Colias philodice eriphyle*. Evolution 37: 150.

Tabashnik, B. E. 1983b. Oviposition specificity in single *vs*. cluster egg-laying butterflies: a discrimination phase in *Colias eurytheme*? Oecologia 58:278.

Tabashnik, B. E., H. Wheelock, J. D. Rainbolt, and W. B. Watt. 1981. Individual variation in oviposition preference in the butterfly, *Colias eurytheme*. Oecologia 50: 225.

Waldbauer, G. P. 1968. The consumption and utilization of food by insects. Adv. Insect Physiol. 5: 229.

Wasserman, S. S., and D. J. Futuyma. 1981. Evolution of host plant utilization in laboratory populations of the southern cowpea weevil, *Callosobruchus maculatus* Fabricius (Coleoptera: Bruchidae). Evolution 35: 605.

Watt, W. B., D. Han, and B. E. Tabashnik. 1979. Population structure of pierid butterflies. II. A "native" population of *Colias philodice eriphyle* in Colorado. Oecologia 44: 44.

Wiklund, C. 1975. The evolutionary relationship between adult oviposition preference and larval host range in *Papilio machaon* L. Oecologia 18: 185.

Quantitative Genetic Analysis of Feeding and Oviposition Behavior in the Polyphagous Leafminer *Liriomyza sativae*

Sara Via*

Department of Zoology
Duke University
Durham, North Carolina 27706

Introduction

In the absence of cultural transmission of behavior, behavioral evolution must be the result of the same population genetic factors that produce evolutionary change in morphological and physiological traits: mutation, selection and random genetic drift. Over the past five decades, theoretical efforts in population and quantitative genetics have established that the evolutionary trajectory of any character is a function of the magnitude of its genetic variation, the pattern of its genetic correlations with other characters, and the selection intensity on each character (Hazel 1943, Dickerson 1955, Fisher 1958, Wright 1968, Antonovics 1976, Lande 1979, 1982).

The formal relationships between genetic variation and evolutionary change that are provided by population genetics make the estimation of intraspecific genetic variation a useful tool in the study of behavioral evolution. Knowledge of extant patterns of genetic variation and covariation in behaviors like habitat selection may be useful in two ways. First, estimates of genetic variance in behavior within and among populations can be used to formulate hypotheses about how particular patterns of behavior could have arisen (Lande 1976, Arnold 1981a). These hypotheses can then be investigated further by experiment or computer simulation. Secondly, estimates of genetic covariance among characters in present-day populations can reveal genetic constraints on future evolutionary trajectories. The study of genetic variation within species can thus provide a view of the evolutionary potential of particular behaviors or life history characters under selection in various ecological circumstances.

In the experimental study of evolution, the estimation of intraspecific genetic variation and correlation complements the study of differences among closely related species. Interspecific or interracial comparisons may be quite useful to estimate the magnitude or type of genetic differentiation after divergence (e.g., Huettel and Bush 1972, Arnold 1981b). However, comparisons or crosses between closely related species do not permit the determination of the genetic mechanisms underlying evolutionary divergence because in any present-day study, the genetic changes that may have been involved in speciation are confounded with additional differentiation that has occurred since that time.

*Present address: Department of Zoology, University of Iowa, Iowa City, Iowa 52242

Here, the methods of expermental quantitative genetics (see Falconer 1981 for general treatment) were used to determine experimentally the causal components of variation in host plant preference behavior by females of a polyphagous herbivore, *Liriomyza sativae* Blanchard (Diptera: Agromyzidae). The choice of a host plant for feeding and oviposition is of central importance in herbivore populations. Such choice clearly determines the ecological circumstances for larval growth. Also, in insects that mate on the plant (reviewed in Gilbert 1978), variation in host plant preference can lead to *de facto* assortative mating. Intraspecific variation in the choice of host plants can thus have a large effect on the ecological and genetic structure of herbivore populations (Fox and Morrow 1981).

Variation in host plant selection by adults may result from (1) "induction" during larval exposure to particular host plants (*sensu* Jermy et al. 1968), (2) conditioning by early adult experience, or (3) congenital genetic variation among individuals in host plant preference. Conditioning of preference by experience in adult stages does appear to occur for some herbivores (Prokopy et al. 1982, Rausher 1983). However, adult experience appears to have little effect on host plant preferences in *Liriomyza* (Via, unpublished) and so it will not be considered further here. The effects of larval experience on adult host plant preference have long been controversial, and a mixture of positive and negative results have been obtained in various experimental studies (reviewed in Fox and Morrow 1981). In most previous studies, however, the roles of genetics and experience as determinants of preference have been assumed to be independent (e.g., Hovantiz 1969, Wicklund 1974, but see Rausher 1983 and Lofdahl, this volume). The experiment discussed here was designed specifically to test this assumption.

To date, there have been few studies of the genetic basis of host plant preference (Huettel and Bush 1972, Tabashnik et al. 1981 and this volume, Carson and Ohta 1981, Lofdahl, this volume). Using lepidopteran herbivores, the hypothesis of larval induction has been refuted in some cases by demonstrating that there is no effect of larval food plant on the *mean* oviposition behavior of groups of unrelated individuals without keeping track of the effects on individual genotypes (e.g., Tabashnik et al. 1981). However, variation in the responses of individuals to the inducing stimuli may cancel out when the mean is estimated. Thus, the simple examination of population means may not always reveal variation occuring within populations.

The experiment described here employs the techniques of quantitative genetics to determine the relative effects of larval rearing environment and genetic constitution on the oviposition behavior of adults. The examination of individual responses illustrates that variation among genotypes in the susceptibility to induction of adult preference by larval experience can occur. The possibility that host plant species can affect genotypes differentially within an herbivore species should therefore be considered in theoretical treatments of the evolutionary relationships between insects and their host plants.

Methods

Quantitative Genetic Analysis in an Ecological Context

The statistical techniques of quantitative genetics permit phenotypic variation in polygenic characters to be partitioned into components corresponding to the effects of genotype and environment as well as to their interaction. The central feature of quantitative genetics is the use of resemblances among relatives to estimate genetic variation (Fisher 1918, Falconer 1981). When these techniques are applied to family groups that have been generated from controlled matings of field-collected parents, considerable information can be obtained concerning the ecological and genetic structure of natural populations.

The experimental design used here was a two-way mixed model ANOVA in which families (a random effect) were tested in different environments (host plants, a fixed effect). In this analysis, the family "main effect" tests whether families varied in preferences of sibs reared in different environments were averaged over rearing environments. This variance among families allows estimation of the genetic main effect, that is, the variation among genotypes when averaged over environments (sometimes also called the "overall" genetic variance). The environmental main effect tests the effect of larval rearing environment on the preference of the average genotype. The interaction term, known as the "genotype-environment interaction," reveals whether responses to the environment are genotype-specific (Comstock and Moll 1963), that is, whether genotypes vary in the relativepreferences of siblings reared on different hosts. Via and Laude (1985) illustrate how genotype-environment interaction provides evolutionary biologists with a statistic that can be used to predict divergent evolutionary change and specialization in populations exposed to different environments.

The difference between genotype-environment interaction and "overall" genetic variation can be visualized in Figure 1. When the mean character state of each family in environment 1 is plotted on the left axis, and the mean in another environment plotted on the right axis, then the slope of the line connecting each pair of means illustrates the responses of the families to a change in the environment. In Figure 1A, the parallelism of the lines means that all families experience change in the environment in a similar way: the relative rankings of families are maintained across environments. The fact that families differ in their average character value over environments illustrates "overall" genetic variation. The decrease in the mean value in environment 2 illustrates the environmental component of variance. In contrast, in Figure 1B, families vary in their responses to the environment, despite the same magnitude of variation within environments. The nonparallelism of the lines on a graph of this type is thus a visual indication of genotype-environment interaction. Moreover, the change in rank of genotypes in different environments indicates the potential for genetic divergence of populations under selection, because different genotypes would be in the selected group in each environment.

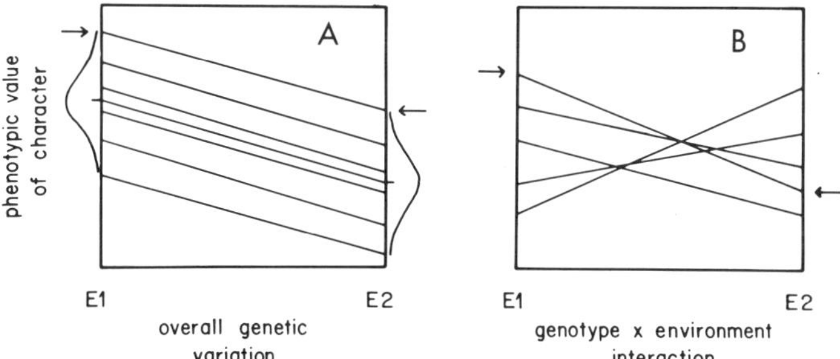

Fig. 1. Graphical representation of the difference between overall genetic variation and genotype-environment interaction. The mean character state for a given family in environment 1 (E1) is plotted on the left axis, the mean in environment 2 (E2) is plotted on the right axis, and the two family means are connected by a line. The slopes of the lines denote the phenotypic response of each family to a change in the environment. Within environments, variation among families is expected to be normally distributed. A. All families respond similarly to a change in the environment: the lines are parallel. Overall genetic variation means that some families achieve a higher value of the character than others when averaged over environments. No changes in

genotypic rank occur. B. The variation among families within environments is the same as in A, but crossing of the lines indicates that families vary in their responses to the environment. Arrows marking certain families show that changes in rank can occur.

Organism and Study Site

Because the leafminer *Liriomyza sativae* is an economically important pest of vegetable crops, its basic biology has been studied extensively by agriculturalists (Oatman 1959, Oatman and Michelbarger 1959, Musgrave et al. 1975). In Sampson Co., North Carolina, where the flies used in this experiment were collected, *L. sativae* can be found on a wide range of crops and wild plants in the families Solanaceae, Cucurbitaceae, and Leguminoseae. For this study, I chose two crop species from families that appear to differ in chemical and mechanical defenses against insects (Gibbs 1974): tomato (var. "UC82," Solanaceae) and cowpea (var. "Dixie Lee," Leguminoseae).

The rationale for studying variance among individuals in a collection from different crops within a local area should be noted. Variation in host plant utilization has most usually been studied in geographically separated populations. In such cases, different host plants may predominate due to ecological or climatic differences among locations (Hsaio 1978, Tabashnik et al. 1981). The focus on geographical rather than local variation found in most previous studies may be due partially to the fact that if one uses the mean phenotype as a measure of response to variation in the host, only populations with very little gene flow will be able to be resolved as distinct (Mitter et al. 1979). In contrast, the techniques of quantitative genetics allow variation both within and among populations to be estimated because individual family members are monitored. This means that variation within populations that may not differ when genotypes are averaged can be resolved. Techniques focused on individuals thus improve detection of evolutionarily important variation and permit the study of insect-host plant relations without the confounding influences of nonhost plant related variation among populations in different geographical areas.

Experimental Design

The central feature of this experiment was the generation of families with siblings reared in each of the two environments, cowpea and tomato. Parents were collected as larvae from two pairs of adjacent cowpea and tomato fields. A nested mating design was performed among individuals from the same population: several females were mated to each male (Comstock and Robinson 1952).

The test siblings were produced by allowing each female to oviposit on both crop species in sequence. The growth and development of each larva was monitored and the genetic variation and correlations in larval characters were estimated (Via, 1984a, b). The significant relationship between the oviposition preference of the female parent and the pupal weight of female offspring is described elsewhere (Via, 1986). One day after eclosion, each female sib was mated individually to a male chosen at random from another family. Then several females from each family with representatives from the two larval environments were subjected individually to a choice test to determine feeding and oviposition preference. Each trial lasted 24 h, after which feeding punctures were counted. The number of incipient larval mines were counted as an estimate of oviposition; the tiny eggs, which are inserted into the leaf tissue, cannot be counted directly with accuracy unless elaborate staining procedures are used (Tavormina 1982, Parella and Robb 1982). Although differences in average egg mortality among females could bias the main effects in the oviposition analysis, both a plant- and genotype-specific egg mortality would be required to bias the interaction term; this seems unlikely.

Statistical Analysis

Because the two plant species in a choice test are not independent, the appropriate dependent variables for this analysis are the percentage of an individual's feeding or oviposition punctures which occurred on one of the two plants, in this case, cowpea. The basic model is a two-way ANOVA, with family and larval rearing plants as crossed factors. Because the data are unbalanced and the interactions are of interest, the analyses were performed using the Type IV sums of squares in the General Linear Models procedure of SAS (Freund and Littell 1981). In the two-way ANOVA, small sample size precluded resolution of the sire and dam components of the half-sib design, as well as specific analysis of variation among populations. The data were thus collapsed to produce an analysis of full sib families with families from all populations pooled. Because the families were collected in such a way as to provide a broad sample of genotypes from each field, and "family" was taken as a random effect, the results of the analysis of the sampled genotypes can be extended to make inferences about the source populations.

Results

Feeding Behavior

The analysis of variance for the effect of larval experience on adult feeding preferences of individuals from full sib families (Table 1) illustrates no significant main effects of either larval rearing plant ("Rplant," $p > .90$) or family ($p > .50$) on the percentage of an adult's feeding that was performed on cowpea. In other words, when averaged over all genotypes, larval experience had no effect on adult feeding preference, nor was there significant variation among genotypes when averaged over both larval rearing plants. However, the presence of a significant interaction term ("Rplant*Family," $p < .05$) suggests that responses do occur to the larval host plant as an inducing stimulus, but that they are genotype-specific. In other words, families appear to vary in the susceptibility to larval induction of adult preference. Moreover, this genotype-environment interaction provides an experimental demonstration that the effects of genotype and the larval rearing plant are not independent in the organisms tested here.

Table 1. Analysis of variance on feeding preference (arcsin $\sqrt{\%}$ feeding on pea). Rplant = larval rearing plant, Family = full sib family.

Source	DF	SS	F	p > F	R^2
Model	40	9.73	1.67	.018	.37
Rplant	1	.00	.01	>.90	
Family	23	6.07	1.05	>.50	
Rplant*Family	16	4.03	1.73	.05	
Error	115	16.78			
Total	155	26.51			

Genotype-environment interaction in which families change rank in the different environments means that no genotype in the experimental population has the highest value of a given character in all environments (Fig. 1B). For example, that would mean here that the genotypes that fed most on pea when reared on pea were not the ones that showed the highest preference for pea when reared on tomato. This change of rank of genotypes in different environments is what Haldane called "crossing" genotype-environment interaction (Haldane 1946), and it suggests the potential for genetic differentiation among populations under directional selection in different environments (Via and Laude, 1985).

If the mean values for the feeding preference of each family in the two larval rearing environments are plotted and connected by lines, then the slope of each line provides a visual representation of the phenotypic response of that family to a change in the environment (Fig. 2). Nonparallelism of the lines indicates variation in environmental response, i.e., genotype-environment interaction. The homogeneous subsets of families within environments are indicated by the bars beside each axis. Figure 2 illustrates that the families which preferred pea the most or the least when reared on pea were not so extreme when reared on tomato. The Spearman rank correlation of family means for feeding preference in the two larval environments is not significantly different from zero ($r_s = -0.04$, $p > .86$, $n = 17$). This low correlation is consistent with the genotype-environment interaction estimated in the ANOVA (Table 1), because it indicates that the preference of a genotype reared in tomato cannot be predicted by knowledge of its preference when reared in cowpea.

The genetic variation within environments can be used to determine the potential for response to directional selection within either crop to increase preference for that plant species. The magnitude of the genetic variation in feeding preference within larval rearing plant species was therefore estimated using two separate one-way ANOVAs. It is important to note that because genetic and environmental effects are not independent here (Table 1), genetic variance and genotype-environment interaction are confounded in the separate analyses. Moreover, because the "broad sense" heritability estimated from variation among full sib families is a combination of additive genetic variance, dominance genetic variance and maternal effects, it must be used with some caution as a predictor of response to selection. Heritabilities are thus presented here primarily to compare variation expressed in the two larval rearing environments.

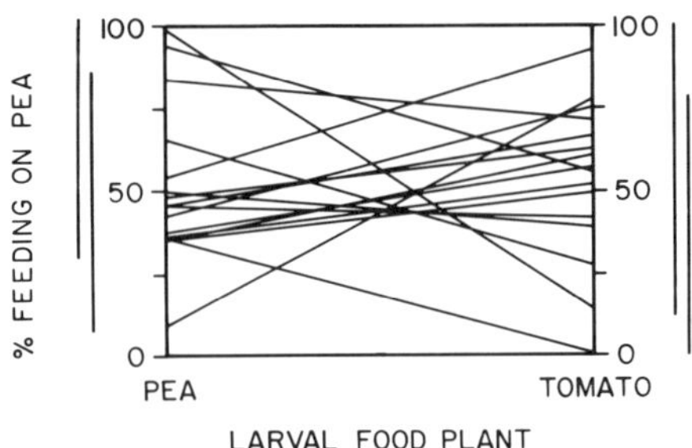

Fig. 2. Variation in the proportion of an individual's feeding which occured on cowpea for siblings reared on either cowpea or tomato. Means for each full sib family are plotted on the vertical axes and connected by lines. Non-parallelism of lines indicates genotype-environment interaction. Bars beside each axis are the homogeneous subsets calculated by the Tukey studentized range procedure.

Table 2. Variance components within environments for feeding preference (arcsin $\sqrt{\%}$ feeding on pea). V_P = phenotypic variance, V_F = variance among full sib families, V_E = error variance, h_b^2 = "broad sense" heritability (see text).

Rearing plant	V_P	V_F	V_E	h_b^2
Cowpea	.1841	.0434	.1407	.47 ± .25
Tomato	.1623	.0092	.1531	.11 ± .23

Variance components from the analyses within host plant species (Table 2) indicate that error variances in the two environments were equivalent, but the heritability estimate for full sib families ($h^2 = V_F/V_p$, Falconer 1981) is only different from zero for larvae reared in cowpeas. No significant differences in feeding preference could be detected among families reared in tomato by an F test, which is more conservative in certain cases (Scheffe 1959) than is the Tukey studentized range test shown in Figure 2.

Oviposition Preference

The results for oviposition behavior (Table 3) are qualitatively similar to those for feeding preference. Again, there is no average effect of larval experience on the host preferred for oviposition ("Rplant," $p > .40$). There is also no overall genetic (main-effect) variation in oviposition preference ("Family," $p > .75$). That is, when sibs were averaged over larval rearing environments, families did not differ in the magnitude of their oviposition preference for pea. A weak genotype-environment interaction is suggested in the "Rplant*Family" term ($p < .067$), with the Spearman rank correlation for mean oviposition of families in the two rearing environments slightly negative but not significantly different from zero ($r = -0.11$, $p > .65$, $n = 17$).

Table 3. Analysis of variance on oviposition preference (arcsin $\sqrt{\%}$ oviposition on pea). Rplant = larval rearing plant, Family = full sib family.

Source	DF	SS	F	p > F	R^2
Model	40	9.62	1.26	.169	.31
Rplant	1	.24	.78	>.40	
Family	23	5.71	.79	>.75	
Rplant*Family	16	5.02	1.65	.067	
Error	115	21.89			
Total	155	31.50			

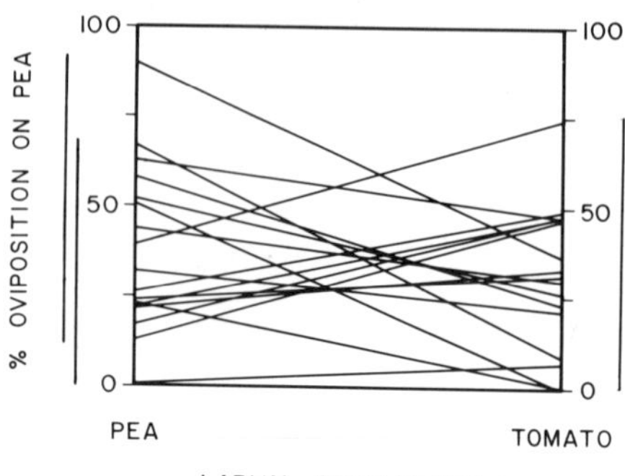

Fig. 3. Variation in the proportion of an individual's oviposition which occurred on cowpea for siblings reared on either cowpea or tomato. Means for each full sib family are plotted on the vertical axes and connected by lines. Bars beside each axis are the homogeneous subsets calculated by the Tukey studentized range procedure.

When the mean oviposition preferences for families in the two larval rearing environments are plotted, it can again be seen that the families which preferred pea the most when reared on pea were not so extreme when reared on tomato (Fig. 3). In oviposition behavior, as in feeding preference, no significant variation can be detected for larvae reared on tomato, even by the more sensitive Tukey procedure; significant variation among families in oviposition preference can only be detected when sibs were reared on cowpea (Table 4).

Taking an increase in feeding punctures as an indication of preference for a particular plant, the Spearman rank correlations between feeding and oviposition preference were estimated for individuals within each larval rearing environment. For sibs reared in cowpea, the correlation between feeding and oviposition preference for larvae reared in pea was $r_{pea} = 0.45$ ($p < .04$) and for siblings reared in tomato, the correlation was $r_{tom} = 0.53$ ($p < .01$). The 95% confidence limits estimated for these correlations by the *z*-transformation (Snedecor and Cochran 1967) are close to zero but do not overlap 1 ($-0.04 < r_{pea} < 0.77$; $0.07 < r_{tom} <$

Table 4. Variance components within environments for oviposition preference (arcsin $\sqrt{\%}$ oviposition on pea). V_P = phenotypic variance, V_F = variance among full sib families, V_E = error variance, h_b^2 = "broad sense" heritability (see text).

Rearing plant	V_P	V_F	V_E	h_b^2
Cowpea	.2125	.0420	.1705	.40 ± .25
Tomato	.2113* .1669**	−.0446	.2113	0*

* V_F treated as 0

** V_F −.0446

0.81). The lack of perfect correlation among family means indicates some genetic uncoupling between feeding and oviposition behavior. In other words, regardless of larval rearing plant, some families that preferred cowpea for feeding preferred tomato for oviposition. This suggests that the taste of the plant is not the only criterion involved in the selection of a suitable oviposition site by *L. sativae*.

It is possible that the genotype-specific responses to rearing plant species evidenced in Tables 1 and 3 could be an artifact of phenotypic variation among individual plants within plant species. For example, apparent interaction variation at the family level could result if, by chance, some families experienced several "good" test plants of one of the species, which induced preference for that host, and several "bad" plants of the other host, which induced aversion for that species. This possibility was tested by including a factor in the model for replicate test plant nested within the "Family*Rplant" term. This element, "Repl(Family*Rplant)," was not significant in either the feeding analysis ($F = 0.78, p > .80$) or in the oviposition analysis ($F = 0.96, p > .55$), and so it was pooled with the error to produce the analyses seen in Tables 1 and 3.

Discussion

The genotype-environment interactions demonstrated here have several ecological and evolutionary implications:

(1) The genotype-environment interaction between family and larval rearing environment indicates that genetic variation and induction of preference by larval experience are not independent in this system. In other words, genotypes varied in their susceptibility to the effects of larval experience on adult preference behavior. This means that the potential for genotype-specific responses to the environment must be considered even when addressing what appears to be a very straightforward ecological question, that is, the effect of larval environment on adult behavior.

(2) Unless the variation estimated among full sib families within the cowpea environment is entirely due to nongenetic maternal effects, this experiment suggests that selection could operate to increase the preference for pea. It is unclear at present why the estimated variation among families was less when siblings were reared on tomato than when they grew up on pea.

(3) If directional selection for increased preference were to occur within these two larval rearing environments, we would expect different genotypes to be represented in the selected groups. Thus, populations could diverge in host plant preference, at least temporarily, by processes outlined in Via and Laude (1985).

The fact that genotype-environment interaction can result in genetic differentiation of populations under selection in different environments is of considerable importance in the study of genetic substructuring within species; it underscores the idea that individuals within a species may not be ecologically identical. Moreover, genetic differences among individuals in response to the environment are central to the determination of the potential of a given species to evolve in response to a change in resource type or abundance. This possibility should be considered in any general model of the evolution of insect-host plant relationships.

(4) If we observe a response to selection for increased preference when larvae are reared in only a single environment, then we cannot discriminate the following two alternatives. First, the observed response may involve an increase in frequency of genotypes that have congenital (i.e., genetically determined) preference and that would maintain their preference even if reared in the alternate environment. This is the expectation if "main-effect" variation among families were present in the base population, that is, if some families always preferred pea more than others, regardless of rearing environment. Second, the response could involve an

increase in frequency of individuals that are susceptible to the induction of preference by larval experience. In this case, individuals could not be expected to maintain their preference when reared on the other host. This is the expected outcome if selection were to operate on the genotype-environment interaction variation illustrated here.

The confounding of these alternatives results from the fact that genetic variation and genotype-environment interaction variation are both included in the variance among families in an analysis performed in a single environment (Comstock and Moll 1963). Because "main-effect" and "interaction" variation have very different implications evolutionarily, the possibility of confounding them must be taken seriously, not only in the interpretation of selection experiments and sib analyses (see Lofdahl, this volume), but also in the interpretation of natural patterns of insect-host plant relationships.

For example, if response to selection for increased preference were mediated by the proliferation of genotypes that could be induced more effectively than others to a particular host plant, then this preference might be expected to break down if the resource base and, thus, the predominant larval rearing environment were to change. Such a breakdown due to the absence of a genetically fixed preference could provide a certain degree of flexibility in host plant use. However, one must understand that the evolution of preference in any given case is opportunistic, that is, that the mechanism that spreads is the one that contributes the most to the selected group, even though "flexibility" might be advantageous in the future. Only experimental tests in several environments will resolve the underlying causes of the responses to selection that are observed in particular circumstances.

In conclusion, the data presented here illustrate the existence of genetic variation in the susceptibility to induction of adult host plant preference by larval experience. Because genotype-environment interaction may be as widespread in natural systems as in agricultural ones (Dickerson 1962, Allard and Bradshaw 1964), the possibility of genotype-specific responses in behavior and life history to different host plant species must be considered in the formulation of realistic models of the evolution of insect-host plant relationships.

Acknowledgments

I am grateful to Drs. Janis Antonovics, Mark Rausher and Ruth Shaw for discussion of genetic approaches to ecological problems, and to Drs. Donald Burdick and Robert Moll for consultations on statistics and quantitative genetics, respectively. I thank Drs. Douglas Futuyma, Montgomery Slatkin, Guy Bush, Robert Rockwell, Joseph Hegmann, Milton Huettel, Katharine Lofdahl, and Fred Gould for comments on these ideas during the symposium. Support was provided by NIH Training Grant (5 T32 GM 07754) to the Duke University Program in Genetics; the Department of Zoology, Duke University, and NSF grant DEB 8016414 to M. Rausher.

Literature Cited

Allard, R. W., and A. D. Bradshaw. 1964. Implications of genotype-environment interaction in applied plant breeding. Crop Sci. 4: 503.

Antonovics, J. 1976. The nature of limits to natural selection. Ann. Mo. Bot. Garden 63: 224.

Arnold, S. J. 1981a. Behavioral variation in natural populations I. Phenotypic, genetic and environmental correlations between chemoreceptive responses to prey in the garter snake *Thamnophis elegans*. Evolution 35: 489.

Arnold, S. J. 1981b. Behavioral variation in natural populations. II. Inheritance of a feeding response in crosses between geographic races of the garter snake *Thamnophis elegans*. Evolution 35: 510.

Carson, H. L., and A. T. Ohta 1981. Origin of the genetic basis of colonizing ability. *In* G.G.E. Scudder and J. L. Reveal, eds. Evolution Today. Proc. of 2nd Int. Congr. of Syst. and Evol. Biol.

Comstock, R. E., and R. H. Moll. 1963. Genotype-environment interactions. *In* W. D. Hanson and H. F. Robinson, eds. Statistical Genetics and Plant Breeding. NSA-NRC Publ. -982.

Comstock, R. E., and H. F. Robinson. 1952. Estimation of average dominance of genes. *In* J. W. Gowen, ed. Heterosis. Iowa State College Press, Ames.

Dickerson, G. E. 1962. Implications of genetic-environmental interaction in animal breeding. Animal Prod. 4: 47.

Falconer, D. S., 1981. Introduction to Quantitative Genetics, 2nd ed. Longman Press, New York.

Fisher, R. A. 1918. The correlation between relatives on the supposition of Mendelian inheritance. Trans. Roy. Soc. Edinburgh 52: 399.

Fisher, R. A. 1958. The Genetical Theory of Natural Selection. Dover Publications, New York.

Fox, L. R., and P. A. Morrow. 1981. Specialization: Species property or local phenomenon? Science 211: 887.

Freund, R. J., and R. C. Littell. 1981. SAS for linear models. SAS Series in Statistical Appl. SAS Institute. Cary, North Carolina.

Gilbert, L. E. 1978. Development of theory in the analysis of insect-plant interactions. *In* Analysis of Ecological Systems 8: 117.

Gibbs, R. D. 1974. Chemotaxonomy of Flowering Plants. McGill-Queens Univ. Press.

Haldane, J.B.S. 1946. The interaction of nature and nurture. Ann. Eugen. (Lond.) 13: 197.

Hazel, L. N. 1943. The genetic basis of constructing selection indices. Genetics 28: 476.

Hovanitz, W. 1969. Inherited and/or conditioned changes in host-plant preference in *Pieris*. Entomol. Exp. Appl. 24: 437.

Huettel, M. D., and G. L. Bush. 1972. The genetics of host selection and its bearing on sympatric speciation in *Procecidochares* (Diptera: Tephritidae). Entomol. Exp. Appl. 15: 465.

Jermy, T., F. E. Hanson, and V. G. Dethier. 1968. Induction of specific food preferences in lepidopterous larvae. Entomol. Exp. Appl. 11: 211.

Lande, R. 1976. Natural selection and random genetic drift in phenotypic evolution. Evolution 30: 314.

Lande, R. 1979. Quantitative genetic analysis of multivariate evolution, applied to brain-body size allometry. Evolution 33: 402.

Lande, R. 1982. A quantitative genetic theory of life history evolution. Ecology 63: 607.

Mitter, C., D. J. Futuyma, J. C. Schneider, and J. D. Hare. 1979. Genetic variation and host plant relations in a parthenogenetic moth. Evolution 33: 777.

Musgrave, C. A., S. L. Poe, and D. R. Bennett. 1975. Leafminer population estimation in polyculture vegetables. Proc. Fla. State Hort. Soc. 88: 155.

Oatman, E. R. 1959. Host range studies of the melon leaf miner *Liriomyza pictella* (Diptera:Agromyzidae). Ann. Entomol. Soc. Am. 52: 739.

Oatman, E. R., and A. E. Michelbarger. 1959. The melon leaf miner, *Liriomyza pictella*. II. Ecological studies. Ann. Entomol. Soc. Am. 52: 83.

Parella, M. P., and K. L. Robb. 1982. Technique for staining eggs of *Liriomyza trifolii* within chrysanthemum, celery and tomato leaves. J. Econ. Entomol. 75: 383.

Prokopy, R. J., A. L. Averill, S. S. Cooley, and C. A. Roitberg. 1982. Associative learning in egg laying site selection by apple maggot flies. Science 218: 76.

Rausher, M. D. 1983. Conditioning and genetic variation as causes of individual variation in the oviposition behavior of the tortoise beetle *Deloyala gutatta*. Anim. Behav. (in press).

Scheffe, H. 1959. The Analysis of Variance. John Wiley and Sons, New York.

Snedecor, G. W., and W. G. Cochran. 1967. Statistical Methods. Iowa State Univ. Press, Ames.

Tabashnik, B. E., H. Wheelock, J. D. Rainbolt, and W. B. Watt. 1981. Individual variation in oviposition preference in the butterfly *Colias eurytheme*. Oecologia 44: 225.

Tavormina, S. 1982. Sympatric genetic divergence in the leaf mining insect *Liriomyza brassicae* (Diptera: Agromyzidae). Evolution 36: 523.

Via, S. 1984a. the quantitative genenics of polyphagy in an insect herbivore. I. Genotype—environment interaction in larval performance on different host plant species. Evolution 38:881.

Via, S. 1984b. The quantitative genenics of polyphagy in an insect herbivore. II. Genetic correlations in larval performance within and among host plants. Evolution 38:896.

Via, S. 1986. Covariance of female host plant preference and larval performance in an insect herbivore (submitted to Evolution).

Via, S. and R. Laude. 1985. Genotype—environment interaction and the evolution of phenotypic plasticity. Evolution 39:505.

Wiklund, C. 1974. Oviposition preference in *Papilio machaon* in relation to the host plants of the larvae. Entomol. Exp. Appl. 17: 189.

Wright, S. 1977. Evolution and the Genetics of Populations. V3: Experimental results and evolutionary deductions. Chicago University Press, Chicago.

Drosophila Larval Foraging Behavior and Correlated Behaviors

Marla B. Sokolowski

Department of Biology
York University
Downsview
Ontario, Canada M3J 1P3

Introduction

Correlated behaviors are important from an evolutionary viewpoint, especially if the correlations are genetic. Phenotypic correlations of behaviors do not necessarily reflect genotypic correlations, however, since they may also be the result of environmental factors or have a large environmental component (Hedrick 1982). It is for this reason that it is important not to use phenotypic correlations as the basis for evolutionary predications (for example, in predicting correlated responses to selection). Genetic analyses of phenotypic correlations are essential to our understanding of evolutionary processes. Two behavior patterns may be correlated genetically because they are influenced by a common set of genes (pleiotropy) or because they are controlled by different genes that are closely linked on the same chromosome (linkage). For example, suppose there exists a positive genotypic correlation between the activity rates of two behaviors. In the case of pleiotropy, there might exist a general activity factor that influences the rates of activity of the two otherwise discrete behaviors. In the case of linkage, two sets of genes would influence activity rates independently in each of the behaviors.

Genetic correlations are relevant from an evolutionary viewpoint since during selection for one trait, changes in the phenotypic value of other traits may occur simultaneously. It is of considerable evolutionary interest to determine whether the behavioral components of habitat selection are interrelated, especially at the genetic level. The present paper identifies the behavioral components of habitat selection in the third instar larva of *Drosophila melanogaster* Meigen. Many of these behaviors are correlated phenotypically, related developmentally and influenced environmentally. Chromosomal analysis has indicated that these phenotypic correlations are influenced significantly by genes on a single pair of autosomes in this species.

Genetic and environmental determinants of habitat selection in *Drosophila* have focused, until recently, on adult behavior, and particularly the effect of certain parameters (Pyle 1976, 1978, Ohnishi 1977) such as the presence of males (Mainardi 1968, 1969, Ayala and Ayala 1969), gregariousness (Del Solar and Palomino 1966, Del Solar 1968, Soliman 1971, Parsons 1978), light (Rockwell and Grossfield 1978), temperature (Fogleman 1979), ethanol (Richmond and Gerking 1979), and the nature of the substrate (Takamura and Fuyama 1980, Fogleman et al. 1981a, Sokolowski, unpublished data) on oviposition site preference. However, there has been a recent surge of interest in the importance of larval behavior in habitat choice (Sewell et al. 1975, Parsons 1977, Burnet et al. 1977, Godoy-Herrera 1977, 1978, Cavener 1979, Ohnishi 1979, Manning and Markow 1981, Parsons 1980, Sokolowski 1980, 1981, and 1982a-1982c, Fogelman et al. 1981b, Green et al. 1983, Sokolowski and Hansell 1983a and 1983b, Sokolowski et al. 1983).

My research focuses on the behavior of third instar larvae of *Drosophila melanogaster*. The present paper will summarize briefly published work from our lab. Recent unpublished findings will be discussed in greater detail. In the first section, a chromosomal analysis of early third instar larval foraging behavior is discussed. Similar genetic control of larval foraging behavior is demonstrated in both laboratory stocks (Sokolowski 1980, 1982, Sokolowski et al. 1983) and isofemale lines derived from a natural population (Bauer and Sokolowski). The relationship between larval age and foraging behavior is discussed in section two. An age-related change from foraging behavior to wandering, a preparation behavior, is outlined. In the third section, late third instar prepupation behavior is discussed. The polygenic nature of pupation heights in laboratory stocks (Sokolowski and Hansell 1983b) and a natural population (Bauer and Sokolowski, 1985) of *D. melanogaster* is described. In section four, the importance of oviposition site preferences in habitat selection is documented briefly. The paper concludes with a discussion of the implications of the correlations between larval foraging patterns, digging behavior, pupation heights and oviposition site preferences.

Early Third Instar Larval Foraging Behavior

The Rover and Sitter Larval Forager Types

Larval foraging behavior can be defined as the relative amounts of feeding and locomotor behavior performed on the feeding substrate during a test period. The number of probes with the mouth hooks (shovels) and the number of muscular contractions passing along the body of the larva (crawls) are discrete measures of feeding and locomotor behavior (Sokolowski 1980). Sewell et al. (1975), Burnet et al. (1977), and Ohnishi (1979) studied larval feeding rate in an aqueous yeast suspension. *Drosophila* larvae also tend to crawl along the feeding substrate while shovelling. Larval feeding and locomotor behavior can be examined simultaneously when a larva is placed in a petri dish covered with a yeast paste so that a moving larva leaves a visible trail or path in the yeast. The length of the foraging trail during a 6-min test period is measured. This measurement is termed the *path length* of the foraging trail. Since path length is strongly and positively correlated (+.9) with crawling behavior, path length can be used to provide a rapid determination of the locomotory component of the foraging behavior phenotype (Sokolowski 1980).

I have called larvae that have long path lengths, high crawling scores and traverse a large area (while foraging in a yeasted dish), *rover* larval foragers. *Sitter* larval foragers have relatively shorter path lengths, low crawling scores and cover a smaller area while foraging (Fig. 1). Two stocks, isogenic for the second and third pairs of chromosomes and designated W_2W_3 and E_2E_3 were used in this study. A breeding scheme that utilizes the presence of crossover suppressors to permit substitution of intact second or third chromosome pairs from one stock into another (described in Sokolowski 1980) was used to produce a second pair of stocks, E_2W_3 and W_2E_3. E_2W_3 has the same second chromosome pair as E_2E_3 and the third chromosome pair from W_2W_3. E_2E_3 and W_2E_3 both carry recessive alleles for the gene for ebony body color (e^{11}) on the third chromosome. Genetic analyses using the chromosome substituted stocks W_2W_3, W_2E_3, E_2E_3 and E_2W_3 revealed that differences in these forager types could be attributed to the second pair of chromosomes, whereas differences in larval feeding rate were affected by both the second and third chromosomes.

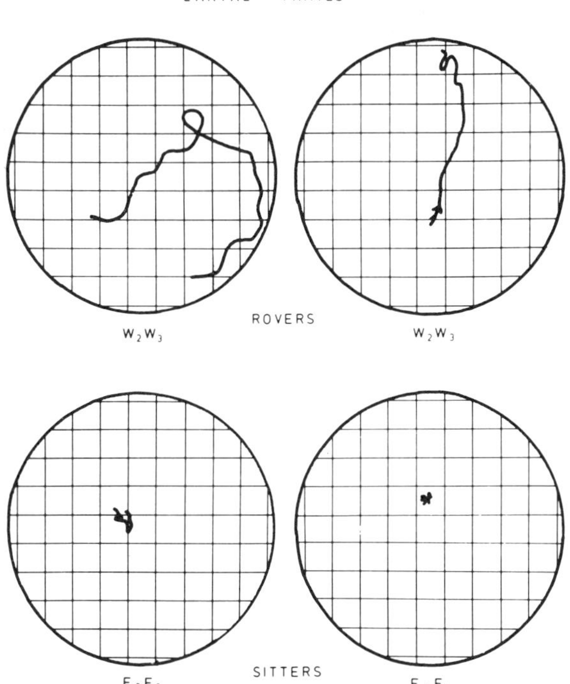

Fig. 1. A random sample of *Drosophila melanogaster* larval trails of W_2W_3 and E_2E_3 shown superimposed on a centimeter grid. The lengths of the trails (path length) and the numbers of squares traversed are clearly different. W_2W_3 shows the *rover* behavioral phenotype whereas E_2E_3 shows the *sitter*.

Figure 2 illustrates the mean crawling and shovelling scores ± S.E. of W_2W_3, E_2E_3, the chromosome substituted stocks W_2E_3 and E_2W_3 and the reciprocal crosses of W_2W_3 and E_2E_3. Stocks with the W_2 homologues showed significantly higher (< 0.001, Mann-Whitney U Test) crawling scores than did those with the E_2 homologues. The heterozygous larvae of both reciprocal crosses showed very high crawling and shovelling scores compared with the scores of the parental stocks. When adults reared from *rover* larvae were mated with adults reared from *sitter* larvae, hybrid progeny of both sexes exhibited high levels of crawling behavior (the *rover* phenotype). Thus a high level of crawling behavior is dominant and the X chromosome does not affect crawling behavior significantly. The differences in shovelling behavior could not be attributed to a single pair of autosomes. The stocks' mean crawling and shovelling scores well represented the temporal changes in foraging behavior during the 6-min test period (Sokolowski 1982c).

Digging Behavior

Drosophila larvae do not only travel in a horizontal plane, they also may move downward through the medium. Polygenic control of digging behavior has been shown by Godoy-Herrera (1977, 1978) who studied variation in digging behavior in strains of *D. melanogaster* and successfully selected for low digging activity. In Sokolowski (1982a), a correlation

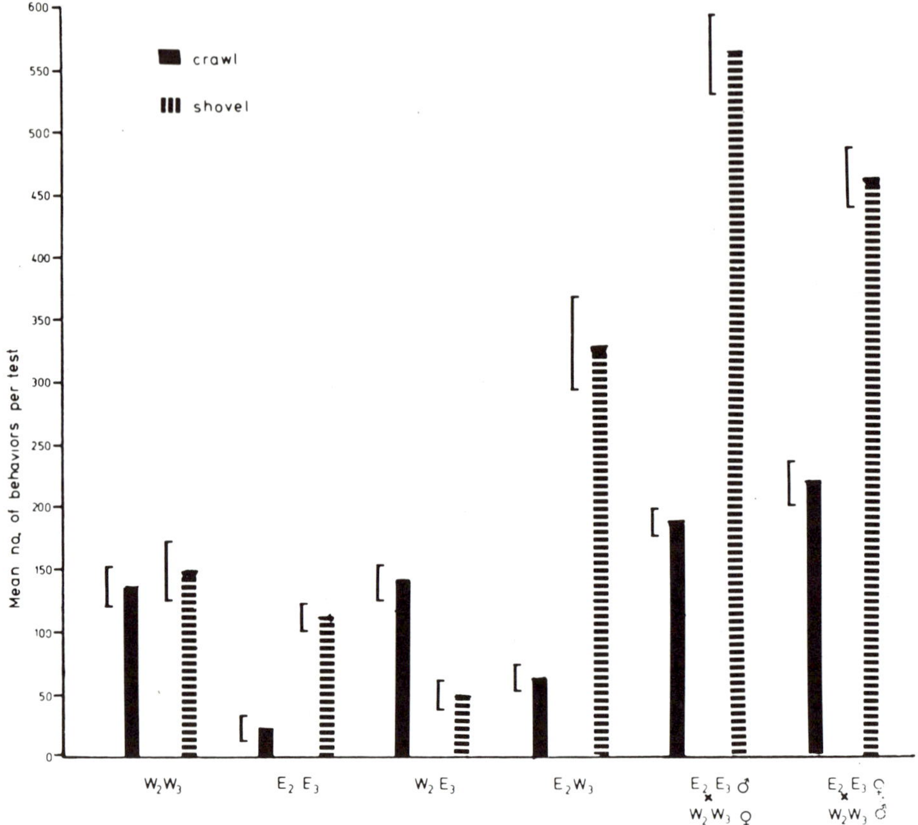

Fig. 2. The mean (±S.E.) crawling (solid) and shovelling (hatched) scores performed by W_2W_3, E_2E_3, W_2E_3, E_2W_3 and the reciprocal crosses of W_2W_3 and E_2E_3. Twenty-five early third instar *Drosophila melanogaster* larvae of each stock and 12 larvae of each of the reciprocal crosses were tested. Part of this graph was taken from Sokolowski (1980).

between crawling and digging behavior is reported. *Rover* larval foragers tended to dig deeper into the medium than did *sitters*. The significant effect of both the second and third chromosomes on digging behavior supported Godoy-Herrera's contention that digging behavior was under polygenic control in *D. melanogaster*.

The Locomotory Component of Larval Foraging Behavior in Isofemale Lines from a Natural Popuation

Studies of natural population variation in larval foraging behavior are rare. Ohnishi (1979) used isofemale lines of the sibling species *D. melanogaster* and *D. simulans* Sturtevant to study the relationship between second instar larval feeding rate (shovelling) and viability. He found significant variation among isofemale lines within a species but not between species. The correlation between feeding rate and egg-to-adult viability reaffirmed the importance of larval feeding behavior to fitness (Bakker 1961, 1969, Nunney 1983, Taylor and Condra 1983).

In the fall of 1982, isofemale lines of *D. melanogaster* were collected in the Toronto area. Four-day-old (posthatching) third instar larvae were tested for a 5-min period in the yeast covered petri dish apparatus described in Sokolowski (1980). Fifty individual larval path lengths were scored for each of 15 isofemale lines (termed B-1, B-2 ... B-15). An ANOVA for path lengths by isofemale line showed significant between-line variation ($F = 18.55$, $p = .0001$, between line mean square = 80.5, df = 14, within-line mean square = 4.3, df = 735). Larvae of the B-15 line showed the longest path lengths (the mean ± S.E. was 86.1 ± 3.4) compared with larvae of the B-1 line which had the shortest path lengths (44.7 ± 2.8) (Fig. 6). Crosses between these lines were performed in order to determine whether the genetic control of path length in a natural population was consistent with that found in the laboratory stocks. The results of these crosses and crosses between lines B-15 and B-8 are presented in Table 1. Crosses between B-15 and B-1 confirmed the results found in the reciprocal crosses of the laboratory stocks (Fig. 2). Longer path lengths (*rovers*) were dominant over shorter path lengths (*sitters*) and there was neither sex linkage nor a maternal effect. The mean path length phenotypes in the F_2 generations (Table 2) were as long as the long path length parental stocks (B-15). The F_1 backcrosses to the short path length parents (B-1 or B-8) had significantly shorter path lengths than did F_1 backcrosses to the long path length parental stock (B-15). These patterns were evident in both sets of crosses presented in Table 1. A 1:1 ratio of the long

Table 1. Path lengths in isofemale lines of *Drosophila melanogaster.*

Lines B-1 and B-15		*Lines B-8 and B-15*	
Parental line	Path length (mean ± S.E.	Parental line	Path length (mean ± S.E.
B-1	44.7 ± 2.8	B-8	60.0 ± 2.3
B-15	86.1 ± 3.4	B-15	86.1 ± 3.4
Reciprocal Crosses		**Reciprocal Crosses**	
1. B-1 ♀ x B-15♂	82.1 ± 3.4	1. B-8 ♀ x B-15♂	82.7 ± 2.8
2. B-15 ♀ x B-1♂	91.9 ± 3.8	2. B-15 ♀ x B-8♂	79.6 ± 3.1
F_2 Crosses		**F_2 Crosses**	
F_2 from **1.** above	100.4 ± 4.2	F_2 from **1.** above	77.6 ± 4.2
		F_2 from **2.** above	89.8 ± 4.5
Backcrosses	**Backcrosses**		
F_1♂ from **2.** above x B-1♀	60.3 ± 2.9	F_1♂ from **1.** above x B-8♀	68.5 ± 2.9
F_1♀ from **2.** above x B-15♂	87.0 ± 3.4	F_1♂ from **2.** above x B-15♀	84.7 ±3.2

Table 2. *Drosophila melanogaster* stock specific OSP for larval inhabited (I) as compared to uninhabited (U) plugs.

Ovipositing adult stock	Larval stock on inhibited plug *(I)*	Inhibited plug preference number of times $I > U$	Uninhabited plug preference number of times $I < U$	Sign test Z score and 2-tailed probability
W_2W_3	none (control)	28	29	n.s.
W_2W_3	W_2W_3	31	29	n.s.
W_2E_3	none (control)	31	29	n.s.
W_2E_3	W_2E_3	39	25	n.s.
E_2E_3	none (control)	33	32	n.s.
E_2E_3	E_2E_3	51	24	$Z=3.3$, $p=.001$
E_2W_3	none (control)	28	31	n.s.
E_2W_3	E_2W_3	38	18	$Z=2.7$, $p=.007$

path length (B-15) phenotype to the short path length (B-1) phenotype was found in the backcross to the short path length parental stock (B-1). A 1:1 phenotypic ratio of *rover* to *sitter* also was found when backcrossing the isogenic laboratory stocks (Sokolowski, unpublished).

The similarity in the genetic control of laboratory and natural populations indicates that the genetic differences in the locomotory component of foraging behavior found in the lab stocks are not artifacts of culturing animals in the laboratory. Further evidence for the existence of genetic variation for path length was evident from the bimodal distribution of larval path lengths of 100 *D. melanogaster* larvae picked from a natural population on a pear (Sokolowski 1982b). Thus, *rover* and *sitter* larval forager types appear to exist in natural populations as well as in laboratory populations.

The Relationship between Third Instar Larval Behavior and Age

Many variables influence larval behavior in the third instar including age (Sewell et al. 1975, Sokolowski et al., 1984), strain (Burnet et al. 1977, Godoy-Herrera 1977, Pruzan and Bush 1977, Ohnishi 1979, Sokolowski 1980, 1982a, Sokolowski and Hansell 1983a,b), type and homogeneity of the medium (Sokolowski et al. 1984) and light (Manning and Markow 1981, Markow 1981). The effects of age are of particular interest, since larvae are searching for a pupation site, rather than foraging, by the end of the third instar. Pupation site choice and rate of development are strain-dependent (Sameoto and Miller 1968, Markow 1979, Sokolowski and Hansell 1983b). Larvae may be expected, therefore, to shift from foraging-related behaviors to prepupation behaviors at some time in the third instar in a strain-dependent manner. Thus, we can ask whether these different behaviors can be separated reliably by assaying for locomotory behavior. Larvae of four strains, W_2W_3, W_2E_3, E_2E_3 and

E_2W_3 (Sokolowski 1980) were tested at three different stages in third instar development (3, 4 and 5 days posthatching) in order to answer this question.

Culture dishes (100 larvae/dish) of closely (± 0.75 h) aged larvae were incubated at 24 ± 1°C, approximately 60% RH and a light cycle of 12 h light followed by 12 h darkness. The lights were turned on at 0800 h. Culture dishes were 8.5 cm in diameter and 1.4 cm high and contained no less than 28 g of a standard nonliving Brewer's yeast agar medium. Replicate culture dishes were prepared to enable the testing of 3-, 4- and 5-day-old (posthatching) larvae. All behavioral tests were performed between 1400 and 1900 h.

Larvae were tested in the paired behavioral testing apparatus (Fig. 3) which was prepared as follows. Two large (13.5-cm diam x 2.2-cm high) petri dishes were filled to a depth of 0.5 cm with hot Difco® agar (prepared by combining 8 gm of agar with 500 ml of distilled H_2O and then boiling the solution for 5 min). The agar in the dishes was then flamed (using a Bunsen burner) to eliminate bubbles from the surface, thereby ensuring a smooth surface texture for larval locomotion. After the agar had cooled, two plugs (2.5-cm diam x 0.5-cm high) were placed on the agar layer in the positions diagrammed in Figure 3. The apparatus on the left (C-T) contained one agar (nonnutritive) plug which will be called the C or control plug and one food plug made from a standard yeast-agar medium darkened with charcoal (Sokolowski 1982a), that will be called the T or treatment plug. The apparatus on the right (T-T) contained two food plugs.

The testing procedure was as follows. All of the larvae were collected from each culture dish and rinsed twice in a small amount of distilled H_2O to remove all food residue. All of the larvae removed from a single culture dish comprised the group from which a random sample of 50 larvae was chosen for the paired behavioral tests. Two sub-groups of 25 larvae were tested simultaneously in the paired behavioral testing apparatus. One sub-group of larvae was placed on the center of the control plug in the C-T apparatus using a paint brush. The second

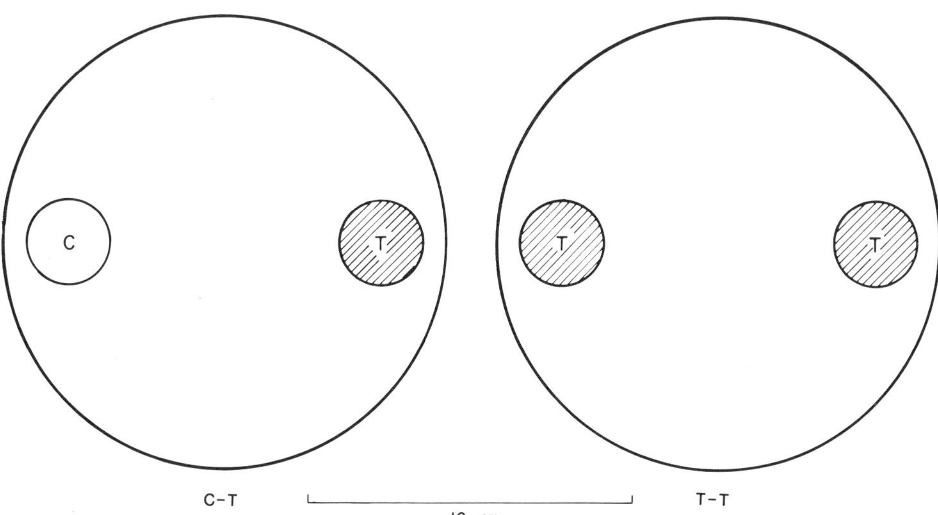

Fig. 3. The paired behavioral testing apparatus. Two plugs were placed on an agar surface. The C-T apparatus contained one agar plug (C) and one food plug (T). The T-T apparatus contained two food plugs positioned as shown.

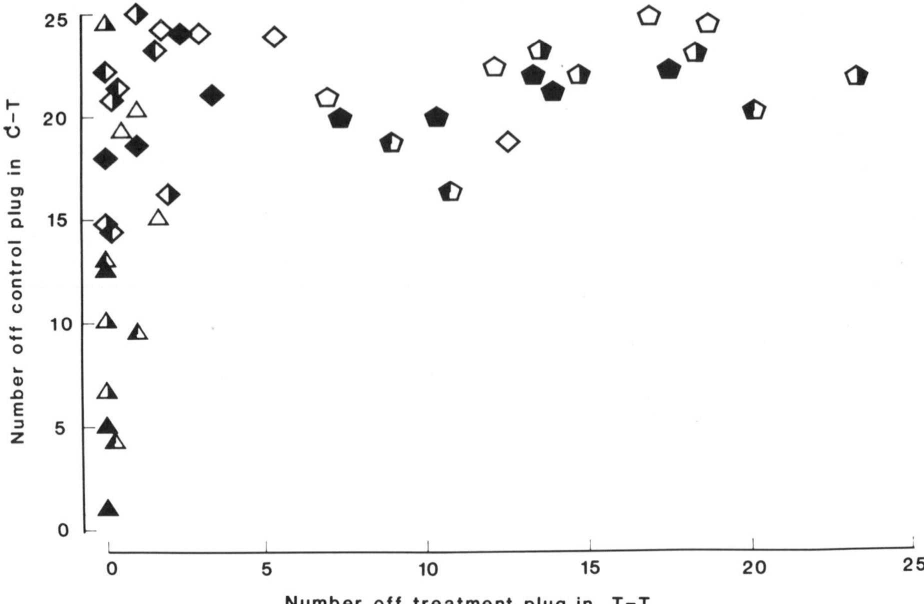

Fig. 4. The average number of *Drosophila melanogaster* larvae out of 25 that left the plug they were originally placed upon (from 20 to 60 minutes). The Y axis shows the number having left the C plug in the C-T apparatus. The X axis shows the number having left the T plug in the T-T apparatus. Triangles show 3-day, diamonds show 4-day, and pentagons show 5-day-old larvae. W_2W_3 symbols are shaded, E_2E_3 are open, W_2E_3 are shaded on the left side whereas E_2W_3 are shaded only on the right hand side of the symbol.

sub-group was placed on the center of the left treatment food plug in the T-T apparatus at the same time (see Fig. 3) after which the lids of the dishes were placed on the apparatus, marking the beginning of the experiment. One investigator observed the entire paired behavioral testing apparatus and recorded the number of larvae on each of the four plugs and the number of larvae off both plugs in C-T and T-T. This information was recorded every minute for the first 10 min of the test, and afterwards at 10-min intervals, up to 60 min. Only the "long-term" results (20-60 min) are discussed.

We calculated the number of larvae that left the plug on which they were placed at each time interval. Since larvae from each culture were placed on the control plugs (nonfood) in the C-T apparatus and on a food plug in the T-T apparatus, the number of larvae that were off each plug at a given time formed a matched pair of observations. By averaging these values for the observations at times 20 through 60 min, we obtained a single pair of numbers for each culture. This (paired) observation represents the mean long-term tendency to leave a plug. Each symbol in Figure 4 represents results from one culture.

There was a *simple* relationship between the behaviors observed in the two parts (C-T and T-T) of the paired behavioral test apparatus (Fig. 4). (1) Three-day-old larvae (triangles) remained on food plugs. Larvae on the control plugs moved slowly to food plugs at a rate that is strain dependent. (2) Four-day-old larvae (diamonds) tended to remain on the food plugs but less so than 3-day-old larvae. The major difference in 4-day-old larvae (compared with 3-day-old larvae) is that they left the C plug and moved to the food plug more rapidly. (3) Five-day-old larvae (pentagons) left the C plug as readily as 4-day-old larvae. However, 5-day-old larvae also left the food plug.

Two distinct developmentally related behavior patterns can be identified in third instar larvae of all stocks, "foraging" and "wandering." These behavior patterns are performed sequentially. "Foraging" involves remaining on the food plug in T-T whereas in C-T larvae leave the nonfood plug (C), subsequently move toward the food plug (T) and remain on it. The frequency at which larvae leave the C plug is low at 3 days and reaches a maximum at 4 days. "Wandering" involves leaving food plugs. As larvae get older, they leave food and nonfood plugs equally (Fig. 5). In our apparatus, the second behavior (wandering) is performed only after the first behavior (foraging) is expressed maximally.

We are presently repeating these experiments with other strains and are encouraged to believe that this sequence of behaviors may be representative of those found in most well-fed, third instar *D. melanogaster* larvae. We feel justified in labeling wandering a prepupation behavior since the tendency to remain off plugs in the C-T and T-T apparatus anteceded puparium formation.

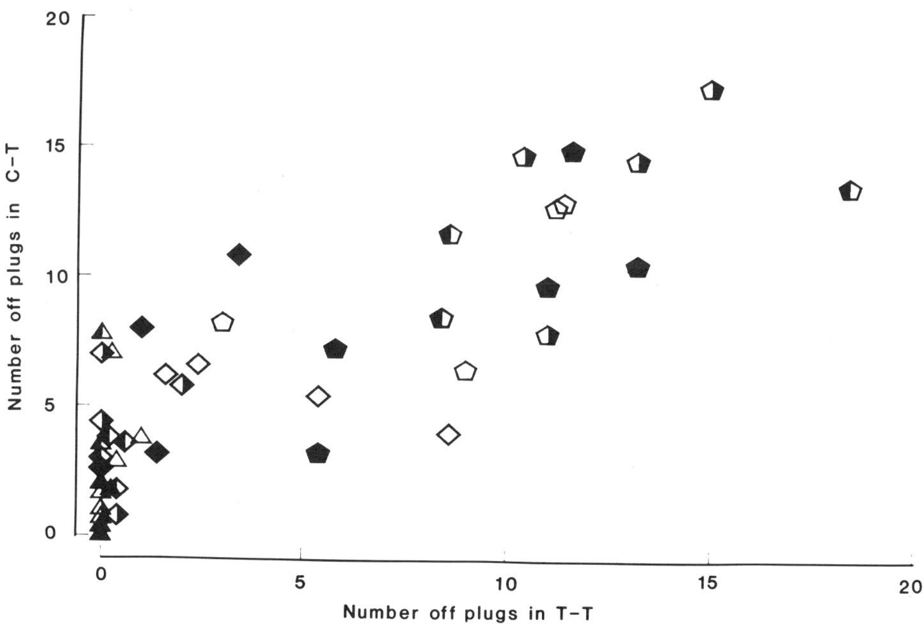

Fig. 5. By calculating the mean numbers of *D. melanogaster* larvae off the plugs during times 20 to 60 minutes, for C-T and T-T, a second pair of numbers was found for each culture. These (paired) observations are plotted in this figure. Symbols are the same as for Figure 4. Three and 4-day larvae tended to be found on either plug in both C-T and T-T. In contrast, 5-day larvae were more likely to be found off all plugs.

Foraging Behavior on Nutritive and Non-Nutritive Medium

The behavioral test used by Sokolowski (1980) (larval locomotion during a 6-min period in yeast paste) reliably distinguished early third instar larvae of different strains. The yeast paste medium has at least three characteristics important to foraging: it is homogenous, moist, and is a concentrated food source. How do larvae behave when presented with heterogenous medium and with choices between food and nonfood medium (Fig. 3)?

Differences in 3-day-old larval behavior in the paired behavioral test apparatus were strain dependent and attributable to the second pair of chromosomes. This can be seen in Figure 4 where W_2 stocks (triangles with shaded left halves) fall lower than E_2 stocks. Stocks that shared E_2 chromosomes (E_2E_3 and E_2W_3) showed a higher propensity to leave a nonfood (C) plug than W_2 (W_2W_3 and W_2E_3) stocks. Three-way analysis of variance was performed using the number off the control plug as the dependent variable for 3-day-old larvae. The independent variables were: (1) second chromosome, (2) 3rd chromosome, and (3) age of the culture. The third variable was included to determine the effect of small differences ($\pm$ 2 h) in the age of cultures. The F values for variables (d.f. = 1, 8) were (1) 6.83 (p = .03), (2) 1.24 (NS), and (3) 1.78 (NS). This indicates that neither age of the culture nor the third chromosomes had a significant effect on leaving nonfood plugs at 3 days; the effect of the 2nd chromosomes *is* significant. No significant strain differences were found in the behavior of 4- and 5-day-old larvae.

Three-day-old larvae showed some novel behaviors that are of general interest. Many larvae, after being placed on top of an agar plug (nonfood), remained in contact with the side of the plug and/or burrowed beneath it. The presence of any physical inhomogeneities or "edges" in the medium affected larval behavior profoundly in a strain-dependent manner. This is of importance in experimental apparati with punctures, holes or edges in otherwise uniform medium.

There is a significant effect of the second chromosomes on the likelihood of leaving a C plug at 3 days. However, of those larvae that left the C plug, there was no significant strain difference in the number that reached the food plug within an hour. The strain-dependent manner in which 3-day-old larvae reached the food plug in the C-T apparatus was a function of differences in "edge-leaving" behavior, rather than the rate of locomotion on agar.

Do any of the strain differences in the 3-day-old larvae reflect the *rover* and *sitter* behaviors described by Sokolowski (1980)? The 6-min locomotory tests were done in a moist, nutrient-rich, paste-like yeast slurry. The present longer-term experiment was performed using a heterogeneous environment and required that larvae traverse a dry, nonnutritive, firm, smooth (agar) surface, in order to reach a food source. The *sitter* larvae reached the food as quickly (if not quicker) than the *rover* larvae. This implies that the *rover* and *sitter* forager types described by Sokolowski (1980) do not reflect differences in general activity levels. Rather, *roving* and *sitting* are specific behaviors, dependent on the moist, nutrient-rich test apparatus.

Prepupation Behavior

Third instar larval behavior changes from *foraging* in 3- and 4-day-old larvae to *wandering* in 5-day-old larvae. Evidence for the genetic control of the locomotory component of foraging behavior in 3- and 4-day-old larvae has been presented for both laboratory and natural populations. Five-day-old larvae tended to leave food plugs and wander on the surface of the agar (and the walls or lid) of the paired behavioral test apparatus. The significant differences in path length described in early third instar larvae (first section) were no longer evident in older (5.5-day-old) larvae of the four chromosomally substituted laboratory stocks. Prior to pupation, larvae of all stocks displayed long path lengths while traversing the yeasted petri dish. The strain-dependent differences observed in early third instar larvae in both the yeasted petri dish and the paired behavioral test apparatus were no longer evident in the late third instar. The average number of hours until pupation, counted from 3 days (72 h) posthatching, was 87.3 for E_2E_3, 71.4 for E_2W_3, 64.5 for W_2W_3 and 73.2 for W_2E_3. Puparia are formed between approximately 5.5 and 6.5 days posthatching.

Pupation site choice is an important component of fitness since it influences the success of adult emergence (see Grossfield 1978, for a detailed review). During the immobile pupal stage, animals are subject to dessication, predation and disease. When larvae have reached their minimum weight for pupation (late in the third instar) they begin to search for a pupation site (Bakker 1961). In the laboratory, the larvae often utilize the walls of culture bottles as pupation sites. In a natural population, McCoy (1962) observed that larvae of *D. melanogaster* pupated on the dry skins of fruits, on the soil surface, or approximately 1.25 cm below the soil surface.

The results of pupation site studies vary with biotic factors such as density (Barker 1971, Ringo and Wood 1983, Sokolowski and Hansell 1983b), age structure (Sokolowski, unpublished), and the genetic architecture of the population or species studied (Sameoto and Miller 1968). Sokal et al. (1960) concluded that there were significant developmental (as well as gene-environment) interactions affecting pupation site choice and they failed to find any correlation between pupation site choice and larval behavior patterns. Pupation behavior appears to be under polygenic control in many *Drosophila* species (Sokolowski and Hansell 1983b). de Souza et al. (1970) found that a single major gene was responsible for larval choice of pupation site in *D. willistoni* Sturtevant. Abiotic factors that affect pupation height include moisture (Sameoto and Miller 1968), temperature (Mensua 1967) and light (Manning and Markow 1981, Markow 1981). The results of selection experiments and other genetic analyses have led to differing views of the genetic basis for the choice of pupation site. Models include polygenic and single gene inheritance, as well as little or no genetic control of this behavior. Certainly, these results are dependent upon experimental conditions.

Sokolowski and Hansell (1983b) showed that: (1) there were strong second chromosome effects on pupation height; (2) pupation height increased with larval density; and (3) there was a strong correlation between larval foraging patterns and pupation site choice in the chromosomally substituted stocks, indicating that *rover* larvae pupate higher than *sitter* larvae.

The isofemale lines described earlier in this paper were tested in order to study variation in pupation height in a natural population. Pupation height was measured as the distance (mm) from the surface of the medium to the spot between the spiracles of each pupa. The density of larvae per vial was kept constant by placing ten freshly hatched larvae in each (9.5-cm high x 2-cm diam). All other methods were as described in Sokolowski and Hansell (1983b).

The mean ± S.E. pupation heights and the larval path length scores for each of the 15 isofemale lines are graphed in Figure 6. Significant between-line variation was found for pupation height ($F = 19.55$, d.f. $= 14, 863$, $p = .0001$). Although there was a strong positive correlation between larval path length and pupation height in the laboratory stocks (Sokolowski and Hansell 1983b), there was only a weak positive correlation (+.5) between these behaviors in the isofemale lines. Crosses were performed between lines with extreme pupal heights (B-15, high line and B-8 low line) in order to obtain evidence about the dominance relations and the influence of the X chromosomes on pupation height and path length, respectively.

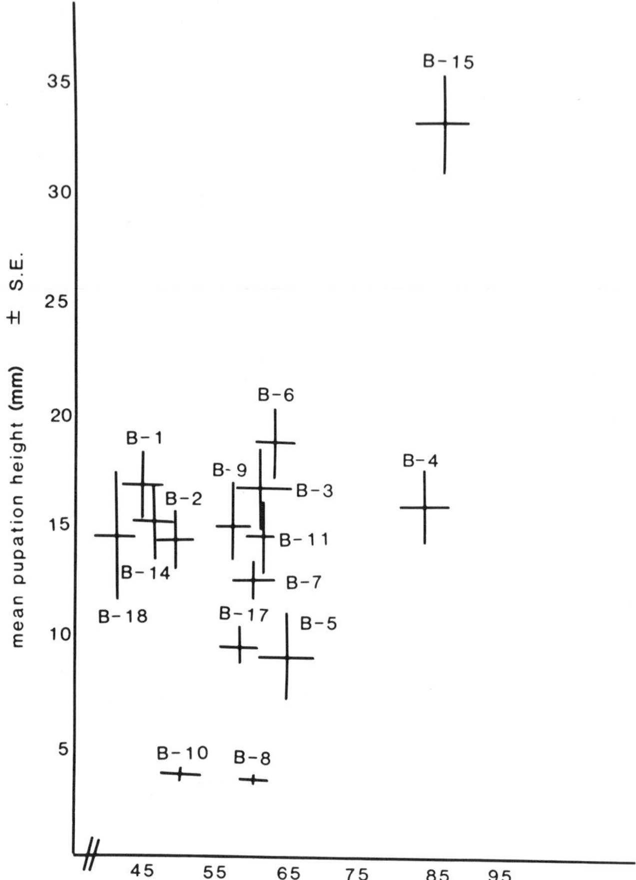

Fig. 6. The mean pupal height (mm ± S.E.) is plotted against the mean larval path length (mm ± S.E.) for each of the 15 isofemale lines of *Drosophila melanogaster*. The path lengths of 50 larvae and the pupation heights of approximately 50 pupae were measured for each line.

Even though pupation height is a quantitative character, it was possible to conclude that the high pupal height phenotype (B-15, X ± S.E. = 33.0 ± 2.2, N = 69) had considerable dominance over the low pupal height phenotype (B-8, X ± S.E. ± 3.8 ±0.4, N ± 64). Further genetic analysis and the construction of chromosome substituted stocks from B-15 and B-8 will help to elucidate the polygenic nature of pupation height.

Oviposition Site Preferences: Adult Behavior

Females from stocks known to have genetic differences in preadult behavior (larval foraging behaviors) also have differences in oviposition site preferences (OSP) when given a choice between uninhabited oviposition sites (U) and sites inhabited by larvae (I). The measure of OSP used in the present study was described initially by Del Solar and Palomino (1966).

The oviposition preference apparatus consisted of a petri dish (13.5-cm diam x 2.0-cm height) that contained four pairs of oviposition sites positioned around the periphery of the dish (Sokolowski 1981). A pair of oviposition sites consisted of two plugs of medium (2.5 cm diam x .75 high) designated I and U plugs. Thirty 5-day-old flies (15 females and 15 males) from one of the four stocks (W_2W_3, E_2E_3, W_2E_3 and E_2W_3) were placed into the center of the oviposition preference apparatus. Adults were left to oviposit for 24 hours (starting between 1300 and 1500 h) under conditions of constant illumination, 23 ± 1 °C and approximately 60% R.H. After the oviposition period, the flies were removed and the number of eggs laid on each of the plugs was counted.

Each of the four stocks was tested under the following two testing conditions: (1) control—*I* and *U* plugs contained no larvae; (2) treatment—each *I* plug was colonized with 15/second instar (48 h after hatching) larvae of the same stock as the ovipositing adults, the *U* plugs were uninhabited. All adults were placed in the OSP apparatus 2 h after the addition of the larvae.

Significant differences in OSP for inhabited as compared to uninhabited plugs were found in the treatment experiment. *Sitter* females of the stocks sharing an E_2 pair of second chromosomes (E_2E_3, E_2W_3) showed a significant preference for plugs inhabited by larvae as compared to uninhabited plugs. *Rover* females demonstrated no significant OSP for inhabited (*I*) sites as compared with uninhabited (*U*) sites (Table 2). In order to test the statistical significance of OSP's, the sign tests were used to determine the number of times the *I* plugs contained more eggs ($I > U$) as compared to the number of times the *U* plugs had more eggs oviposited upon in ($I > U$). The test statistic used was $Z = (X\text{-}0.5N)/(0.5\sqrt{N})$ (Siegel 1956) where N is the total number of $I - U$ pairs which had an unequal number of eggs oviposited on them.

The correlation between larval foraging behavior and OSP is not surprising since the capacity of an adult female to select a suitable oviposition site determines strongly the resources available to the preadult *Drosophila*; preimaginal forms have low mobility compared with adults. Details of these studies will be published elsewhere.

Conclusions: Correlated Behaviors

The relationship between larval behavior and age was studied by measuring the tendency to forage (move toward and remain on food) or to wander (move away from food) in a heterogeneous environment. A quantifiable change in behavior from *foraging* to *wandering* (a prepupation behavior) was observed in all stocks. Further evidence indicates that this sequence of events was probably characteristic of most well-fed, third instar larvae.

Changes in larval foraging and wandering behavior may co-occur with known developmental events such as ecdysone secretion or the attainment of minimal pupation weight (Bakker 1961). Further studies of third instar larval behaviors may help to establish the relationship between behaviors controlled genetically and developmental events.

By examining different aspects of larval foraging behavior in laboratory and natural populations, it was possible to investigate the influence of both genes and the environment on the behavior of the developing larva. A behavioral polymorphism in early third instar larval foraging behavior was quantified in the yeasted petri dish (a homogeneous, rich, moist food supply). Differences in the behavior of *rover* and *sitter* larvae were attributed to the second pair of chromosomes (Sokolowski 1980). *Rover* larvae not only had longer foraging path lengths, they also dug deeper into a (relatively) homogeneous yeast-agar medium than did *sitter* larvae. Both the second and third pairs of autosomes were found to have a significant effect on digging behavior (Sokolowski 1982a). *Rover* larvae also tended to burrow into inhomogeneities in the medium (both food and nonfood plugs), when food was distributed unevenly (as in the C-T and T-T apparatus). Although *sitter* larvae had low locomotory scores on a rich, homogeneous food supply, their activity level on a nonnutritive, dry food source (agar) was as high as that of *rover* larvae. When placed on a nonnutritive food source, *sitters* and *rovers* were equally able to locate, move toward and remain feeding on a nutritive food source. This indicated that *sitter* larvae were not simply low-activity animals (short path lengths, little digging and low pupal height), rather, *roving* and *sitting* represented actual locomotory components of foraging behaviors conditional on the environment (i.e., the nature of the feeding substrate).

When tracing the chromosomal contribution to the development of behavior in third instar larvae of *D. melanogaster* we found: (1) no significant X chromosome, Y chromosome or maternal effects on any quantified behavior; (2) complete dominance of the long path length phenotype (*rover*) over the short path length (*sitter*) in both the laboratory and a natural population; (3) a strong effect of chromosome number two on third instar larval behaviors (i.e., crawling, "edge effect," path length and digging); (4) disappearance of all strain-dependent differences found in the early third instar by the late third instar coinciding with the change in behavior from *foraging* to *wandering*; (5) strain-dependent differences in pupation height also were attributable to the second pair of chromosomes with *rovers* pupating higher than *sitters*; (6) significant variation between isofemale lines for pupation height, and larval foraging path length in a natural population; and (7) a correlation between larval behavior and adult OSP, with *sitter* females preferring to oviposit on sites inhabited by larvae of their own strain compared with uninhabited sites and *rover* females showing no OSP for inhabited sites.

A group of correlated, developmentally related, third instar larval behavior patterns has been identified. Many of these behavior patterns are strain dependent; all of them are dependent on the environment. These correlations reflect some joint genetic influences on these behaviors that are to some extent attributable to the second pair of chromosomes. We hope to use compound autosomes and deletion mapping to determine the location of the second chromosome genes responsible for the suite of behaviors described in this paper. Research that demonstrates how these behaviors respond to different selection regimes will help elucidate the importance of correlated behaviors in evolutionary processes.

Why should genetic variability for larval foraging behavior exist in nature? The observed differences in path length between isofemale lines may be related to the larval environment. Food acquisition by a larval population depends on the kind of foraging behaviors present and the distribution of food available to it (Sokolowski 1980). Sokolowski (1980) hypothesized that a discontinuous food supply may give an advantage to larvae with longer path lengths, whereas larvae with shorter path lengths may be at an advantage in a continuous food supply. In terms of energy budgets, a long path length larva may waste energy in locomotor activity in a continuous food environment. Another hypothesis that may help elucidate why genetic variability for path length is maintained in nature may be related to *Drosophila* larval parasitoids. Alphen (1982) has demonstrated that the frequency of larval parasitism is related to larval movement. In future studies, we plan to investigate whether the proportion of *rover* and *sitter* larval foragers in a natural population is related to the search images used by *Drosophila* larval parasitoids.

Acknowledgments

I would like to thank Johathan Wong, Clement Kent, Roger Hansell and Allen Sokolowski for many stimulating hours of discussion. Sharon Bauer, Joel Berger, Daniela Rotin and Rhonda Taylor supplied excellent technical assistance. I would especially like to thank J. Wong, C. Kent and S. Bauer for encouraging me to refer to our collaborative unpublished work. Krystyna Tarkowski expertly typed the manuscript. John Ringo, Robert Rockwell, Spencer Johnston, Sharon Bauer and especially Milton Huettel generously helped with the editing of this manuscript. This work was supported by an ASERC University Research Fellowship to M. B. Sokolowski.

Literature Cited

Alphen, J. van. 1982. Host selection by *Asobara tabida* Nees (Braconidae: Alysiinae) a larval parasitoid of fruit inhabiting *Drosophila* species. Neth. J. Zool. 32: 215.

Ayala, F. J., and Ayala, M. 1969. Oviposition preferences in *D. melanogaster*. Dros. Infor. Serv. 44: 120.

Bakker, K. 1961. An analysis of factors which determine success in competition for food among larvae of *Drosophila melanogaster*. Arch. Neerl. de Zool. 14: 200.

Bakker, K. 1969. Selection for rate of growth and its influence on competitive ability in *Drosophila melanogaster*. Neth. J. Zool. 19: 541.

Barker, J. S. F. 1971. Ecological differences and competitive interaction between *Drosophila melanogaster* and *Drosophila simulans* in small laboratory populations. Oecologia 8: 139.

Bauer, S. J. and M. B. Sokolowski. 1984. Larval foraging behavior in isofemale lines of *Drosophila melamogaster* and *D. pseudoobscura*. J. Heredity. 75:131.

Bauer, S. J. and M. B. Sokolowski. 1985. A gentic analysis of path length and pupation height in a natural population of *Drosophila melanogaster*. Can. J. Genet. Cytol. 27:334.

Burnett, B., D. Sewell, and M. Bos. 1977. Genetic analysis of larval feeding behavior in *Drosophila melanogaster*. II. Growth relations and competition between selected lines. Genet. Res. Camb. 30: 149.

Cavener, D. 1979. Preference for ethanol in *Drosophila melanogaster* associated with the alcohol dehydrogenase polymorphism. Behav. Genet. 9: 359.

Del Solar, E. 1968. Selection for and against gregariousness in the choice of oviposition sites by *Drosophila pseudoobscura*. Genetics 58: 275.

Del Solar, E., and H. Palomino. 1966. Choice of oviposition in *Drosophila melanogaster*. Amer. Natur. 100: 127.

Fogelman, J. C. 1979. Oviposition site preference for substrate temperature in *Drosophila melanogaster*. Behav. Genet. 9: 407.

Fogelman, J. C., K. R. Hackbarth, and W. B. Heed. 1981a. Behavioral differentiation between two species of cactophilic *Drosophila*. III. Oviposition site preference. Amer. Natur. 118: 541.

Fogelman, J. K., W. B. Heed, and W. T. Starmer. 1981b. Utilization of food resources by *Drosophila* larvae. Proc. Nat. Acad. Sci. USA. 78: 4435.

Godoy-Herrera, R. 1977. Inter- and intra-populational variation in digging in *Drosophila melanogaster* larvae. Behav. Genet. 7: 433.

Godoy-Herrera, R. 1978. Selection for digging behavior in *Drosophila melanogaster* larvae. Behav. Genet. 8: 475.

Green, C. H., B. Burnet, and K. J. Connolloy. 1983. Organization and patterns of inter- and intraspecific variation in the behavior of *Drosophila* larvae. Anim. Behav. 31: 282.

Grossfield, J. 1978. Non-sexual behavior of *Drosophila*. *In* M. Ashburner and T. R. F. Wright, eds. The Genetics and Biology of *Drosophila*. Vol. 2b. Academic Press, New York.

Hedrick, P. W. 1982. Genetics of Populations. Science Books International, Boston.

Mainardi, M. 1968. Gregarious oviposition and pheromones in *Drosophila melanogaster*. Boll. Zool. 35: 135.

Mainardi, M. 1969. Oviposition preferences in *Drosophila melanogaster* and *Drosophila simulans*. Boll. Zool. 36: 101.

Manning, M., and T. A. Markow. 1981. Light-dependent pupation site preferences in *Drosophila*. II. *Drosophila melanogaster* and *D. simulans*. Behav. Genet. 11: 557.

Markow, T. A. 1979. A survey of intra- and interspecific variation for pupation height in *Drosophila*. Behav. Genet. 9: 209.

Markow, T. A. 1981. Light-dependent pupation site preference in *Drosophila*: Behavior of adult visual mutants. Behav. Neurol. Biol. 31: 348.

Mensua, J. L. 1967. Some factors affecting pupation height of *Drosophila*. Dros. Info. Serv. 42: 76.

McCoy, C. E. 1962. Population ecology of the common species of *Drosophila* in Indiana. J. Econ. Entomol. 55: 978.

Nunney, L. 1983. Sex differences in larval competition in *Drosophila melanogaster*: The testing of a competition model and its relevance to frequency-dependent selection. Amer. Natur. 121: 67.

Ohnishi, S. 1977. Oviposition pattern of several *Drosophila* species under various light environments. J. Insect. Physiol. 23: 1157.

Ohnishi, S. 1979. Relationship between larval feeding behavior and viability in *Drosophila melanogaster* and *Drosophila simulans*. Behav. Genet. 9: 129.

Parsons, P. A. 1977. Larval reaction to alcohol as an indicator of resource utilization differences between *Drosophila melanogaster* and *Drosophila simulans*. Oecologia 30: 141.

Parsons, P. A. 1978. Habitat selection and evolutionary strategies in *Drosophila*: An invited address. Behav. Genet. 8: 511.

Parsons, P. A. 1980. Larval responses to environmental ethanol in *Drosophila melanogaster*: Variation within and among populations. Behav. Genet. 10: 183.

Pruzan, A., and G. Bush. 1977. Genotypic differences in larval olfactory discrimination in two *Drosophila melanogaster* strains. Behav. Genet. 7: 457.

Pyle, D. W. 1976. Oviposition site difference in strains of *Drosophila melanogaster* selected for divergent geotactic maze behavior. Amer. Natur. 110: 181.

Pyle, D. W. 1978. Correlated responses to selection for a behavioral trait in *Drosophila melanogaster*. Behav. Genet. 8: 333.

Richmond, C. R., and J. L. Gerking. 1979. Oviposition site preferences in *Drosophila*. Behav. Genet. 9: 233.

Ringo, J., and D. Wood. 1983. Pupation site selection in *Drosophila simulans*. Behav. Genet. 13: 17.

Rockwell, R. F., and J. Grossfield. 1978. *Drosophila*: Behavioral cues for oviposition. Amer. Midl. Natur. 99: 361.

Sameoto, D. D., and R. S. Miller. 1968. Selection of pupation site by *Drosophila melanogaster* and *D. simulans*. Ecology 49: 177.

Seigel, S. 1956. Nonparametric Statistics for the Behavioral Sciences. McGraw-Hill, New York.

Sewell, D., B. Burnett, and K. Connolly. 1975. Genetic analysis of larval feeding behavior in *Drosophila melanogaster*. Genet. Res. Camb. 24: 163.

Sokal, R. R., P. R. Ehrlich, P. E. Hunter, and G. Schlager. 1960. Some factors affecting pupation site of *Drosophila*. Ann. Entomol. Soc. Amer. 53: 174.

Sokolowski, M. B. 1980. Foraging strategies of *Drosophila melanogaster*: A chromosomal analysis. Behav. Genet. 10: 291.

Sokolowski, M. B. 1981. Evolutionary Strategies in *Drosophila*: Genetic Analyses. Ph. D. Thesis. University of Toronto, Toronto, Ontario.

Sokolowski, M. B. 1982a. *Drosophila* larval foraging behavior: Digging. Anim. Behav. 30: 1252.

Sokolowski, M. B. 1982b. *Rover* and *sitter* larval foraging patterns in a natural population of *D. melanogaster*. Dros. Infor. Serv. 58: 138.

Sokolowski, M. B. 1982c. Temporal patterning of foraging behavior in *D. melanogaster* larvae. Dros. Infor. Serv. 58: 139.

Sokolowski, M. B., and R. I. C. Hansell. 1983a. *Drosophila* larval foraging behavior: I. The sibling species, *Drosophila melanogaster* and *D. simulans*. Behav. Genet. 13: 159.

Sokolowski, M. B., and R. I. C. Hansell. 1983b. Elucidating the behavioral phenotype of *Drosophila melanogaster* larvae: Correlations between larval foraging strategies and pupation height. Behav. Genet. 13: 267.

Sokolowski, M. B., R. I. C. Hansell, and D. Rotin. 1983. *Drosophila* larval foraging behavior: II. Selection in the sibling species *Drosophila melanogaster* and *D. simulans*. Behav. Genet. 13: 267.

Sokolowski, M. B., C. Kent, and J. L. Wong. 1984. *Drosophila* larval foraging behavior: Developmental stages. Anim. Behav. (32:645).

Soliman, M. H. 1971. Selection of oviposition by *Drosophila melanogaster* and *D. simulans*. Amer. Midl. Natur. 86: 487.

de Souza, H. M. L., A. B. da Cunha, and E. P. dos Santos. 1970. Adaptive polymorphism of behavior evolved in laboratory populations of *Drosophila willistoni*. Amer. Natur. 104: 175.

Takamura, T., and Y. Fuyama. 1980. Behavior genetics of choice of oviposition sites in *Drosophila melanogaster*. I. Genetic variability and analysis of behavior. Behav. Genet. 10: 105.

Taylor, C. E., and C. Condra. 1983. Resource partitioning among genotypes of *Drosophila pseudoobscura*. Evolution 37: 135.

Neurogenetics of *Drosophila* Circadian Rhythms

Ronald Konopka*

Division of Biology 216-76
California Institute of Technology
Pasadena, California 91125

Introduction

The isolation of single-gene mutations has proved to be a useful tool in investigating the molecular basis of cellular processes. Although the discovery that living organisms possess endogenous oscillators with a period of about a day (circadian) was made more than two centuries ago, the molecular nature of these oscillators is still unknown. In an effort to understand the genetic control of circadian rhythmicity, as well as to provide a means of identifying a molecule that may be a component of the underlying oscillator, my laboratory is studying the genetics, physiology, and behavior of *Drosophila* that bear chemically induced mutations which alter the periodicity of the circadian eclosion and adult locomotor activity rhythms. This paper summarizes some of the results of these investigations.

Table 1 lists the available mutations at three genetic loci on the X chromosome of *Drosophila melanogaster* Meigen, as well as their map locations and the average periods of the adult locomotor activity rhythms measured in constant environmental conditions. The most drastic effect on periodicity is produced by mutations at the *per* (period) locus in the 3B1-2 region of the chromosome. This is the only locus, studied by us, for which more than one allele is available. In this case there are two long-period alleles, two arrhythmic alleles, and one short-period allele. The *and* (andante) mutation, located at 10E3, close to the miniature-dusky region, lengthens the period of the eclosion and locomotor activity rhythms by about two hours. (Periodicity of eclosion is the average interval between the medians of each emergence peak. Normally, this interval is 24 hours for wild type, but is different in the mutants.) The clk^{KO6} (clock^{KO6}) mutation shortens the period of the activity rhythm by about 1.5 hours. Its eclosion rhythm has not yet been measured. This locus has not been mapped precisely but appears to be located near the *white* locus. It is not an allele of the *per* locus or the *and* locus.

Genetics of the *per* Locus

The *per* locus is conveniently situated in the zeste-white region, intensively studied by Judd and co-workers (Judd et al. 1972). As a result, there are several chromosomal aberra-

* Present Address: Ronald Konopka, Department of Biology, Clarkson University, Potsdam, New York 13676.

Table 1. Locomotor activity rhythm periods of X-linked clock mutants measured in constant infrared and constant temperature (22°C for per^0 and per^{02}). Data shown are for adult males.

Strain	Chromosome Band Location	Period (h) ± Std. Dev.	N
Wild type		23.9 ± 0.3	27
per^s	3B1-2	19.0 ± 0.3	95
per^l	,,	29.6 ± 0.7	67
per^{l2}	,,	29.2 ± 1.6	18
per^{01}	,,	arrhythmic	18
per^{02}	,,	arrhythmic	10
and	10E3	26.0 ± 0.8	72
clk^{KO6}	unknown	22.6 ± 0.4	104

tions available with breakpoints near or at the *per* locus. The rhythm phenotypes influenced by these chromosomal aberrations were studied by Young and Judd (1978) and by Smith and Konopka (1981). With one exception, the results of the two studies were similar. Deficiencies (deletions) with a breakpoint at 3B1-2, in combination with per^0 (arrhythmic) or with each other, lead to either an arrhythmic or wild type phenotype. The one exception was a combination of two deficiencies which Young and Judd (1978) reported as producing a long-period phenotype, but which Smith and Konopka (1981) found to result in a wild type phenotype for both eclosion and activity rhythms. These results suggested that the deficiencies studied either had breakpoints within the *per* locus that caused an arrhythmic phenotype, or outside the *per* locus, leaving it intact and resulting in a wild type rhythm.

The only available translocation with a 3B1-2 breakpoint, known as JC43, produced an interesting and unusual phenotype in combination with per^0 and other deficiencies (Smith and Konopka 1981). Although most individuals heterozygous for JC43 and per^0 or another deficiency had activity profiles that were arrhythmic or showed transient long-period cycles that graded into arrhythmicity, 27% of these individuals had well-defined rhythms, some persisting for more than nine days, with periods in the range 32 to 36 h, much longer than those produced by per^l and per^{l2}.

Dosage Analysis of the *per* Locus

Additional insight into the nature of the *per* locus was provided by a study of dosage alterations at this locus (Smith and Konopka 1982). The *per* locus, like other X-linked loci in *Drosophila*, is dosage compensated in males. Thus, one dose of per^+ in a male results in the same period as two doses in a female. However, an extra dose of per^+ in a male shortens the period of both eclosion and activity rhythms by about an hour, and a reduction of one dose in a female lengthens the period by about an hour. This result, considered with the results of the chromosomal aberration studies, suggests that the per^l and per^{l2} alleles exhibit lenghened rhythms because of reduced gene or gene product activity; per^0 and per^{02} are arrhythmic because of a lack of gene or gene product activity; and per^s produces short rhythms because of increased gene or gene product activity. Since the period of per^s rhythms is significantly shorter than the period of males carrying a total of five doses of per$^+$, the nature of the per^s mutation may lie in increased activity of the *per* gene product and not merely an overproduction of per^+ product.

Partial Dominance of Clock Mutants

Of the five *per* alleles, per^s exhibits the greatest amount of partial dominance over wild type; $per^s/+$ heterozygotes have a period about two hours shorter than wild type and about two hours longer than per^s homozygotes (Konopka and Benzer 1971, Smith and Konopka 1981). On the other hand, $per^0/+$ and $per^l/+$ heterozygotes have periods only about half an hour longer than wild type.

As shown in Table 2, both the *and* and clk^{KO6} mutations exhibit partial dominance over wild type. The periods of heterozygotes are intermediate between the periods of homozygotes and wild type.

Mosaic Analysis and Physiology of Drosophila Rhythms

The locomotor activity of the fly is produced by the output of motorneurons located in the thoracic ganglia. The clock controlling the locomotor activity rhythm could also be located in the thoracic nervous system or, alternatively, in the brain. By the use of genetic mosaics possessing both male (mutant) and female (heterozygous mutant) tissue, it is possible to construct a two-dimensional, blastoderm fate-map of external cuticle structures, upon which a focus for the site of action of a mutation affecting behavior can be located (Hotta and Benzer 1972, Hall 1978). Construction of mosaics is accomplished by the use of a ring-X chromosome which is unstable, or by use of a parental loss (*pal*) mutation. Both of these methods result in the loss of an X chromosome early in development. Consequently, male tissue is XO (not XY), and female tissue is XX. This is described in greater detail in Konopka et al. (1983). Mutations that affect the brain map close to head cuticle markers, while mutations that affect the thoracic nervous system map close to (and usually ventral to) thoracic structures. When the focus for the per^s mutation was determined with the aid of a least-squares computer program (Flanagan 1976, 1977), the fate-map location was close to the head cuticle markers and distant from thoracic cuticle markers, consistent with a brain location for the driving oscillator (Konopka et al. 1983). Some mosaic flies, especially those with mosaic heads, showed both males (mutant) and female (heterozygous mutant) components, suggesting that there was an oscillator of different genotype on each side of the brain, each capable of operating independently and contributing to the fly's activity profile.

Table. 2. Partial dominance of the andante and clock^{KO6} mutations. Rhythms were measured at 25°C in constant infrared.

Genotype	N	Period (h) ± Std. Dev.
+/+	7	24.2 ± 0.3
and/+	9	24.8 ± 0.2
and/*and*	10	25.7 ± 0.3
$clk^{KO6}/+$	19	23.2 ± 0.3
clk^{KO6}/clk^{KO6}	63	22.5 ± 0.6

The mechanism by which the brain clock controls the thoracic nervous system could be electrical or humoral. Handler and Konopka (1979) obtained evidence that the coupling between brain and thoracic ganglia is humoral by implanting short-period (*per*^{S}) brains in the abdomens of genetically arrhythmic (*per*^{0}) hosts and demonstrating that some of these hosts had short-period rhythms for several cycles. This result also suggests that the output of the brain clock might be produced by a group of brain neurosecretory cells. It is interesting in this context that Konopka and Wells (1980) found a posterior neurosecretory cell group in the brain whose morphology was affected by the *per*^{0} mutation. Normally these cells appear as a cluster of four cells on each side of the brain about halfway between the top and bottom edges of the brain. Occasionally, cells of this group can be found located in an abnormal position near the top edge of the brain. The percentage of abnormally located cells was 39.6% for *per*^{0} as opposed to only 16.8% for wild type. A similar increase in the percentage of abnormally located cells was found by Konopka and Wells (1980) in arrhythmic mutant strains isolated by Fox and Pittendrigh in *Drosophila pseudoobscura* Frolova (Pittendrigh 1974). It is possible that this cell group may be involved in the brain clock system. This question may be answered by the use of internally marked mosaics for which the genotype of cells in the nervous system can be determined directly (see Hall, 1978 for review).

Membrane Model for the *Drosophila* Clock

Konopka and Orr (1980) have proposed a model for the *Drosophila* oscillator. In this model, the oscillation consists of a buildup of a gradient across a membrane during the subjective day, and a dissipation of the gradient during the subjective night. The gradient is established by the operation of a light insensitive ion pump. When the gradient reaches its maximum value, the pump shuts down and light-sensitive ion gates open to allow the gradient to dissipate. Light pulses during the subjective night cause the ion gates to close and the pump to restart, thereby producing advances and delays in the rhythm.

By measuring the subjective day and night portions of the light response curve of the *per*^{s} mutant (Konopka 1979) and by measuring the active and inactive portions of the *per*^{s} activity cycle (Konopka and Orr 1980), it was found that the *per*^{s} mutation acts by shortening the subjective day, when the fly is active. There was no significant effect on the duration of the subjective night. This result suggests that there are separate molecular processes corresponding to the subjective day and subjective night, consistent with the above model. According to the model, then, the *per* gene product would be part of the ion pump that establishes the ion gradient during the subjective day. The genetic evidence discussed above that indicates increased gene product activity in *per*^{s} also fits well into the model; if pump activity is increased, the gradient should be established more rapidly, and the subjective day should be shortened, which is what is observed.

Short-Term Rhythms in Courtship Song

While studying the male's courtship song, Kyriacou and Hall (1980) discovered that there is a short-term oscillation in the interpulse interval of the song, with a period on the order of a minute. Surprisingly, the *per*^{s} mutation shortens the period of this oscillation, the *per*^{l} mutation lengthens it, and the *per*^{0} mutation abolishes it. These results indicate an unexpected link between the mechanism of short-term rhythms and that of circadian rhythms. It is possible that the period of a cellular oscillator might be regulated by the amount of *per* product in that cell. It is significant that both the courtship song rhythm and circadian rhythm are temperature compensated, so that the variation of period with temperature is small. Additional information concerning courtship song rhythms may be found in the paper by Hall in this volume.

Longevity of *per* Mutants

There is evidence in the literature that the life span of organisms is decreased in environmental cycles with periods greatly different from their genetically determined periods (Went 1959, Pittendrigh and Minis 1972, Saint Paul and Aschoff 1978). If this is true in *Drosophila*, one would expect wild type flies to have maximum longevity in 24-hour cycles, per^s in 20-hour cycles, and per^l in 28-hour cycles. The mean longevities of adult male flies of each genotype are given in Table 3. The longevities of wild type and per^l are very similar in each of the three cycles, with the maximum longevity occurring in the 20-hour cycle. However, the differeces in longevities among the three cycles are not statistically significant for these genotypes. In the case of per^s, there is more variability among the three environmental cycles. The maximum longevity, however, is in the 28-hour cycle, not the 20-hour cycle that is closest to its genetically determined period. These data, therefore, provide no support for the hypothesis that longevity is greatest in environmental cycles with periods close to the genetically determined periods.

The longevities of each of these three genotypes was also determined in combined light and temperature cycles having the same phase and having reversed phase. The mean longevities under these conditions are given in Table 4. Again, there is no significant difference in longevity under the two conditions for each genotype. The results do not support the hypothesis that longevity should be decreased in light and temperature cycles of opposite phase.

Table 3. Mean lifespans of adult male *Drosophila* in light cycles of various durations at 26 °C. The number of flies is given in parentheses.

Genotype	Light-Dark (12 h)	Light-Dark (10 h)	Light-Dark (14 h)
Wild type	41.5 ± 9.9 d (58)	45.7 ± 12.1 d (87)	40.7 ± 10.4 d (72)
per^l	45.6 ± 7.7 d (97)	48.3 ± 7.3 d (80)	43.7 ± 8.2 d (73)
per^s	48.3 ± 18.2 d (100)	41.4 ± 26.5 d (96)	59.0 ± 13.3 d (91)

Table 4. Mean lifespans of adult male *Drosophila* in light and temperature cycles. The number of flies is given in parentheses.

Genotype	12 h Light - 12 h Dark 26 °C Day - 21 °C Night	12 h Light - 12 h Dark 21 °C Day - 26 °C Night
Wild type	52.4 ± 9.6 d (95)	51.0 ± 9.0 d (80)
per^l	53.4 ± 12.0 d (95)	58.2 ± 4.6 d (78)
per^s	69.8 ± 14.8 d (99)	68.5 ± 18.2 d (94)

Future Prospects

With the available techniques in molecular biology, it will be a straightforward process to obtain a molecular clone of the *per* gene. By comparing the sequence organization of wild type and mutant *per* genes, the nature of the genetic alterations resulting in the short-period, long-period, and arrhythmic phenotypes can be determined. Eventually, the gene can be transcribed and translated, and the gene product identified. This type of analysis should ultimately lead to identification of the molecules involved in the mechanism of the circadian oscillator in *Drosophila*.

Acknowledgments

I thank Dominic Orr, Randall Smith, and Steven Wells for aid in data collection and analysis. This research was supported by grants from the Whitehall Foundation and the USPHS (GM 22227 and AG 01844).

Literature Cited

Flanagan, J. R. 1976. A computer program automating construction of fate maps of *Drosophila*. Dev. Biol. 53: 142.

Flanagan, J. R. 1977. A method for fate mapping the foci of lethal and behavioral mutants in *Drosophila melanogaster*. Genetics 85: 587.

Hall, J. C. 1978. Behavioral analysis in *Drosophila* mosaics. *In* W. Gehring, ed. Genetic Mosaics and Cell Differentiation. Springer-Verlag, New York. pp. 259-305.

Handler, A. M., and R. J. Konopka. 1979. Transplantation of a circadian pacemaker in *Drosophila*. Nature 279: 236.

Hotta, Y. and S. Benzer. 1972. Mapping of behavior in *Drosophila* mosaics. Nature 240: 527.

Judd, B., M. Shen, and T. Kaufman. 1972. The anatomy and function of a segment of the X-chromosome of *Drosophila melanogaster*. Genetics 71: 139.

Konopka, R., and S. Benzer. 1971. Clock mutants of *Drosophila melanogaster*. Proc. Natl. Acad. Sci. USA 68: 2112.

Konopka, R. 1979. Genetic dissection of the *Drosophila* circadian system. Fed. Proc. 38: 2602.

Konopka, R., and D. Orr. 1980. Effects of a clock mutation on the subjective day: Implications for a membrane model of the *Drosophila* circadian clock. *In* O. Siddiqi, P. Babu, L. M. Hall, and J. C. Hall, eds. Development and Neurobiology of Drosophila. Plenum Press. New York. pp. 409-416.

Konopka, R. J., and S. Wells. 1980. *Drosophila* clock mutations affect the morphology of a brain neurosecretory cell group. J. Neurobiol. 11: 411.

Konopka, R., S. Wells, and T. Lee. 1983. Mosaic analysis of a *Drosophila* clock mutant. Mol. Gen. Genet. 190: 284.

Kyriacou, C. P., and J. C. Hall. 1980. Circadian rhythm mutations in *Drosophila* affect short-term fluctuations in the male's courtship song. Proc. Natl. Acad. Sci. USA 77: 6729.

Pittendrigh, C. S. 1974. Circadian oscillations in cells and the circadian organization of multicellular systems. *In* F. Schmitt, and F. Worden, eds. The Neurosciences: Third Study Program. MIT Press, Cambridge. pp.437-458.

Pittendrigh, C. S., and D. H. Minis. *1972*. Circadian systems: longevity as a function of circadian resonance in *Drosophila melanogaster*. Proc. Natl. Acad. Sci. USA 69: 1537.

Saint Paul, U. v., and J. Aschoff. 1978. Longevity among blowflies *Phormia terraenovae* R.D. kept in non-24-hour light-dark cycles. J. Comp. Physiol. 127: 191.

Smith, R. F., and R. J. Konopka. 1981. Circadian clock phenotypes of chromosome aberrations with a breakpoint at the *per* locus. Mol. Gen. Genet. 183: 243.

Smith, R. F., and R. J. Konopka. 1982. Effects of dosage alterations at the *per* locus on the period of the circadian clock of *Drosophila*. Mol. Gen. Genet. 185: 30.

Went, F. W. 1959. The periodic aspect of photoperiodism and thermoperiodicity. *In* R. B. Withrow, ed. Photoperiodism. Amer. Assoc. for the Advance. Sci., Washington, D. C. pp. 551-564.

Young, M., and B. Judd. 1978. Nonessential sequences, genes, and the polytene chromosome bands of *Drosophila melanogaster*. Genetics 88: 723.

The Behavioral Effects of a Carboxylesterase in Drosophila

Rollin C. Richmond

Suresh D. Mane

Department of Biology
Indiana University
Bloomington, Indiana 47405

Laurie Tompkins

Department of Biology
Temple University
Philadelphia, Pennsylvania 19122

Introduction

The analysis of the genetic bases of behavior and its adaptive significance is complex. Until recently, the genetic analysis of behavior relied largely on quantitative genetic methodology. While this approach has been successful in identifying the degree to which genetics influence the variability in behavioral traits, it has been unable to elucidate the mechanistic bases for adaptive behavior or to reveal the possible evolutionary histories of behavioral adaptations. Work in our laboratory has been concerned principally with an understanding of the evolutionary significance of polymorphisms at enzyme-coding loci in natural populations of *Drosophila*. Our work has concentrated on the function and adaptive significance of a carboxylesterase polymorphism—Esterase-6 (EST-6)—in *Drosophila melanogaster* Meigen (Richmond et al. 1980). This work has revealed that EST-6 has several effects on the reproductive biology of *Drosophila*. Among these effects is a direct influence on the sexual attractiveness of mated *Drosophila* females. This work reveals the complexity of interactions which a single locus, presumably coding for the structure of EST-6, can have on several aspects of reproductive behavior. Our findings further demonstrate the importance of a multi-disciplinary approach to problems of behavior genetics (Hirsch and McGuire 1982).

In the discussion that follows, we use the nomenclature EST-6 to refer to the enzyme itself and *Est-6* to refer to the genetic locus. Superscripts indicate electrophoretic variants of EST-6 or alleles of *Est-6*.

Biology of the EST-6 System

EST-6 is a non-specific carboxylesterase (E.C. No. 3.1.1.1) localized on the left arm of chromosome 3 in *Drosophila melanogaster*. Two major electrophoretic variants of this enzyme—EST-6^S and EST-6^F—are found in the vast majority of natural populations

(Oakeshott et al. 1981). The frequencies of the electrophoretic variants of EST-6 exhibit clinal variation in natural populations on three continents. EST-6 variants have been described which have no activity in a variety of biochemical tests (Richmond et al. 1980). Such null variants (EST-6°) have not been found in natural populations (Langley et al. 1981). These observations suggest strongly that the polymorphism is subject to some sort of selective process.

Studies of the ontogenetic expression of the EST-6 locus have revealed that gene expression begins very early after fertilization and is maintained at a low level throughout the larval and pupal stages (Sheehan et al. 1979). At approximately 24 h after adult eclosion from the pupal case, the level of activity in adult males begins to rise and by 3-5 days after eclosion has reached a level 2-10 times that found in adult females. The sex-limited differences in the activity of this enzyme led us to determine its tissue distribution in adult males. The enzyme is highly concentrated in the reproductive system where it is localized to the anterior ejaculatory duct, a secretory organ in *D. melanogaster*. We have shown recently that the enzyme is synthesized in this duct tissue (Stein et al. 1984.) EST-6 is transferred to females during the first 2-3 min of the approximate 20 min copulation. In the female, enzyme activity can be detected for up to 2 h after mating (Richmond and Senior 1981). Our finding that EST-6 is localized to the male reproductive tract and is transferred to females early during copulation, suggested to us that the enzyme might be involved in some mechanism to influence male or female mating behavior. To test this hypothesis, we have completed a series of behavioral tests utilizing strains of flies that carry active or null alleles. These data demonstrate that EST-6 can have significant effects on reproductive behavior.

Long-Term Behavioral Effects of EST-6

Manning (1967) showed in an elegant study that the sexual receptivity of mated females was controlled by two factors: an effect associated with copulation and a long-term effect associated with sperm transfer. We (Gilbert et al. 1981) have investigated in detail the possible effects of EST-6 on the timing of remating by females inseminated by males carrying either an active or null allele of EST-6. Females inseminated by *Est-*6^{S} males remate significantly sooner than females inseminated by *Est-*6° males.

Manning (1962) and Gilbert (see Gromko et al. 1984) have shown that the timing of female remating is at least in part controlled by the number of sperm a female has in her storage organs. In order to partition the effects of sperm storage and seminal-fluid EST-6 on the timing of female remating, we performed a remating experiment in which the first mating of a female was interrupted 3-4 min after the initiation of copulation. This procedure allows the transfer of EST-6 but prevents the transfer of sperm. The results of this experiment show that the EST-6 type of the first male has no effect on the timing of female remating if the first mating is interrupted. We tentatively concluded that the timing of remating effect noted in our earlier remating experiments (Gilbert et al. 1981) arose from an interaction between EST-6 and other factors (most likely sperm) transferred to the female.

Sterile Male Remating Experiments

The interrupted mating-remating experiment reviewed above is open to criticism from at least two points. The rather drastic method used to terminate copulation (vortexing of mating vials) may have damaged the female's reproductive system or sensory physiology such that she did not respond normally thereafter. Secondly, the EST-6 effect on remating might result from an interaction between the enzyme and some other component of the seminal fluid (other than sperm) which is transferred to females later in copulation. In order to account for these possibilities, we devised a scheme to permit the generation of sterile XO males which are either EST-6^S or EST-6°. XO males do not produce or transfer motile sperm (Kiefer 1966) but do transfer EST-6 normally to females (our unpublished observations). We constructed two attached-X stocks which carry no free Y chromosome and which are homozygous for either *Est-*6° or *Est-*6^S. When a female from the attached-X, *Est-*6° stock is crossed to an XY, *Est-*6° male, the male offspring are all XO, *Est-*6°, sterile males. *Est-*6^S, sterile males are obtained from a similar cross using the *Est-*6^S attached-X female and *Est-*6^S, XY male.

We have used the sterile XO males to determine if the timing of remating in females is affected by the presence or absence of EST-6 in the seminal fluid of the first mate of a female. The experiment was done at both 18 and 25 °C. Since Manning (1967) found a rapid return to receptivity in females mated to XO males, we opted to use a lower temperature which might act to exaggerate differences in the timing of remating. The design of the experiment was as follows: Ore R (a standard wild type strain) females were mated first to either *Est-*6°, XO or *Est-*6^S, XO males; subsequent to female insemination by the XO males (whose duration of copulation is normal, i.e., ~ 20 min), females were tested individually for remating (2-h test period) with Oregon R males at 4, 24, 48, and 72 h (Gilbert et al. 1981). The results are given in Table 1.

These data confirm the findings of Manning (1967). Females inseminated by sterile XO males are refractory to remating for about 24 h and recover full receptivity by 48 h after their first mating. There is a clear temperature effect. A return to receptivity occurs more rapidly at 25 ° C. These data also confirm the results of our interrupted mating experiments described above. The EST-6 type (*Est-*6^S or *Est-*6°) of the first mate of a female does not significantly influence the timing of her remating if no sperm are transferred. Since we know that XO males transfer EST-6 to females, we must conclude that the timing of remating effect observed with normal XY males which differ in EST-6 type is due to an interaction between the sperm and EST-6 in the female's reproductive tract. We can eliminate the possibility that *Est-*6^S and *Est-*6° males transfer different numbers of sperm since we (Gilbert et al. 1981) have shown that the productivity of females inseminated by these two male types is virtually identical, and Gilbert (1981) has shown that the number of sperm stored by females mated to these two male types is not significantly different.

Table 1. Cummulative proportion of females that remated after insemination by *Est-*6° or *Est-*6^S males at 18 and 25 °C. A total of 50 females in each group was tested for remating over a 2-h period at four times after their initial mating.

Hours post	18 °		25 °	
first mating	*XO-*6°	*XO-*6^S	*XO-*6°	*XO-*6^S
4	10	16	16	20
24	58	46	86	80
48	98	96	98	96
72	——	——	100	100

Timing of Remating with Olfactory Insensitive Males

Our results on the relationship between the EST-6 type of a female's first mate, the rate of sperm loss from females, and the timing of remating suggest that EST-6 interacts with sperm to influence the female's remating behavior. If this latter hypothesis accounts fully for the effect of EST-6 on the timing of remating, one would predict that the olfactory capabilities of a female's potential second mate should not alter the time at which she remates. In other words, if the EST-6 effect on female remating behavior is mediated solely by an interaction between sperm and EST-6 rather than some pheromonal mechanism, using a second male type (in remating tests) which is insensitive to a female's pheromonal cues should produce the same result as we obtained previously: *Est-6^S* inseminated females remate sooner than *Est-6°* inseminated females (Gilbert et al. 1981).

We have completed a remating experiment using as second males flies expressing the (*sbl*) smellblind mutation (Aceves-Pina and Quinn 1979). This mutant does not respond to organic compounds or sex pheromones produced by normal virgin females (Tompkins et al. 1980). As we discussed above, we predict that the timing of remating in two groups of females initially inseminated by *Est-6^S* or *Est-6°* males should not be changed from previous experiments when *sbl* males are used as potential second mates. The design of this experiment incorporated 50 females in each group and was as follows: Ore R females were mated first to either *Est-6^S* or *Est-6°* males; subsequent to insemination by the *Est-6* males, females were tested individually for remating with *sbl* males at 4, 12, and 24 h and at 24 h intervals. A plot of the cumulative proportion of remated females in both groups as a function of hours after the first mating (Fig. 1) reveals that the two groups behave identically. In contrast to experiments employing wild type males as second males, *sbl* males are apparently unable to distinguish between females inseminated by *Est-6°* or *Est-6^S* males. The results are surprising and do not accord with the prediction made above. Rather they suggest that EST-6 may be involved in how a male perceives a previously mated female.

Short-Term Behavioral Effects of EST-6

Manning (1976) showed that the sexual receptivity of mated *Drosophila* females was influenced by an event occurring at copulation. Our results employing the *sbl* mutant reviewed above suggest that EST-6 may be involved in a mechanism to affect the sexual attractiveness of previously mated females. This possibility is strengthened by the observations of Tompkins and Hall (1981) who noted that even a very short period of copulation (2-4 min) rendered a female less attractive to males later given an opportunity to court her. It seemed unlikely to us that EST-6 could be the only component of the male ejaculate which was responsible for this phenomenon. Butterworth (1969) and Brieger and Butterworth (1970) discovered that male flies synthesize a male lipid in the ejaculatory bulb of their reproductive system and transfer this substance to females at copulation. The formula for the male lipid (called *cis*-vaccenyl acetate) is

$$CH_3(CH2)_5CH = CH(CH_2)_9CH\text{-}O\text{-}COCH_3$$

This compound is an acetate ester and had been identified previously as an aphrodisiac pheromone in other insects (Pliske and Eisner 1969). Moreover Jallon et al. (1981) have shown that approximately 70% of the male lipid disappears from the female reproductive tract within 6 h after mating. Thus, the male lipid is a potential in vivo substrate for EST-6. The products of the EST-6 catalyzed hydrolysis of the male lipid might be involved in the effects on remating behavior revealed in our tests utilizing the *sbl* male, and those of Tompkins and Hall (1981) employing females that had copulated with males for short periods of time before being given an opportunity to remate.

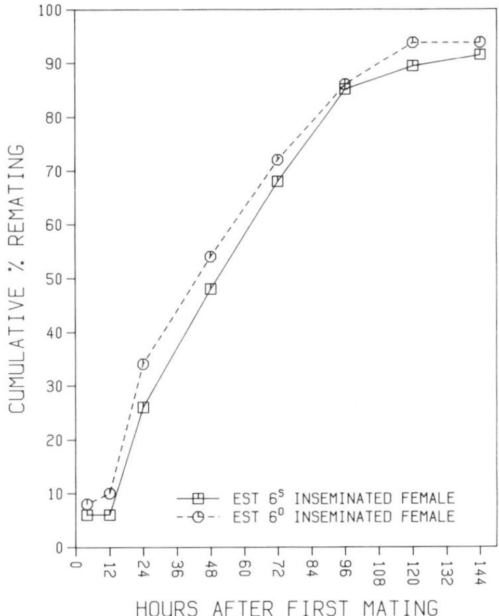

Fig. 1. Cummulative proportion of remating in females initially inseminated by *Est-6°* or *Est-6^S* males. Fifty females were initially tested for remating with wildtype Ore-R males for two hours at each time period.

EST-6 Catalyzes the Hydrolysis of the Male Lipid

We tested the possibility that the male lipid might be a substrate for EST-6 by incubating purified enzyme (Mañe et al. 1983a,b) with ^{14}C-labelled male lipid. The labelled atom was present in the acetate portion of the lipid and was detected by its solubility in the aqueous phase of a chloroform extract of the reaction mixture. Figure 2 (Mane et al. 1983b) reveals that the release of ^{14}C-labelled acetate is a function of both enzyme concentration and incubation time. The products of a 3 h incubation of purified EST-6 with male lipid were extracted in chloroform and subjected to gas-liquid chromatography. This analysis revealed that an alcohol (*cis*-vaccenyl alcohol, cVOH) was the product of the cleavage of the male lipid as predicted.

Behavioral Effects of the Male Lipid and cVOH

Initial tests of the behavioral effects of these compounds employed topical application of the male lipid and cVOH to the abdomens of virgin females. After a recovery period of 30 min, test females were introduced into a chamber with a virgin male and the number of pairs mating in a 1-h period was determined. Controls consisted of virgin females and females treated with acetone—the solvent for the two test compounds. The results were clear. Either singly or in combination, the male lipid or cVOH were equivalently effective in reducing the number of females which mated in 1 h. The two control groups showed high and equivalent levels of mating.

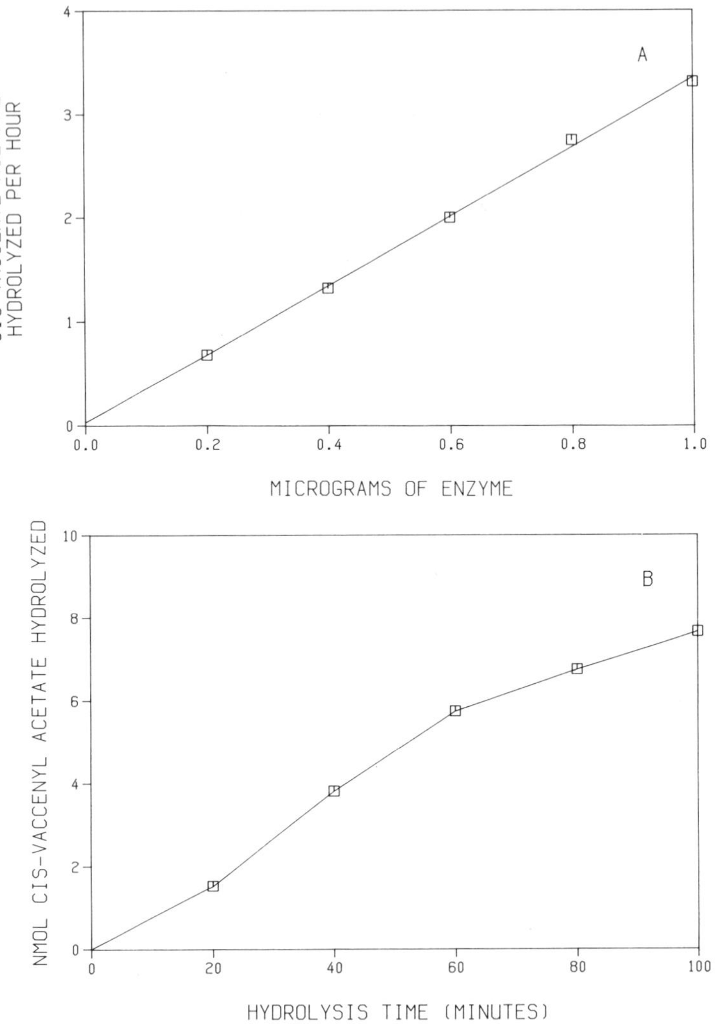

Fig. 2. Hydrolysis of *cis*-vaccenylacetate by purified EST-6. In panel A, purified EST-6 was incubated with ^{14}C-*cis*-vaccenylacetate following the procedures of Mane et al. (1983b). In panel B, purified EST-6 was incubated with ^{14}C-*cis*-vaccenylacetate and aliquots of the reaction mixture were removed at 20 min intervals and assayed for the liberation of labelled acetate.

This behavioral assay suffers from two difficulties: the dose of the presumed pheromones may be outside the biologically relevant range and the only measurement of the effect of these compounds on courtship is an end point—copulation. We utilized the *Est-6*^{O} and *Est-6*^{S} strains to control the production of cVOH in inseminated females genetically. Females inseminated by *Est-6*^{O} males will receive the male lipid at mating but will be unable to convert it to cVOH as will occur in *Est-6*^{S} inseminated females. We compared the ability of females to elicit male courtship either 10 min or 6 h after they copulated with an *Est-6*^{S} or *Est-6*^{O} male. Previously inseminated females were tested with Canton S (CS), virgin wild type males using the technique of Tompkins et al. (1980). This procedure involves recording the cumulative time that a male performs courtship in response to a female during a 10 min period and calculating a courtship index (CI) which is the percentage of the observation period that a male spent courting. The results (Fig. 3) show that 10 min after mating, females elicit significantly reduced levels of courtship. There is no significant effect of the EST-6 type of the females' first mate. However, by 6 h after mating, females who received active EST-6 from their first mate elicit significantly less courtship than females initially mated to *Est-6*^{O} males. These data strongly suggest that males courting mated females are able to detect the presence of cVOH or that cVOH alters the "behavior" of females thereby rendering them less attractive to males.

If the ability of males to detect chemical signals from mated females is involved in the recognition of previously mated females, then males which are unable to detect acetates but are capable of recognizing alcohols should be unable to differentiate recently mated females from virgins. However such males should be able to differentiate females which are producing cVOH from those which do not. The mutant *olfC* (olfactory C) (Rodrigues 1980), can detect some alcohols but is unable to detect acetates (Rodrigues and Siddiqi 1978). Males that carry this mutant were used in the same experimental design discussed immediately above (Fig. 4).

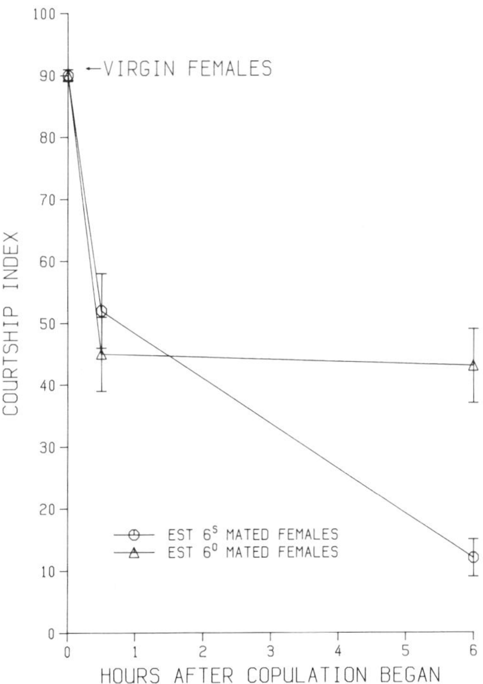

Fig. 3. Courtship indices (± SEM) of wild type Canton S males tested with virgin females or females inseminated by *Est-6*^{O} or *Est-6*^{S} males 10 min or 6 h previously.

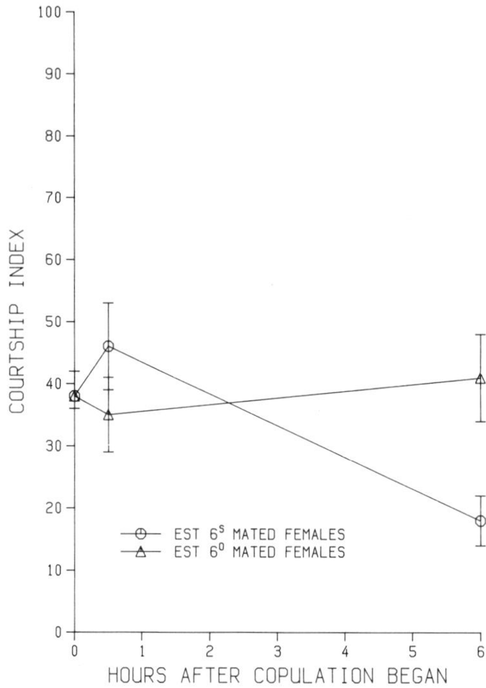

Fig. 4. Courtship indices (±SEM) of olfactory deficient (olfactory C) males tested with virgin females or females inseminated by *Est-6*^{O} or *Est-6*^{S} males 10 min or 6 h previously.

As predicted, *olfC* males are unable to differentiate recently mated females from virgins, but they are able to differentiate *Est-6*^{S} from *Est-6*^{O} mated females 6 h after insemination. These results suggest that cVOH is acting directly on males as an antiaphrodisiac or that this compound is inducing females to produce a pheromone which is not detected by *olfC* males.

Other Behavioral Effects Associated with EST-6

In a series of preliminary experiments, we have tested the possibility that EST-6 might be involved in other courtship behaviors. D. G. Gilbert (unpublished MS), working in our lab, used a diallele cross design to examine the effect of EST type (*Est-6*^{S} and *Est-6*^{F}) on several parameters including mating speed. He found that the average latency to copulation at 25 °C in *Est-6*^{F} males is 5.7 min while *Est-6*^{S} males require an average of 10.3 min to achieve copulation. Gilbert and Richmond (1982) have investigated the latency to copulation for *Est-6*^{S} and *Est-6*^{O} male types at 18 °C. Although the time required to achieve mating at 18 °C is increased over that required at 25 °C, *Est-6*^{S} males mated an average of 12.5 min after introduction to the female while *Est-6*^{O} males require an average of 19.7 min to mate. Our results confirm the experiments of Aslund and Rasmuson (1976) who used entirely different strains and measured only the number of copulations occurring after mixing 50 pairs of *Est-6*^{F} and *Est-6*^{S} flies. Recent results of Tompkins et al. (1982) suggest that latency to copulation is at least in part controlled by a male-emitted pheromone which induces females to become stationary just prior to copulation.

Table 2. Proportion of females mating with their own or a different *Est-6* genotype male in choice tests. The x^2 value in this table tests the hypothesis that females show no mating preference.

Female genotype	No. Females	Male Genotypes	% Females Mating with Est 6°	x^2
Est-6°	38	Est-6° + Est-6^S	45	0.42
Est-6^S	46	Est-6° + Est-6^S	54	0.35

Ehrman and her colleagues (Petit 1958, Ehrman 1966, 1968, Ehrman and Spiess 1969) have shown repeatedly that *Drosophila* females appear to exhibit a "preference" for mates when several male genotypes are present. Accordingly, we have tested the possibility that females might show a mating preference for *Est-6°* or *Est-6^S* males by allowing a female to "choose" between two males, one of each type, in a 8-dram vial. The results are given in Table 2.

Pruzan (1976) and Ehrman and Probber (1978) have shown that prior mating experience alters the mating choice of females in subsequent encounters with males. We tested the possibility that EST-6 might be involved in this "conditioning" of female mating preference by initially mating females (2 days old) to either an *Est-6^S* or *Est-6°* male and then giving the female (4 days old) a choice between the two male types. The results are summarized in Table 3. These data provide no evidence that the EST-6 type of a female's first mate influences her choice of subsequent mates.

Table 3. Effect of the EST-6 type of female's first mate on her preference for a subsequent mate. Females were mated intially to either an *Est-6^S* or *Est-6°* male and then given a choice between these two male types two days later.

Female	1st Male	No. 1st Matings	No. 2nd Matings	% EST-6° in 2nd matings	x^2
Est-6°	Est-6°	119	56	41	1.79
Est-6°	Est-6^S	119	63	54	0.18
Est-6^S	Est-6°	117	51	61	2.42
Est-6^S	Est-6^S	114	45	49	0.02

^{2*} Tests deviation of female preference from random choice.

Summary of EST-6 Effects on Behavior

The EST-6 type of a female's first mate influences the timing of her remating. Females inseminated by males having active enzyme remate sooner than females inseminated by males lacking active enzyme. The effect of EST-6 on sperm use may partially mediate this long term remating effect. This effect is abolished by interrupting courtship prior to sperm transfer, initially mating females with sterile males, or using olfactory-insensitive males as second mates. The latency to copulation is also affected by the EST-6 type of the male. *Est-6*^{F} males mate approximately twice as fast a *Est-6*^{S} males which mate faster than *Est-6*^{O} males. Females show no mating preference for *Est-6*^{O} vs. *Est-6*^{S} males and female "choice" in subsequent matings is unaffected by the EST-6 type of her first mate.

Discussion

A gene which codes for the structure of a non-vital hydrolytic enzyme would scarcely seem likely to have major effects on reproductive behavior. Our investigations of the physiological role of EST-6 in *Drosophila melanogaster* have revealed that this locus can have important effects on the reproductive behaviors of both sexes. The presence of this enzyme in the male ejaculate provided one of the initial clues that EST-6 was involved in reproduction. The application of analytic chemical, biochemical, physiological, and population biological analyses to this system have suggested an explanation for the presence of EST-6 in the male ejaculate. Gromko and Pyle (1978), working in our laboratory, showed that the sperm of a female's most recent mate take precedence over that of a previous mate in fertilizing the female's eggs (see also Lefevre and Jonsson 1962, Boorman and Parker 1976, Prout and Bundgaard 1977). Thus, males should experience selection for mechanisms which will act to reduce the sexual attractiveness and receptivity of their mates. EST-6 appears to be involved in systems by which males influence the short-term sexual attractiveness of a female and her long-term receptivity to remating. This is accomplished by transferring to the female in the seminal fluid a substance which is converted in the female's reproductive tract into an antiaphrodisiac. The conversion is catalysed by an enzyme—EST-6—which is also present in the seminal fluid. This is the first example of a pheromone whose final synthetic step is accomplished in one sex using substrate and enzyme provided by the other. The effects of EST-6 on sperm storage and use (Gilbert 1981) likely mediate the long-term effect on female receptivity.

The EST-6 system has a recently described parallel in the cricket, *Teleogryllus commodus* Walker (Loher et al. 1981, Stanley-Samuelson and Loher 1983). In this species, the male transfers a precursor of a prostaglandin to the female in the spermatophore. Also included is an enzyme complex which converts the precursor into prostaglandin E_2. This substance stimulates the female to ovipost. The seminal fluid of *Drosophila* has been demonstrated to contain a variety of components including compounds which induce oviposition (Gromko et al. 1983, Heinstra and Thorig 1982). These and many other examples lend strong support to the hypothesis that selection for male-derived, seminal-fluid compounds which affect female reproductive physiology and behavior has been intense (Davey 1980). Although we are far from an understanding of the selective forces which may act to maintain the EST-6 polymorphism, the involvement of the enzyme in reproduction, the stability of allele frequencies on a global scale, and the absence of null variants of EST-6 in natural populations (Voelker et al. 1980, Langley et al. 1981) suggest that the polymorphism is actively maintained by selection.

Many investigators have obtained evidence that *Drosophila* courtship is modulated by chemical cues (Averhoff and Richardson 1974, Jallon et al. 1981, Tompkins and Hall 1981). A series of elegant studies of the evolutionary history and reproductive behavior of the unparalleled Hawaiian *Drosophila* species implicate changes in behavioral and chemical repertoires as an important prelude to speciation (Kaneshiro 1980, Carson 1981). Averhoff and Richardson (1976) suggested that airborne chemical cues were responsible for the failure of reproduction in highly inbred strains of *Drosophila*. These studies all suggest that genetic variability for the production of pheromones exists in natural populations and may form an important substrate for the elaboration of reproductive isolation. Detailed biochemical, genetic and behavioral studies of systems such as EST-6 are likely to lead to a better understanding of the synthetic pathways involved in pheromone production, the regulation of these systems and their possible role in the evolution of new species.

The hypothesis that EST 6 modulates the sexual receptivity of the female by hydrolyzing *cis*-vaccenyl acetate to *cis*-vaccenyl alcohol in the female reproductive tract has been disproved by us (Vander Meer, Obin, Zawistowski, Sheehan, and Richmond, 1986. J. Insect Physiol., in press). We have shown that *cis*-vaccenyl acetate is lost from a mated female reproductive tract independently of the Est-6 genotype of her mate. Bartelt et al. (1985. J. Chem. Ecol. 11:1747) have shown that *cis*-vaccenyl acetate is deposited by mated females into their feeding vials within 6 *hours* after mating. Their analysis indicates that *cis*-vaccenyl acetate acts as an aggregation pheromone. These observations account for the loss of *cis*-vaccenyl acetate from the female in a manner which is independent of the Est-6 genotype of her mate.

Acknowledgments

We are particularly grateful for insightful and careful review of this paper by Drs. M. H. Gromko, J. C. Hall and M. D. Huettel. Kathy Sheehan and Alayne Senior provided excellent technical assistance. The research was supported by grants from The National Science Foundation and The National Institutes of Health.

Literature Cited

Aceves-Pina, E. O. and W. G. Quinn. 1979. Learning in normal and mutant *Drosophila* larvae. Science 206: 93.

Aslund, S. E. and M. Rasmuson. 1976. Mating behavior as a fitness component in maintaining allozyme polymorphisms in *Drosophila melanogaster*. Hereditas 82: 175.

Averhoff, W. W. and R. H. Richardson. 1974. Pheromonal control of mating patterns in *Drosophila melanogaster*. Behav. Genet. 4: 207.

Averhoff, W. W. and R. H. Richardson. 1976. Multiple pheromone systems controlling mating in *Drosophila melanogaster*. Proc. Natl. Acad. Sci. USA 73: 591.

Boorman, E. and G. A. Parker. 1976. Sperm (ejaculate) competition in *Drosophila melanogaster*, and the reproductive value of females to males in relation to female age and mating status. Ecol. Entomol. 1: 145.

Brieger, G. and F. M. Butterworth. 1970. *Drosophila melanogaster*: Identity of male lipid in reproductive system. Science 167: 1262.

Butterworth, F. M. 1969. Lipids of *Drosophila*: A newly detected lipid in the male. Science 163: 1356.

Carson, H. L. 1981. Chromosomal tracing of evolution in a phylad of species related to *Drosophila hawaiiensis*. *In* W. R. Atchley and D. S. Woodruff, eds. Evolution and Speciation. Cambridge University Press, Cambridge.

Davey, K. G. 1980. The physiology of reproduction in *Rhodnius* and other insects: Some questions. *In* M. Locke and D. S. Smith, eds. Academic Press, New York.

Ehrman, L. 1966. Mating success and genotype frequency in *Drosophila*. Anim. Behav. 14: 332.

Ehrman, L. 1968. Frequency dependence of mating success in *Drosophila pseudoobscura*. Genet. Res. 11: 135.

Ehrman, L. and J. Probber. 1978. Rare *Drosophila* males: The mysterious matter of choice. Amer. Sci. 66: 216.

Ehrman, L. and E. B. Spiess. 1969. Rare type mating advantage in *Drosophila*. Amer. Natur. 103: 675.

Gilbert, D. G. 1981. Ejaculate esterase 6 and initial sperm use by female *Drosophila melanogaster*. J. Insect Physiol. 27: 641.

Gilbert, D. G. and R. C. Richmond. 1982. Esterase 6 in *Drosophila melanogaster*: Reproductive function of active and null males at low temperature. Proc. Natl. Acad. Sci. USA. 79: 2962.

Gilbert, D. G., R. C. Richmond, and K. S. Sheehan. 1981. Studies of esterase 6 in *Drosophila melanogaster*. VII. The timing of remating in females inseminated by males having active or null alleles. Behav. Genet. 11: 195.

Gromko, D. G., and D. W. Pyle. 1978. Sperm competition, male fitness, and repeated mating by female *Drosophila melanogaster*. Evolution 32: 588.

Gromko, M. H., D. G. Gilbert, R. C. Richmond. 1984. Sperm transfer and use in the multiple mating system of *Drosophila*. *In* R. L. Smith, ed. Sperm Competition and the Evolution of Animal Mating Systems. Academic Press, New York.

Heinstra, P. W. H. and G. E. Thorig. 1982. Multiple function of pteridines in *Drosophila melanogaster*. J. Insect. Physiol. 28: 847.

Hirsch, J. and T. R. McGuire. 1982. Behavior-Genetic Analysis. Hutchinson Ross Publishing Company, Stroudsburg, Pennsyslvania.

Jallon, J-M, C. Antony, and T. Iwatsubo. 1981. Elements of chemical communication between Drosophilids and their modulation. Taniguchi Symposium on Biophysics, Kyoto.

Kaneshiro, K. Y. 1980. Sexual isolation, speciation and the direction of evolution. Evolution 34: 437.

Kiefer, B. I. 1966. Ultrastructural abnormalities in developing sperm of X/O *Drosophila melanogaster*. Genetics 54: 1441.

Langley, C. H., R. A. Voelker, A. J. Leigh-Brown, S. Ohnishi, B. Dickson, and E. Montgomery. 1981. Null allele frequencies at allozyme loci in natural populations of *Drosophila melanogaster*. Genetics 99: 151.

Lefevre, G. and U. B. Jonsson. 1962. Sperm transfer, storage, displacement and utilization in *Drosophila melanogaster*. Genetics 47: 1719.

Loher, W., I. Ganjian, I. Kubo. D. Stanley-Samuelson, and S. S. Tobe. 1981. Prostaglandins: Their role in egg-laying of the cricket *Teleogryllus commodus*. Proc. Natl. Acad. Sci. USA 78: 7835.

Mane, S., C. S. Tepper, and R. C. Richmond. 1983a. Studies of esterase 6 in *Drosophila melanogaster*. XIII. Purification and characterization of the two major isozymes. Biochem. Genet. 21:1019.

Mane, S., L. Tompkins, and R. C. Richmond. 1983b. A sex pheromone is synthesized in the reproductive tract of *Drosophila melanogaster* females by an enzyme and substrate provided by males. Science. 222:419 .

Manning, A. 1962. A sperm factor affecting the receptivity of *Drosophila melanogaster* females. Nature 194: 252.

Manning, A. 1967. The control of sexual receptivity in female *Drosophila.* Anim. Behav. 15: 239.

Oakeshott, J. G. G. K. Chambers, J. B. Gibson, and D. A. Willcocks. 1981. Latitudinal relationships of esterase-6 and phophoglucomutase gene frequencies in *Drosophila melanogaster.* Heredity 47: 385.

Petit, C. 1958. Le determinisme genetique st psychopsysiologique de la competition sexuelle chez *Drosophila melanogaster.* Bull. Biol. Fr. Belg. 92:1.

Pliske, T. E. and T. Eisner. 1969. Sex pheromone of the queen butterfly: biology. Science 164: 1170.

Prout, T. and J. Bundgaard. 1977. The population genetics of sperm displacement. Genetics 85: 95.

Pruzan, A. 1976. Effects of age, rearing, and mating experiences on frequency dependent sexual selection in *Drosophila melanogaster.* Evolution 30: 130.

Pyle, D. W. and M. H. Gromko. 1978. Repeated mating by *Drosophila melanogaster*: The adaptive importance, Experientia. 34: 449.

Richmond, R. C. and A. S. Senior. 1981. Studies of esterase 6 in *Drosophila melanogaster.* IX. Kinetics of transfer to females, decay in females, and male recovery. J. Insect Physiol. 27: 849.

Richmond, R. C., D. G. Gilbert, K. B. Sheehan, M. H. Gromko, and F. M. Butterworth. 1980. Esterase 6 is involved in the reproduction of *Drosophila melanogaster.* Science 207: 1483.

Rodriques, V. 1980. Olfactory behavior of *Drosophila melanogaster. In* O. Siddiqi, P. Babu, L. Hall, and J. Hall, eds. Develoment and neurobiology of *Drosophila.* Plenum. New York.

Rodriques, V. and O. Siddiqi. 1978. Genetic analysis of chemosensory pathway. Proc. Indian Acad. Sci. 87B: 147.

Sheehan, K. B., R. C. Richmond, and B. J. Cochrane. 1979. Studies of esterase 6 in *Drosophila melanogaster.* III. The developmental pattern and tissue distribution. Insect Biochem. 9: 443.

Stanley-Samuelson, D. W. and W. Loher. 1983. Arachidonic and other long-chain polyunsaturated fatty acids in spermatophores and spermathecae of *Teleogryllus commodus*: Significance in prostaglandin-mediated reproductive behaviour. J. Insect. Physiol. 29:41.

Stein, S. P., C. S. Tepper, N. D. Able and R. C. Richmond. 1984. Studies of esterase 6 in *Drosophila melanogaster.* 2XVI. Synthesis occurs in the male reproductive tract (anterior ejaculatory duct) and is modulated by juvenile hormone. Insect Biochem. 14:527.

Tompkins, L., J. C. Hall, and L. M. Hall. 1980. Courtship-stimulating compounds from normal and mutant *Drosophila.* J. Insect Physiol. 26: 689.

Tompkins, L. and J. C. Hall. 1981. The different effects on courtship of volatile compounds from mated and virgin *Drosophila* females. J. Insect Physiol. 27: 17.

Tompkins, L., A. C. Gross, J. C. Hall, D. A. Gailey, and R. W. Siegel. 1982. The role of female movement in the sexual behavior of *Drosophila melanogaster*. Behav. Genetics 12: 295.

Voelker, R. A., C. H. Langley, A. J. Leigh-Brown, S. Ohnishi, B. Dickson, E. Montgomery, and S. C. Smith. 1980. Enzyme null alleles in natural populations of *Drosophila melanogaster*: Frequencies in a North Carolina population. Proc. Natl. Acad. Sci. USA 77: 1091.

Mutants of Biological Rhythms and Conditioned Behavior in *Drosophila* Courtship

Jeffrey C. Hall

Department of Biology
Brandeis University
Waltham, Massachusetts 02254

Introduction

We use genetic variants to perturb courtship and mating behavior in *Drosophila*. Our purpose is to find out how these complex actions of the flies are controlled—genetically and neurally—and to add to knowledge that is accumulating on the significance of the genes that we and other investigators are manipulating. Most of the pertinent background information, from our work and that of many others, is reviewed in Hall (1981, 1982), Spieth and Ringo (1983), and Quinn and Greenspan (1984).

Circadian Rhythm Mutants and Courtship

We established a few years ago that certain *Drosophila* mutations, isolated on the basis of disrupted circadian rhythms (see review of Konopka 1981), also perturb a short-term rhythm in the male's courtship song. The most conspicuous features of these songs are trains of pulses, produced at a rate of approximately 30 per second in *D. melanogaster* (e.g., Bennet-Clark and Ewing 1969). We found that these interpulse intervals are not constant, but instead vary in a systematic, oscillating manner. The period length of such a "song rhythm" is about one minute, again in *D. melanogaster* (Kyriacou and Hall 1980). Three allelic period (*per*) mutations were shown to make the song clock run too fast, too slowly, or not at all. These disruptions of the song rhythm are strikingly parallel to the aberrations of circadian cycles that are induced by these same mutations (Kyriacou and Hall 1980).

This discovery has been extended somewhat. First, five other mutants, identified orginally by circadian defects, have been tested for their song rhythms: The *psi-2, psi-3,* and gate (*gat*) mutations (Jackson 1983) either lengthen the song cycles, or cause them to be "sloppy" (quasi-arrhythmic), just as they do in regard to circadian periods (C. P. Kyriacou and F. R. Jackson, unpublished). The "clock" (clk^{K06}) and "andante" (*and*) mutations of R. J. Konopka, R. F. Smith, and D. Orr (see Konopka, this volume) shorten or lengthen circadian periods, respectively. Yet, whereas males expressing clk^{K06}) have shorter than normal song periods, *and* males appear to sing with normal oscillations of their wing vibrations (W. Zehring and J. C. Hall, unpublished). Therefore, the genetic control of these two different types of rhythms does not always go hand in hand. Stated another way: mechanisms that underly circadian and song rhythms share certain components (the presumed products of the *per, psi-2, psi-3, gat,* and *clk* genes), but other components (so far exemplified only by *and's* hypothetical product) "operate" only in regard to one of these rhythmic characters.

The control of circadian and song rhythms also is "divergent" in regard to the neural tissues exerting primary control of these cycling behaviors. This tentative statement stems from a preliminary analysis of genetic mosiacs (J. C. Hall, R. J. Konopka, and C. P. Kyriacou, unpublished). Each such mosiac was part *per*-mutant, and part normal, with different distributions of mutant *vs.* normal tissue present in separate individual adults. The mosaics were tested, first for circadian rhythms of locomotor activity, then for courtship singing. This was followed by tissue sectioning and histochemistry to determine the distribution of mutant *vs.* normal tissue in each animal (see Tompkins and Hall, 1983, for an example of the cell-marking strategy). Such an analysis has led to the following suggestion: the *per* gene exerts a direct influence on brain cells in its control of circadian cycles, but this gene also is expressed in thoracic neurons in its regulation of song cycles. An example of a result allowing this preliminary conclusion is that one commonly finds a mosaic that has a normal circadian rhythm (and a brain that is at least partly normal, genotypically), but a mutant song rhythm (and a thoracic nervous system that is at least partly mutant).

Other neurogenetic experiments on the fly's courtship song rhythms have involved "stopping the clock" with mutations that block membrane excitability. We have found that the two temperature-sensitive (*ts*) paralytic mutations in Drosophila which eliminate neural action potentials at high temperature (Wu and Ganetzky 1980) each have the following effects on the song rhythms (Kyriacou and Hall 1985): after a courting, mutant male (expressing the no-action-potentials-*ts* or the paralytic-*ts* variant) is paralyzed by high temperature, he begins singing again soon after the temperature is lowered and he recovers normal mobility. Yet his song rhythm "sine wave" (see Kyriacou and Hall 1980) must be shifted along the time axis so that the peaks and troughs can be displayed in a proper "phase relationship" with the rhythm established during the courtship which occurred before the mutation- and temperature-induced paralysis. The necessary shift is nearly the same as the amount of time during which the gene or its product was "turned off." Controls that involve high temperature treatment of wild type males or paralysis of such males elicited by cold shocks or CO^2 caused only minimal phase shifts of the song rhythms.

The characteristics of the song rhythms are species-specific. *Drosophila simulans* males court with oscillating songs having a 35- to 40-sec period (Kyriacou and Hall 1980). In hybrid males—that is, the progeny of reciprocal crosses between *D. simulans* females or males and *D. melanogaster* males or females—the song difference between the two species maps solely to the X chromosome (Kyriacou and Hall, (1986). The usually inviable hybrid males with a *D. melanogaster* X were made to live by use of the lethal-hybrid-rescue factor of T. S. Watanabe, (e.g., Kawanishi and Watanabe, 1981, who also report that the basic interpulse intervals, in the two reciprocal hybrid types, are intermediate between the parental values).

Could the genetic difference controlling the song rhythm difference be more "narrowly" located than in regard to the whole X chromosome—for instance, at the *per* locus (for which there must be a homologous gene in *D. simulans* as well as in its close relative *D. melanogaster*)? The *clk* locus in *D. melanogaster*, which mutation shortens the song rhythm period as well as circadian cycles, also is X chromosomal. Nevertheless, one could still entertain the possibility that it is only the *per* gene that has evolved to be the major factor specifying a 55-sec song rhythm period in *D. melanogaster vs.* an approximately 35-sec period in *D. simulans*. Yet, in both species, *per*'s function (with respect to the wild type allele) would be to maintain its control of a normal 24 h circadian clock.

This hypothesis has been developed against a background of the following cytogenetic and molecular genetic information—and current experimentation—on the *per* gene. This locus is very well localized to a spot on *D. melanogaster*'s X chromosome (e.g., Smith and Konopka 1981). Many "molecular clones" have been purified from the segment of the X chromosome in and around the *per* locus (Pirrotta et al., 1983), using startling techniques that depend on direct excision of quasi-specific chromosomal sites from the fly's large salivary gland cells (cf. Scalenghe et al. 1981). We obtained, from V. Pirrotta, cloned *D. melanogaster* DNA that corresponds to a site including or very near the *per* locus. We have engineered cloning vectors which allowed molecular transformation (cf. Spradling and Rubin 1982, and Rubin and Spradling 1982) of arrythmic per° hosts by this DNA. Once the transformants were selected with regard to expression of a gratuitous molecular marker (cf. Rubin and Spradling 1982), most were per^{+} phenotypically (Zehring et al. 1984, cf. Bargiello et al. 1984). Thus, we can use the now identified per^{+} DNA from *D. melanogaster* to purify the homologous material from *D. simulans*. Transformation with this DNA, of a per° *D. melanogaster* host, will probably cause the transformants to be rhythmic in their circadian behavior. The main question, then, is: will such transformants, as males, now have a "completely" *D. simulans* song rhythm, though their genetic background is of course exclusively a *D. melanogaster* genome?

The species specificity of the song rhythm parameters suggests that these behavioral phenotypes may be of functional significance in male-female courtship interactions. For instance, the information contained in a *D. melanogaster* rhythm period could be used by a female of this species as part of her method for selecting conspecific males as her usual or near-exclusive mating partners. To look into these matters, we have performed a series of experiments with electronically simulated courtship songs (cf. Bennet-Clark and Ewing 1969, and Schilcher 1976a,b). We found that the sounds "worked" best, in partially correcting the poor mating performance of wingless males (courting normal females), when the simulator was specially programmed to generate a rhythmic song (Kyriacou and Hall 1982). That is, instead of setting the machine to generate "pulses of tone" (the most conspicuous feature of courtship song in the *D. melanogaster* subgroup, e.g., Cowling and Burnet [1981]) occurring at invariant intervals, we set the simulator to produce properly oscillating interpulse intervals and achieved a better enhancement of mating kinetics.

The basic interpulse interval, about which the oscillations occur, is different in the songs of *D. melanogaster* and *D. simulans* males, i.e., ca. 35 ms in the former and ca. 45-50 ms in the latter (e.g. Bennet-Clark and Ewing 1969, Cowling and Burnet 1981). When we considered the matter of species specificity in our song simulator experiments, we found that wingless *D. melanogaster* males, plus normal conspecific females, prefer artificial songs with both the correct average interpulse interval (35 ms) and the correct rhythm period (55 sec); whereas the corresponding courting *D. simulans* pairs "want" to hear their own appropriate characteristics of interpulse interval oscillations (35 sec) programmed into the simulated songs along with their normal interpulse intervals *per se* (48 ms) (Kyriacou and Hall 1982).

Another previously demonstrated effect of artificially generated courtship songs is that prestimulation of *D. melanogaster* females with a certain component of these sounds enhances their subsequent mating kinetics. That is, Schilcher (1976b) showed that the ca. 160 Hz hums produced by courting *D. melanogaster* males along with their trains of pulses, cause females to mate faster with "untreated" males which are put with such females just after the latter are exposed to the hums (produced by the song simulator). We reproduced this finding (Kyriacou and Hall 1984), and we have extended it in two ways. We showed that prestimulation of females with oscillating pulse songs also exerts an enhancing aftereffect on mating kinetics

(Kyriacou and Hall 1984), whereas pulse trains generated by the machine at invariant intervals are ineffective. The second result here once more confirms a previous one (Schilcher 1976b).

Conditioning Mutants and Courtship

We also have examined the possibility that these "song aftereffects" bear some relationship to mechanisms of learning and memory in *Drosophila*—which introduces the general subject of conditioned courtship in this organism. Studies of learning and memory in *Drosophila* have been reviewed recently. Mutants that could not learn or remember for very long, were isolated in tests using electric shocks and artificial odorants (Quinn and Greenspan 1984). We have used some of the conditioning variants in experiments on courtship aftereffects exerted by song prestimulation (Kyriacou and Hall 1984). We showed that females homozygous for either of two mutations in the dunce (*dnc*) gene can have no salutary aftereffects induced, following prestimulation by the hums or the rhythmic pulse songs. We also examined the effects of a memory mutation, amnesiac (in a separate gene called *amn*). Whereas the aftereffects of song prestimulation last 3-5 min in experiments using normal females, females homozygous for *amn* exhibit faster mating only when males are introduced to them within 1 min after the songs are played to them (Kyriacou and Hall 1984).

Conditioning mutations such as *amn* and *dnc* also influence other aspects of experience-dependent courtship behavior in *Drosophila*. Mature, normal males exposed to either mated females (but not virgin females) or immature 1-day-old males (but not other mature 3-5-day-old males) exhibit low levels of courtship behavior for a few hours afterward. In the first case, courtship performed with a subsequent virgin female (Siegel and Hall 1979) or another mated female (D. A. Gailey, unpublished) is "weak." In the second case, courtship directed at a subsequent young male is weak (Gailey et al. 1982). These two types of aftereffects do not "cross react." Males trained by immature males will still court virgin females vigorously (Gailey et al. 1982); and males trained by mated females subsequently exhibit their normal high levels of courtship in the presence of young males (D. A. Gailey, unpublished). That courtship is elicited by immature males seems bizarre, but it is observed routinely in this species (Cook and Cook 1975, Jallon and Hotta 1979). It almost certainly has to do with special pheromones that are transiently associated with young adult males; that is, these substances are no longer produced by males older than one day posteclosion (Tompkins et al. 1980). Mature males in the presence of the young males appear to habituate to such males or even to their "aphrodisiac traces" (Gailey et al. 1982).

Let us pause momentarily to ask why a young male fruit fly should want to produce odors that make him a stimulating object for courtship during his first day or so of adult life. Is it conceivable that, during such courtship bouts and exposure to the songs of mature males, the immature fly stores some information from what he hears? This kind of qustion might stem from studies on birds and the songs they must learn as youngsters if they are to emit robust songs with the correct dialect later on (see, for example, the review of Konishi and Gurney 1982). But the animal at hand is supposed to be a "hard-wired lower organism." Consider, though, these old and emerging facts: (1) *Drosophila* males have never been thoroughly tested for influences of prior experience on their eventual singing behavior. (2) Whereas fruit fly males will court, sing, and mate after being stored (posteclosion) in isolation (C. P. Kyriacou and J. C. Hall [unpublished]; plus earlier findings of this kind from *Drosophila* lore, see Speith [1974]), (a) it is not yet known if all the subtleties of his song (including the rhythms) are normal after such deprivation, and (b) wingless *Drosophila* males singing no songs at all can court and eventually mate (see above), so that the mere fertility of males previously kept in isolation is not diagnostic. (3) The brains of *Drosophila keep developing after eclosion*, and this in part depends on sensory input (Technau 1984).

This last point refers to the nerve fibers that are added to the mushroom bodies, during the first several days of adulthood (Technau 1984). These "bodies" are a series of lobes (each a bundle of many axons) that are commonly found in the dorsal brains of many insects (see Heisenberg [1980] for background information of *Drosophila*'s mushroom bodies, including their putative behavioral significance). Such structures have been implicated in the control of sex-specific behaviors in *D. melanogaster*, based on analysis of male-like behavior performed by gynandromorphs, and nerve-cell marking of brain regions (in the vicinity of the mushroom bodies) which are consistently found to be genotypically male in sex mosaics that court as males (Hall 1979, Schilcher and Hall 1979, Greenspan et al. 1980). Subsequent to these behavioral/neural experiments, the fly's mushroom bodies have been found to be sexually dimorphic in nerve fiber number (Technau 1984), just as they are in other insects (e.g., Mobbs 1982)—by analogy to the sexual dimorphisms of structure found in the central nervous systems of many vertebrates, including the brain regions related to sex-specific singing in birds (e.g., Arnold 1980).

We must, then, entertain the possibility that some of the sensory input that can influence mushroom-body development in young adult *Drosophila* comes from nearby songs (elicited by the immature male's special odors); and that this hypothetical modulation of the not-so-hard wiring in the dorsal brain is concerned directly with control of the full, correct species-specific song that can be performed by males after they are partially programmed by prior exposure to the relevant auditory information. Of course, this hypothesis once more is analogous to the picture that has emerged from the many relevant behavioral and anatomical data on birds. And the hypothesis—however bizarre it might be regarded since it now refers to an invertebrate—is at least easily falsifiable. We can readily test this idea's validity by further experiments in the realm of isolated rearing, exposure of males to song sounds during putative "critical periods" early in adulthood (cf. Arnold 1980, Konishi and Gurney 1982), histological analysis of the mushroom bodies so exposed, and possibly even the use of the aforementioned conditioning variants—that is, as expressed in males that one is attempting to program with auditory information, some of which is to be retrieved a few days later.

The aftereffects exerted by exposure of males to mated females are strongly related to the olfactory control of courtship. In this case, the odors are special pheromones associated with fertilized females. The source of all such inhibitory pheromones may be the ejaculatory fluid sent in by the male during copulation (see Richmond et al. [this volume] for evidence, arguments, and background literature). In any event, mated female materials (extracted from them) must be present simultaneously with another fly (such as a virgin female, or even a paralyzed male) in order that a mature courting male can be trained by such exposure (Tompkins et al. 1983). This, then, looks to be a case of associative conditioning.

Mutations, originally isolated on the basis of defects in associative learning or memory (Quinn and Greenspan 1984), impinge on the occurrence or duration of "mated female aftereffects," when such genetic variants are expressed in males undergoing training (Siegel and Hall 1979, Gailey et al. 1984). What about the case of "young-male aftereffects," and its apparent analogy to non-associative habitation (see Gailey et al. 1982)? It has turned out that the associative conditioning mutants, such as *dnc* and *amn*, also affect simple learning, such as habituation to sugar stimuli, in experiments that measure behavioral responses to such stimuli as applied to chemosensory appendages (Duerr and Quinn 1982). The aftereffects of courtship song prestimulation on a female's mating receptivity is not, it would seem, a case of associative conditioning. Instead, this phenomenon resembles sensitization. Yet, again, the *dnc* and *amn* mutations which disturb the effects of song prestimulation (Kyriacou and Hall 1984) turn out to be sensitization variants as well as associative learning and memory mutants (Duerr and Quinn 1982). Some of the effects of learning mutations on experience-dependent courtship are "clean" enough to permit mosaic experiments. That is, mutant *dnc* alleles, expressed in males, do not allow any of these flies to be trained by prior exposure to young males; whereas essentially each individual *dnc*$^{+}$ (wildtype) male is well trained and, hence, courts a subsequent young male poorly (Gailey et al. 1982). Therefore, if a part-*dnc*, part-*dnc*$^{+}$ mosaic male is or is not trained in this kind of experiment, efforts at correlating the distribution of mutant *vs.* normal CNS tissue with the behavioral result will allow meaningful guesses as to which parts of the fly's nervous system (if any) must express *dnc*$^{+}$ to allow for normal conditioning (cf. Booker and Quinn [1981], who developed single-fly learning tests involving electric shocks, for a discussion on the necessity of this kind of strategy in a mosaic analysis of learning, as opposed to an application of the original mass fly, stochastic learning procedures).

We have begun to analyze *dnc*//*dnc*$^{+}$ mosaics (D. A. Gailey and J. C. Hall, unpublished) in courtship learning tests. So far, it appears as if the genotype of the brain is what matters, if a mosaic is trainable or not by exposure to an immature male and its pheromones. This result is to be expected.

We should consider, nevertheless, that these preliminary results already demonstrate a relatively localized effect of the *dnc* gene on learning. This is an important possibility to consider in the context of the fact that the enzyme of cyclic AMP metabolism, which is affected by *dnc* mutations (Byers et al. 1981) and is in fact almost certainly coded for by this gene (Kauvar 1982), has a rather broad tissue distribution not limited to the head (Shotwell 1983). The same kind of logic, and future mosaic experimentation, is pertinent to another "biochemical learning" gene, the dopa-decarboxylase (*Ddc*) locus in *D. melanogaster*. This gene was discovered with regard to the fact that it codes for this enzyme (e.g., Wright et al. 1982). Later, it was shown that certain *ts* mutations in the gene will "turn off" conditioned behavior when the mutants are subjected to high temperature treatments after they have reached adulthood (Tempel et al. 1984). This includes a shutoff of conditioned courtship behavior in heat-treated *Ddc*^{ts} males, which were trained in the presence of mated females (Tempel et al. 1984). We are now attempting to extend this result, to ask if the normal allele need only be expressed in a localized portion of the brain to allow for normal conditioning.

We hope eventually to develop several additional implications of "learning gene" expression in mosaics. Some of the anticipated results may reveal information on putative "conditioning centers" in neural ganglia that would not have been intuitively obvious before the execution of the mosaic experiments. For instance, will all conditioning mutations affect experience-dependent courtship via an influence on the same subset of the central nervous system? Will this subset be a narrowly localizable portion of the brain, or must most of the brain be normal for normal learning or mutant for mutant behavior? Will different types of conditioned courtship be controlled from the same focal or diffuse brain region even though these two different types of conditioning are related to one another by virtue of the mutations that so similarly affect these separate behavioral phenomena?

Conclusions

Neurogenetic analysis of reproduction in *Drosophila* has progressed to the point that the implications of courtship control are considerably more extensive than might have been imagined in the initial stages of these investigations. Thus, the mechanisms underlying biological rhythms are intimately associated with discrete components of reproductive behavior. Some additional phenomena correlated with this conclusion are (1) that the control of short-term behavioral rhythms has strong connections with the regulation of circadian rhythms, and (2) that the short-cycling courtship rhythms have adaptive significance in their stimulation of mating.

Mechanisms that control learning and memory also have important neurogenetic relationships to reproductive behavior. This means, in the first place, that several aspects of interactions between males and females of this insect species can no longer be viewed in the context of fixed action patterns. Second, the ability of these insects to learn and remember would seem to be valuable adaptively. Third, the kinds of experiments using individual flies, in which courtship actions are in part based on the memory of previous experience, provide procedures that allow one to determine portions of the fly's CNS in which the "learning genes" must be active if conditioning is to take place.

The interrelationships among these several features of complex behavior in *Drosophila*, which have been demonstrated neurogenetically, are at present highly formalized connections. Yet, as our investigations proceed into the nerve-cellular analysis of where these genes are acting to control higher functions, and as other studies progress toward molecular understanding of the precise nature of these genes and their products, the network of interconnecting mechanisms threatens to be understood in a concrete sense as well as in abstract terms.

Acknowledgments

The experiments reviewed here were performed in collaboration with my colleagues Bambos Kyriacou, Michael Rosbash, Pranhitha Reddy, Will Zehring, David Wheeler, Ron Konopka, Laurie Tompkins, Dick Siegel, and Don Gailey. I am extremely appreciative of the energy and insight they have brought to bear on these investigations. The work has been supported by grants from the NIH (GM-21473, GM-332305). I also thank John Ringo and Martin Heisenberg and John Hildebrand for helpful comments on the manuscript.

Literature Cited

Arnold, A. P. 1980. Sexual differentiation of the brain. Amer.Sci. 68: 165.

Bargiello, T. A., F. R. Jackson and M. W. Young. 1984. Restoration of circadian behavioral rhythms by gene transfer in *Drosophila*. Nature 312: 752.

Bennet-Clark, H. C., and A. W. Ewing. 1969. Pulse interval as a critical parameter in the courtship song of *Drosophila melanogaster*. Anim. Behav. 17: 755.

Booker, R., and W. G. Quinn. 1981. Conditioning of leg position in normal and mutant *Drosophila*. Proc. Natl. Acad. Sci. USA 78: 3940.

Byers, D., R. Davis, and J. Kiger, Jr. 1981. Defect in cyclic AMP phosphodiesterase due to the *dunce* mutation of learning in *Drosophila melanogaster*. Nature 289: 79.

Cook, R., and A. Cook. 1975. The attractiveness to males of female *Drosophila melanogaster*: effects of mating, age, and diet. Anim. Behav. 23: 521.

Cowling, D. E., and B. Burnet. 1981. Courtship songs and genetic control of their acoustical characteristics in sibling species of *Drosophila melanogaster*. Anim. Behav. 29: 924.

Duerr, J. S. and W. G. Quinn. 1982. Three *Drosophila* mutations that block associative learning also affect habituation and sensitization. Proc. Natl. Acad. Sci. USA 79: 3646.

Gailey, D. A., F. R. Jackson, and R. W. Siegel. 1982. Male courtship in Drosophila: The conditioned response to immature males and its genetic control. Genetics 102: 771.

Gailey, D. A., F. R. Jackson and R. W. Siegel. 1984. Conditioning mutations in *Drosophila melanogaster* affect an experience dependent behavioral modification in courting males. Genetics 106: 613.

Greenspan, R. J., J. A. Finn, Jr., and J. C. Hall. 1980. Acetylcholinesterase mutants in Drosophila and their effects on the structure and function of the central nervous system. J. Comp. Physiol. 189: 741.

Hall, J. C. 1979. Control of male reproductive behavior by the central nervous system of Drosophila: dissection of a courtship pathway by genetic mosaics. Genetics 92: 437.

Hall, J. C. 1981. Sex behavior mutants in *Drosophila*. BioScience 31: 125.

Hall, J. C. 1982. Genetics of the nervous system in *Drosophila*. Quart. Rev. Biophys. 15: 223.

Heisenberg, M. 1980. Mutants of brain structure and function: what is the significance of the mushroom bodies for behavior? *In* O. Siddiqi, P. Babu, L. M. Hall, and J. C. Hall, eds. Development and Neurobiology of *Drosophila*. Plenum Press, New York.

Jackson, F. R. 1983. The isolation of biological rhythm mutations on the autosomes of *Drosophila melanogaster*. J. Neurogenet. 1: 3.

Jallon, J.-M., and Y. Hotta. 1979. Genetic and behavioral studies of female sex appeal in *Drosophila*. Behav. Genet. 9: 257.

Kauvar, L. 1982. Defective cyclic adenosine 3′:5′-monophosphate phosphodiesterase in the *Drosophila* memory mutant *dunce*. J. Neurosci. 2: 1347.

Kawanishi, M., and T. K. Watanabe, 1981. Genes affecting courtship song and mating preference in *Drosophila melanogaster, Drosophila simulans* and their hybrids. Evolution 35: 1128.

Konishi, M., and M. E. Gurney. 1982. Sexual differentiation of brain and behavior. Trends Neurosci. 5: 20.

Konopka, R. J. 1981. Genetics and development of circadian rhythms in invertebrates. *In* J. Aschoff, ed. Handbook of Behavioral Biology. Plenum Press, New York.

Kyriacou, C. P., and J. C. Hall. 1980. Circadian rhythm mutations in *Drosophila melanogster* affect short-term fluctuations in the male's courtship song. Proc. Natl. Acad. Sci. USA 77: 6729.

Kyriacou, C. P., and J. C. Hall. 1982. The function of courtship song rhythms in *Drosophila*. Anim. Behav. 30: 794.

Kyriacou, C. P. and J. C. Hall. 1984. Learning and memory mutations impair acoustic priming of mating behaviour in *Drosophila*. Nature 303: 62.

Kyriacou, C. P. and J. C. Hall. 1985. Action potential mutations stop a biological clock in *Drosophila*. Nature 314: 171.

Kyriacou, C. P. and J. C. Hall. 1986. Interspecific genetic control of courtship song production and reception in *Drosophila*. Science 232: 494.

Mobbs, P. G. 1982. The brain of the honeybee *Apis mellifera*. I. The connections and spatial organization of the mushroom bodies. Phil. Trans. Roy. Soc. B 298: 309.

Pirrotta, V. C. Hadfield, and G. H. J. Pretorius. 1983. Microdissection and cloning of the white locus and the 3B1-3C2 region of the *Drosophila* X chromosome. EMBO J. 2: 927.

Quinn, W. G., and R. J. Greenspan. 1984. Learning and courtship in *Drosophila*: two stories with mutants. Ann. Rev. Neurosci. 7: 67.

Rubin, G. M., and A. C. Spradling. 1982. Genetic transformation of *Drosophila* with transposable element vectors. Science 218: 348.

Scalenghe, F., E. Turco, J. E. Edstrom, V. Pirrotta, and M. Melli. 1981. Microdissection and cloning of DNA from a specific region of *Drosophila melanogaster* polytene chromosomes. Chromosoma 82: 205.

Schilcher, F. v. 1976a. The role of auditory stimuli in the courtship of Drosophila melanogaster. Anim. Behav. 24: 18.

Schilcher, F. V. 1976b. The function of pulse song and sine song in the courtship of *Drosophila melanogaster*, Anim. Behav. 24: 622.

Schilcher, F. V., and J. C. Hall, 1979. Neural topography of courtship song in sex mosaics of *Drosophila melanogaster*. J. Comp. Physiol. 129: 85.

Shotwell, S. L. 1983. Cyclic adenosine 3′:5′-monophosphate phosphodiesterase and its role in learning in *Drosophila*. J. Neurosci. 3: 739.

Siegel, R. W., and J. C. Hall. 1979. Conditioned responses in courtship behavior of normal and mutant *Drosophila*. Proc. Natl. Acad. Sci. USA 76: 3430.

Smith, R. F., and R. J. Konopka. 1981. Circadian clock phenotypes of chromosome aberrations with a breakpoint at the *per* locus. Molec. Gen. Genet. 183: 243.

Spieth, H. T. 1974. Courtship behavior in *Drosophila*. Ann. Rev. Entomol. 19: 385.

Spieth, H. T., and J. M. Ringo. 1983. Mating behavior and sexual isolation in Drosophila. *In* M. Ashburner, H. L. Carson, and J. N. Thompson, eds. The Genetics and Biology of Drosophila Vol 3c. Academic Press, London.

Spradling, A. C., and G. M. Rubin. 1982. Transposition of cloned P elements into *Drosophila* germ line chromosomes. Science 228: 341.

Technau, G. 1984. Fiber number in the mushroom bodies of adult *Drosophila melanogaster* depends on age, sex and experience. J. Neurogenet. 1: 113.

Tempel, B. L., M. S. Livingstone, and W. G. Quinn. 1984. Mutations in the dopa-decarboxylase gene affect learning in *Drosophila* Proc. Natl. Acad. Sa USA. 81: 3577.

Tompkins, L., and J. C. Hall. 1983. Identification of brain sites controlling female receptivity in mosaics of *Drosophila melanogaster*. Genetics 103: 179.

Tompkins, L., J. C. Hall, and L. M. Hall. 1980. Courtship-stimulating volatile compounds from normal and mutant *Drosophila*. J. Insect Physiol. 26: 689.

Tompkins, L., R. W. Siegel, D. A. Gailey, and J. C. Hall. 1983. Conditioned courtship in *Drosophila* and its mediation by association of chemical cues. Behav. Genet. 13: 565.

Wright, T. R. F., B. C. Black, C. P. Bishop, L. March, E. S. Pentz, R. Steward, and E. T. Wright. 1982. The genetics of dopa decarboxylase in *Drosophila melanogaster* V. *Ddc* and *1(2)amd* alleles: isolation, characterization and intragenic complementation. Molec. Gen. Genet. 188: 18.

Wu, C.-F., and B. Ganetzky. 1980. Genetic alteration of nerve membrane excitability in temperature-sensitive paralytic mutants of *Drosophila melanogaster*. Nature 286: 814.

Zehring, W. A., D. A. Wheeler, P. Reddy, R. J. Konopka, C. P. Kyriacou, M. Rosbash and J. C. Hall. 1984. P-element transformation with *period* locus DNA restores rhythmicity to mutant, arrhythmic *Drosophila melanogaster*. Cell 39: 369.

Physiological Tolerance and Behavioral Avoidance of Alcohol in *Drosophila*: Coadaptation or Pleiotropy?

John F. McDonald

Department of Molecular and Population Genetics
University of Georgia
Athens, Georgia 30602

The Problem

The presumed function of natural selection is to enhance the "fit" between organisms and their environments. This process of adaptation is often complex and may occur simultaneously on multiple levels of biological organization. Although a good deal of research has been directed toward elucidating the genetic basis of specific adaptations, relatively little is known concerning the genetic origins of those suites of traits that have apparently coevolved to serve the same adaptive need.

The phenomenon of coadaptation is apparent on all levels of biological organization, but nowhere is it more pronounced than in the relationships that exist between certain adaptive behavior patterns and coordinately adaptive modifications in an organism's morphological structure and/or physiological function. Take, for example, the classic case of industrial melanism in the peppered moth *Biston betularia* L. (Kettlewell 1955a). Every student of evolution is aware of the adaptive significance of the major melanic forms of these moths in providing camouflage protection against bird predators. Less well known, however, is the coordinately adaptive behavior of light and dark colored moths to rest preferentially on backgrounds that match their own color (Kettlewell 1955b).

Two alternative genetic hypotheses have been offered to explain the emergence of this coadaptive complex. According to the hypothesis advocated by Sargent (1969), the morphological and behavioral traits are under separate genetic control, and the genetic variants coding for each trait were selected for separately and, thus, the adaptations evolved independently. Since the adaptive advantage of where a moth chooses to rest is wholly dependent upon the decision being coordinated appropriately with its color, this hypothesis has as a corollary the likely establishment of some form of genetic linkage between the complementary color-coding and behavioral-coding alleles. The alternative hypothesis, favored by Kettlewell (1955b), postulates that moths "self-inspect" their color and move around until the contrast between their own body color and the background reaches some acceptable minimum. The gene or genes coding for such a behavior would not need to be genetically linked to the color-coding locus, for the very nature of the postulated behavioral response allows it to work automatically in conjunction with any color morph, including those newly arisen by mutational or migrational events. Although this model assumes the pre-existence of the moth's "self-inspective" tendencies, the specific adaptive behavior of choosing backgrounds that match body color will arise spontaneously as a simple pleiotropic or secondary effect of the color coding locus *per se*.

This paper, however, is not about *Biston betularia* nor will I present the data which have been offered in support of one or the other of the above hypotheses. The purpose of this example is to introduce two alternate ways of viewing the evolution of specific adaptive behaviors that are part of some larger coadaptive whole. The "coevolution model" views the evolution of adaptive behaviors as arising independently of adaptively related morphologies or functions and thus being under separate, although possibly "linked," genetic control. The "pleiotropic model," on the other hand, views specific adaptive behaviors arising automatically as indirect or pleiotropic effects of changes at loci encoding morphology or function. The two models are certainly not mutually exclusive, and I do not pretend to claim that all behavioral coadaptations must have evolved by one or the other of these two pathways. Indeed, it is possible that many behavioral coadaptations have evolved by a mixture of these strategies. The models are, nevertheless, useful as working alternatives in attempts to dissect the genetic basis of the phenomenon of behavioral coadaptation.

Alcohol Adaptation in *Drosophila*

Background

Adaptation is a complex phenomenon that requires for its understanding an appreciation of the interactions that exist between genotype, phenotype and environment (McDonald 1983). There are many experimental systems that allow analysis of the relationships between genes and their respective structural, functional or behavioral effects on the phenotype. In addition, many other systems are suitable for the study of the adaptive significance of specific phenotypes in particular environments. In contrast, however, relatively few experimental systems permit detailed investigation of the interactions that exist between all of these levels simultaneously. One such system is alcohol adaptation in *Drosophila*.

For the past several years our laboratory has been investigating the genetic basis of alcohol adaptation in *Drosophila* as a model for the study of adaptive evolution (e.g., McDonald and Avise 1969, McDonald and Ayala 1978, McDonald et al. 1977, McDonald et al. 1980, Anderson and McDonald 1983). Most of our effort has focused on elucidating the biochemical and molecular-genetic basis of alcohol tolerance in *Drosophila melanogaster* Meigen, a species that is especially successful in exploiting alcohol-containing natural habitats. We have recently found that the enzyme alcohol dehydrogenase (ADH) plays a key role in the ability of *D. melanogaster* to adapt, both physiologically and behaviorally, to alcohol environments but in a somewhat more complex manner than many workers had originally thought.

It has been recognized for many years that flies lacking ADH activity, i.e., those homozygous for Adh^{null} alleles, succumb to the toxic effects of environmental alcohol significantly sooner than ADH positive flies (Sofer and Hatkoff 1972). Moreover, the relative survivorship of flies in alcohol stress environments was generally found to correlate with their respective ADH activity levels (e.g., Ainsley and Kitto 1975). In addition, early studies on the behavioral preferences of *Drosophila* for alcohol-treated media suggested that flies (both larvae and adults) with relatively high ADH activity levels were attracted to alcohol-supplemented media significantly more often than lower ADH activity flies (McKenzie and Parsons 1974, Parsons 1977, King et al. 1976, Parsons and King 1977, Cavener 1979, Richmond and Gerking 1979). Collectively, these laboratory results suggested that natural selection should favor the fixation of high activity genotypes in natural populations of this alcohol-adapted species.

The unfortunate and (for those of us partial to the idea that what we do in the laboratory has some real world relevance) somewhat uncomfortable reality, however, was that a stable, world-wide genetic polymorphism exists in *D. melanogaster* between high activity and low activity alcohol dehydrogenase-coding alleles (Chambers 1981). Indeed, the low activity Adh^S allele has repeatedly been shown to be less "fit" selectively than the (2-3X) higher activity Adh^F allele in population cage experiments containing alcohol-supplemented food (e.g., Bijlsma-Meeles and van Delden 1974, Cavener and Clegg 1981). Yet the Adh^S allele is consistently maintained at about 50% frequency in even those wild populations that are breeding in obviously high concentration alcohol environments like vineyards and wine cellars (e.g., McKenzie and McKechnie 1978). [Note: The designations "slow" (*S*) and "fast" (*F*) refer to the relative mobilities of the respective allelic products on electrophoretic gels.] These facts, coupled with doubts brought on by some occasionally contradictory laboratory results on the relationship between ADH activity and alcohol tolerance (e.g., Gibson et al. 1979), prompted us to re-examine, in a more systematic fashion, the role of the ADH system on *Drosophila*'s ability to survive within and behaviorally discriminate between alcohol-containing environments. The results have been informative and, I think, relevant to the phenomenon of biochemical-behavioral coadaptation.

Alcohol Tolerance and the ADH System

Faradeih Shadravan and I have recently completed two independent studies that demonstrate that the relative viabilities of different ADH activity genotypes in alcohol environments is a concentration-dependent phenomenon. From a series of adult survivorship studies we have found that: (1) low levels of alcohol enhance the longevity of flies relative to controls when alcohol is the sole source of carbon; (2) at moderate alcohol concentrations, the ability to utilize ethanol is negated by its toxicity; (3) Adh^S (low activity) flies survive longer than Adh^F flies at low alcohol concentrations but less well at moderate concentrations. We had hypothesized previously (Anderson et al. 1981) that this is due to the fact that although Adh^S flies have 2-3X less ADH enzyme than Adh^F flies, ADH^S enzyme has higher affinity for ethanol (i.e., lower Km) and is thus kinetically more efficient than ADH^F enzyme at low ethanol concentrations; (4) at still higher ethanol concentrations, a second reversal in survivorship occurs. We believe that this second concentration-dependent switch is best explained by a "secondary product threshold model" in which the relative fitness of flies in ethanol environments is positively correlated with ADH activity so long as the rate of toxic acetaldehyde (the immediate product of ADH-mediated alcohol catabolism) accumulation is less than the rate at which it can be eliminated by an as yet unspecified oxidase (David et al. 1978). If and when this rate threshold is surpassed, ADH activity will correlate negatively with alcohol tolerance.

In order to test this hypothesis, we chemically induced partial ADH null phenocopies (~90% reduction in comparison with control ADH activity levels) via 2-propanol pretreatment (Anderson and McDonald 1981) and retested relative survivorships under high ethanol stress. We predicted that the Adh^F genotypes should have improved survivorships due to the induced decrease in their ADH activities and the consequent drop in accumulating acetaldehyde levels. In contrast, we predicted that the Adh^S flies should display decreased survivorships due to a reduced ability to catabolize toxic levels of ethanol. Our results were consistent with both predictions. Also consistent with our hypothesis were the results of a series of selection experiments for increased tolerance to 12% ethanol in which we observed ADH activity to increase concomitantly with alcohol tolerance in those populations having relatively

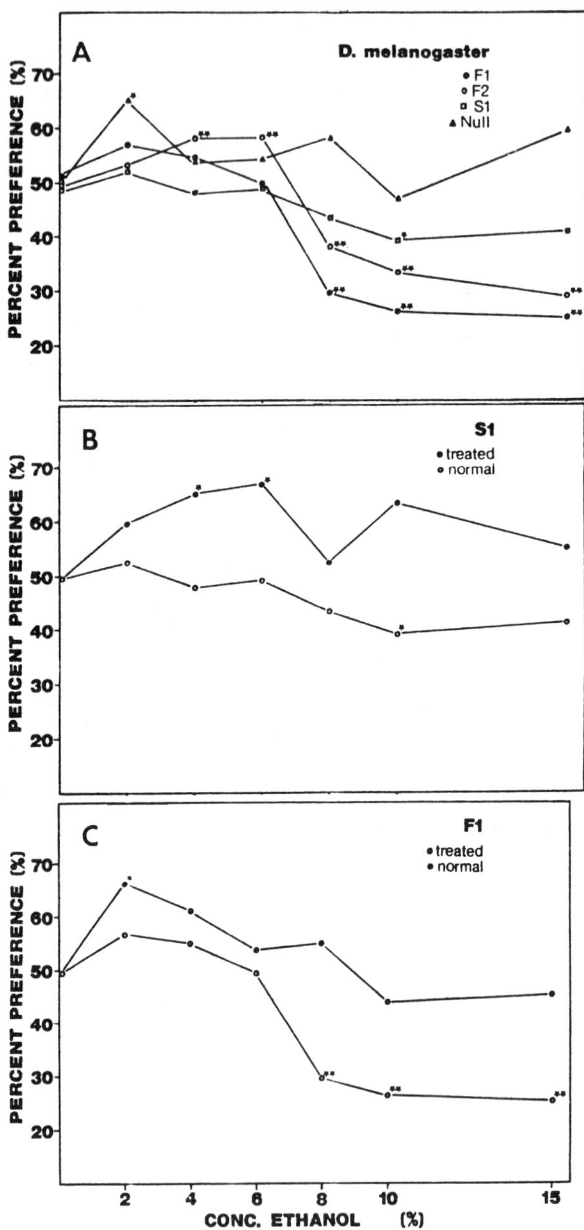

Fig. 1(a). Mean percentage of larvae preferring alcohol over a range of ethanol concentrations for strains of *Drosophila melanogaster*. Note that F1 larvae display greater preference than S1 larvae for alcohol treated media at all alcohol concentrations. (* p [value = control value] < 0.05; ** p [value = control value] < 0.01).

(b). Mean percentage of larvae preferring alcohol over a range of ethanol concentra-

low starting ADH activity levels but to decrease (as tolerance increased) in those populations having relatively high starting ADH activity levels.

In general, then, we believe that the relative fitnesses of different *Adh* genotypes will vary depending upon the concentration of environmental alcohol to which they are exposed. The higher activity Adh^F genotypes are expected to be more fit generally than the lower activity Adh^S genotypes in moderately stressful alcohol environments. In highly stressful alcohol environments, on the other hand, a reversal in relative fitnesses is expected to occur. This model is capable of explaining the existence of a stable *Adh* polymorphism worldwide as well as those "contradictory" laboratory results in which ADH activity was found to have no apparent correlation with alcohol tolerance (e.g., Oakeshott and Gibson 1981).

Alcohol Preference Behavior and the ADH System

As alluded to above, several early studies suggested that *Drosophila*'s relative alcohol preference behavior correlates positively with their relative ADH activity levels. At the time, this behavior pattern was interpreted as being coadaptive with ADH activity-mediated alcohol tolerance. As we have seen, this is a correct conclusion but only in moderately high alcohol environments. In higher concentration alcohol environments, however, the alcohol preference behavior of high ADH activity genotypes would, in fact, be nonadaptive. Lewis Gelfand and I explored the relationship between ADH activity and alcohol preference behavior in more detail by observing the preference of *Drosophila* larvae for a series of ethanol concentrations ranging from 0 to 15% (Gelfand and McDonald 1980, 1983).

The details of the following studies may be found in Gelfand and McDonald (1980). Briefly, each test consisted of placing 15 early third-instar larvae (90-100 h postoviposition) on petri plates in which half of the 1.5% agar medium is supplemented with controlled amounts of ethanol (0-15%) and half supplemented with equivalent volumes of H^2O. Tests were run in an enclosed temperature controlled (22 °C) chamber with overhead flourescent lighting: 15-20 replicate tests were run for each strain at each concentration. The results of the analyses of two Adh^{fast}, one Adh^{slow}, and one Adh^{null} strain are summarized graphically in Figure 1a.

Consistent with the earlier published reports, we observed that the Adh^F strains displayed consistently greater (but statistically insignificant, [$0.10 < p < 0.05$ ANOVA]) preference behavior for media containing low levels of ethanol than the Adh^S strain. Both the Adh^F and Adh^S strains displayed increasing avoidance to alcohol media as the ethanol concentrations were increased. Interestingly, the relative alcohol preferences existing between the Adh^F *and* Adh^S gentoypes were reversed at higher ethanol concentrations (>10%). This switch in relative alcohol preference behavior between the *Adh* genotypes mimics the switch in their relative alcohol tolerances (see above).

---

tions for strain S1 and 2-propanol-induced ADH null phenocopies of strain S1 (*, ** as defined above) Graph showing the alcohol preference behavior of 2-propanol-induced ADH null phenocopies of S1 larvae relative to untreated controls.

(c). Mean percentage of larvae preferring alcohol over a range of ethanol concentrations for strain F1 and 2-propanol-induced ADH null phenocopy of strain F1 (*, ** as defined above) Graph showing the alcohol preference behavior of 2-propanol-induced ADH null phenocopies of F1 iarvae relative to untreated controls.

If the biochemical basis of the switch in behavioral response is analogous to what we believe to be causing the switch in alcohol tolerances (i.e., differential rates of accumulation of "threshold" levels of acetaldehyde), two experimental predictions follow directly. First, Adh^{null} larvae, having no (or substantially reduced) capabilities to convert ethanol to acetaldehyde should not display avoidance to even high concentrations of ethanol. This prediction was confirmed (Fig. 1b). Second, 2-propanol induced partial ADH null phenocopies of both Adh^F and Adh^S genotypes should display reduced avoidance toward ethanol-treated media relative to untreated larvae. Again, this prediction was confirmed (Figs. 1b, 1c).

Conclusions

In general, alcohol avoidance behavior seems to be triggered by accumulation of acetaldehyde levels above a certain critical threshold. As was postulated within the context of our survivorship model, this threshold is reached relatively sooner by the high activity Adh^F larvae when present in high concentration alcohol environments.

The alcohol avoidance response is clearly a pleiotropic or indirect effect of the *Adh* locus and not due to the direct action of a separately evolved "alcohol avoidance gene." As was the case for "color matching" in *B. betularia*, the pleiotropic model assumes the pre-existence of a generalized behavioral response; in this instance an aversion to environments which result in the internal accumulation of noxious substances. In other words, the model assumes the pre-existence of a generalized mechanism that tells larvae to leave environments that make them feel "sick." Presumably, such generalized responses would be acquired very early in the evolution of mobile organisms. Under the pleiotropic model, newly arising ADH activity variants should "automatically" possess adaptive alcohol avoidance behaviors. This fact is demonstrated dramatically in the behavioral responses of the induced ADH null phenocopies (Figs. 1c, 1d).

Finally, there is evidence that ADH activity-mediated alcohol avoidance behavior is working antagonistically with some sort of *Adh*-independent attraction of larvae for ethanol-containing media. This suspicion is supported by the three observations. First, Adh^{null} larvae displayed generally consistent preference for ethanol media. Second, 2-propanol induced reductions in ADH activity in either Adh^F or Adh^S larvae resulted in uniform increases in preference for ethanol-treated media. Third, previously observed preference behavior for alcohol and other volatile compounds in *Drosophila* (Fuyama 1976) has been mapped to a position far removed from the *Adh* locus (Fuyama 1978).

In general, the overall significance of the coadaptive complex reported in this paper is to behaviorally position *Drosophila* larvae into those alcohol containing micro-habitats to which they are most physiologically and biochemically adapted. The avoidance/preference behaviors of insects (and other organisms) to specific environmental substrates is frequently observed to be coordinated with the organism's relative physiological ability to detoxify and/or utilize these substances. Our studies on the genetic and biochemical bases of alcohol adaptation in *Drosophila* suggest that many chemically induced taxes may not have evolved independently but may be the simple pleiotropic or indirect effects of genes coding for the coadaptive ability to catabolize the chemicals.

Acknowledgments

This work was supported by NSF grant DEB-82-00965.

Literature Cited

Ainsley, R., and G. Kitto. 1975. Selection mechanisms maintaining alcohol dehydrogenase polymorphisms in *Drosophila melanogaster*. *In* C. Markert, ed. Isozymes II. Academic Press, New York.

Anderson, S., and J. McDonald. 1981. Effect of environment alcohol on *in vivo* properties of *Drosophila* alcohol dehydrogenase. Biochem. Genet. 19: 421.

Anderson, S., and J. McDonald. 1983. Biochemical and molecular analysis of naturally occurring *Adh* variants in *Drosophila melanogaster*. Proc. Natl. Acad. Sci. USA 80: 4798.

Anderson, S., J. McDonald, and M. Santos. 1981. Selection at the *Adh* locus in *Drosophila melanogaster*: adult survivorship-mortality in response to ethanol. Experientia 37: 463.

Bijlsma-Meeles, E., and W. van Delden. 1974. Intra- and interpopulation selection concerning the alcohol dehydrogenase locus in *Drosophila melanogaster*. Nature 247: 369.

Cavener, D. 1979. Preference for ethanol in *Drosophila melanogaster* associated with the alcohol dehydrogenase polymorphism. Behav. Genet. 9: 359.

Cavener, D., and M. Clegg. 1981. Multigenic response to ethanol in *Drosophila melanogaster*. Evolution 35: 1.

Chambers, G. 1981. Biochemistry of alcohol dehydrogenase variation in *D. melanogaster*. *In* J. Gibson and J. Oakeshott, eds. Genetic Studies of *Drosophila* Populations. Australian National University Press, Canberra.

David, J., C. Bocquet, J. Van Herrewege, P. Fouillet, and M. Arens. 1978. Alcohol metabolism in *Drosophila melanogaster*: uselessness of the most active aldehyde oxidase produced by the Aldox locus. Biochem. Genet. 16: 203.

Fuyama, Y. 1976. Behavior genetics of olfactory responses in *Drosophila*. I. Olfactometry and strain differences in *Drosophila melanogaster*. Behav. Genet. 6: 407.

Fuyama, Y. 1978. Behavior genetics of olfactory responses in *Drosophila*. II. An odorant-specific variant in a natural population of *Drosophila melanogaster*. Behav. Genet. 8: 399.

Gelfand, L., and J. McDonald. 1980. Relationship between ADH activity and behavioral response to environmental alcohol in *Drosophila*. Behav. Genet. 10: 237.

Gelfand, L., and J. McDonald. 1983. Relationship between ADH activity and behavioral response to environmental alcohol in five *Drosophila* species. Behav. Genet. 13:281 and McDonald 1983 is p. 77.

Gibson, J., N. Lewis, M. Adena, and S. Wilson. 1979. Selection for ethanol tolerance in two populations of *D. melanogaster* segregating alcohol dehydrogenase allozymes. Aust. J. Biol. Sci. 32: 387.

Kettlewell, H. 1955a. Selection experiments on industrial melanism in Lepidoptera. Heredity 9: 323.

Kettlewell, H. 1955b. Recognition of appropriate backgrounds by the pale and black phases of Lepidoptera. Nature 175: 943.

King, S., R. Rockwell, and J. Grossfield. 1976. Oviposition response to ethanol in *Drosophila melanogaster* and *Drosophila simulans*. Genetics 83: S39.

McDonald, J. 1983. The molecular basis of adaptation: a critical review of relevant ideas and observations. Ann. Rev. Ecol. Syst. 14: (in press).

McDonald, J., and J. Avise. 1976. Evidence for the adaptive significance of enzyme activity levels: interspecific variation in $^{\alpha}$-GPDH and ADH in *Drosophila*. Biochem. Genet. 14: 347.

McDonald, J., and F. Ayala. 1978. Genetic and biochemical basis of enzyme activity variation in natural populations. I. Alcohol dehydrogenase in *Drosophila melanogaster*. Genetics 89: 371.

McDonald, J., G. Chambers, J. David, and F. Ayala. 1977. Adaptive response due to changes in gene regulation: a study with *Drosophila*. Proc. Natl. Acad. Sci. USA 74: 4562.

McDonald, J., S. Anderson, and M. Santos. 1980. Biochemical differences between products of the *Adh* locus in *Drosophila*. Genetics 95: 1013.

McKenzie, J., and P. Parsons. 1974. Microdifferentiation in a natural population of *Drosophila melanogaster* to alcohol in the environment. Genetics 77: 385.

McKenzie, J., and S. McKechnie. 1978. Ethanol tolerance and the *Adh* polymorphism in a natural population of *Drosophila melanogaster*. Nature 272: 75.

Oakeshott, J., and J. Gibson. 1981. Is there selection by environmental ethanol on the alcohol dehydrogenase locus in *D. melanogaster? In* J. Gibson and J. Oakeshott, eds. Genetic Studies of *Drosophila* Populations. Australian National University Press, Canberra.

Parsons, P. 1977. Larval reaction to alcohol as an indicator of resource utilization differences between *Drosophila melanogaster* and *D. simulans*. Oecologia 30: 141.

Parsons, P., and S. King. 1977. Ethanol: larval discrimination between two *Drosophila* sibling species. Experientia 33: 898.

Richmond, R., and J. Gerking. 1979. Oviposition site preference in *Drosophila*. Behav. Genet. 9: 233.

Sargent, T. 1969. Background selections of the pale and melanic forms of the cryptic moth *Phigalia titea*. Nature 222: 585.

Sofer, W., and M. Hatkoff. 1972. Chemical selection of alcohol dehydrogenase negative mutants in *Drosophila*. Genetics 72: 545.

Evolution of Egg Laying Behavior in Aplysia

Linda B. McAllister

Anne C. Mahon

Richard H. Scheller

Department of Biological Sciences
Stanford University
Stanford, California 94305

Introduction

Invertebrates provide simple model systems for the study of the cellular and molecular bases of behavior. The gastropod mollusc, *Aplysia californica* Cooper (Anaspidea: Aplysiidae), displays a number of behavioral repertoires which can be ascribed to the activity of defined circuits composed of identified central neurons (Kandel 1976). These reflex and fixed action patterns are generated in an all-or-none fashion. Such behaviors are precisely inherited by the animal and therefore are encoded by a discrete set of genes (Lorenz 1958, Kandel 1976). Understanding the organization and structure of these genes enables us to discuss their evolution and, consequently, the evolution of the behavior which they direct.

Egg laying behavior in *Aplysia* is a stereotyped fixed action pattern consisting of a cessation of walking and feeding and increased respiratory activity followed by head waving and egg deposition. This behavior is thought to be generated by a central nervous system program which is activated by a battery of neuroactive peptides. The best characterized of these central neurons are the bag cells, two symmetrical clusters of electrically coupled neuroendocrine cells located at the rostral margins of the abdominal ganglion (Kupfermann 1972). The bag cells may act on and/or with a defined set of interneurons which are distributed throughout the central nervous system (McAllister et al. 1983, E. Meyer and B. Rothman, U. C., San Francisco (unpublished). Both cell types synthesize the same or a related collection of peptides, the egg laying hormone (ELH) and a number of other peptides, some with known transmitter activity (McAllister et al. 1983). A third tissue which may be involved in egg laying behavior is the atrial gland. This exocrine gland is situated at the distal end of the large hermaphroditic duct and is contacted directly by the penis during copulation (Arch et al. 1978). The atrial gland synthesizes a unique set of molecules that are related to those in the bag cells. These include the A and B peptides which differ from each other in only four of their 34 amino acids.

The egg laying hormone and related peptides have several target tissues. When A or B peptide is applied to the processes or soma of the bag cells in vitro, an afterdischarge is elicited. Similarly, activation of the bag cells in vivo produces a synchronous, prolonged discharge of the action potentials. This in turn causes release of ELH and its companion peptides into the cleft between the vascularized sheath and the neurons of the abdominal ganglion (Strumwasser et al. 1980). ELH and its related peptides act on nerve cells of the central ganglia to alter their firing patterns (E. Mayeri and B. Rothman, U. C., San Francisco, unpublished results). In addition, ELH is carried through the circulation to the ovotestis where it induces smooth muscle contraction of the follicle and extrusion of the egg strand (Rothman et al. 1983). The multiple sites of ELH peptide release and action may be important in eliciting different components of the egg laying behavioral repertoire.

We have used recombinant DNA methodology to discover that ELH as well as the A and B peptides are encoded by a small gene family (Scheller et al. 1982, 1983). Expression of the ELH gene family is tissue-specific and gives rise to distinct precursor polyproteins. The ELH, A and B peptides exist on separate precursors in combination with several additional biologically active molecules which are liberated by internal proteolytic cleavages at basic residues. As discussed below, structural analysis of the genes that encode these precursor proteins has provided insight into the evolution of the egg laying behavior.

Discussion

The ELH gene family encodes a collection of related, yet functionally distinct molecules that cooperate to generate the physiological and neuronal activity which governs the egg laying behavior. The nucleotide sequence of these genes reveals that differences in the ELH protein precursors are a direct result of nucleotide substitutions, insertions and/or deletions in the genomic DNA (Scheller et al. 1982, 1983). The sequence data further suggest that these precursors descend from a single primordial protein encoded by a core nucleotide sequence.

Evolution of the ELH gene family from this ancestral nucleotide sequence may have occurred in several stages. First, the primordial sequence was amplified within the transcription unit to create a single gene encoding several peptides: namely, a polyprotein. Then, the entire transcription unit was duplicated and the new copy inserted elsewhere in the genome. Subsequent unequal cross-over events and/or further duplications generated a family of at least ten closely related genes. Finally, mechanisms of differential expression of the multiple ELH genes were established. Evidence for these processes and their significance for the evolution of the egg laying behavior are addressed individually below.

The Egg Laying Peptides Are Encoded in a Polyprotein Precursor

A polyprotein precursor is employed in the synthesis of many neuroactive peptides including calcitonin, arginine vasopressin and neurophysin II, ACTH and MSH, the dynorphins, enkephalins, and the endorphins (Herbert and Uhler 1982). Internal duplications are evident in most of these polyproteins. The most striking example is proenkephalin which encodes four related opioid peptides, one of which (Met-enkephalin) is present four times in the precursor. Nucleotide sequence analysis of the proenkephalin gene and of the gene encoding the endorphin-containing precursor pro-opiomelanocortin, indicate that they too probably arose from multiple duplications of a core nucleotide sequence. Similarly, the homology which exists between the different peptides encoded within the ELH gene suggests that internal duplications established this polyprotein (Scheller et al. 1983). The duplicated sequences then diverged so that the polyprotein now contains several functionally distinct peptides (see Fig. 1). For example, three of the ELH gene products (ELH and the α and β bag cell factors) that have sequences in common have been identified in vivo and shown to have distinct neurotransmitter activities.

Several aspects of the polyprotein structure are advantageous, both physiologically and evolutionarily. The polyprotein allows for coordinate regulation of the biosynthesis and, perhaps, release of several different peptide sequences. With respect to egg laying behavior

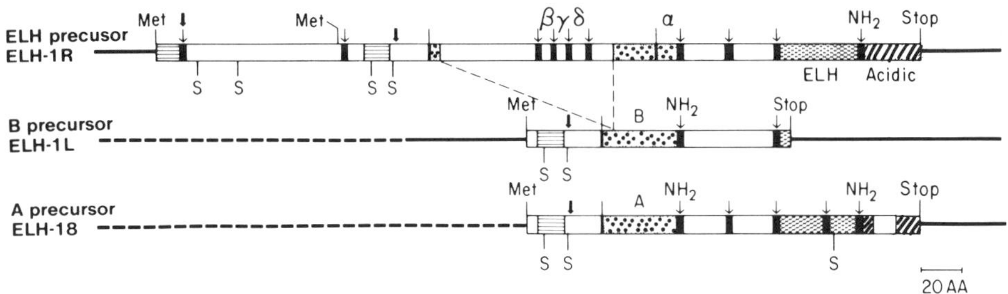

Fig. 1. Comparison of the protein precursors containing the ELH, A and B peptides. Coding regions are derived from the only in-phase amino acid reading frames that match in vitro translation molecular weights. Each of the three proteins is initiated by a methionine followed by a hydrophobic region (///). An S below the coding region indicates the location of a cysteine residue. Thick arrows represent the putative site of cleavage of the signal sequence (↓). Lines above the sequence represent potential cleavages at single arginine residues (|), while arrows represent potential or known cleavages at dibasic, tribasic or tetrabasic residues. If carboxyterminal amidation is believed to occur, an NH_2 appears above the arrow. The A/B peptide homology is represented by stippled boxes. The acidic peptide homology is represented by diagonal hatching enclosed in boxes. Solid lines symbolize sequenced noncoding regions, and dotted lines depict regions not sequenced. (From Scheller et al. 1983.).

this ensures coordination of the behavioral and physiological components of the response and may be necessary for the proper functioning of the small peptide transmitters, some of which are quite labile. In addition, different combinations of peptides might be generated by simply altering the processing of the polyprotein precursor. For example, the pro-opiomelanocortin precursor is processed differently in the anterior and intermediate lobes of the rat pituitary to generate different peptides from a single precursor (Herbert and Uhler 1982). This additional, pleiotropic diversity expands the phenotypic effects of a single gene. Alternate processing schemes could also be used in different tissues or at different times in development.

At the time of its creation by amplification of the ancestral sequence, the polyprotein gene may have been maintained in the population because it provided the organism with additional genetic diversity or because it provided for increased synthesis of the ancestral product. The latter would maintain the functional equivalence of the peptides while the former would allow divergence. The peptides within the ELH polyprotein have diverged so that each precursor now generates several functionally distinct peptides.

The polyprotein organization insures coordinate transcription, translation and, probably, release (by exocytosis) of the entire peptide ensemble. Consequently, any stimulus which evokes the expression of the selected sequence automatically elicits the other sequences. Therefore the additional sequences would be selected for or against based on their ability to enhance the organism's response to the original stimulus. In this way the polyprotein might allow the nervous system to elaborate several new peptides in response to a particular stimulus, thereby promoting the evolution of a more complex behavior.

The alternative to the polyprotein, as a mechanism for the evolution of a collaborative set of peptides, is many separate genes, each encoding a different peptide. If we invoke the same selective pressures which might have acted on the polyprotein, we would suppose that after the gene family was established in the population the "extra" copies of the gene were free to diverge. Each gene would have its own regulatory sequences and these could change as well. However, although new peptides might be created they might no longer be expressed in the original cell type, or at the same time, or at all. Consequently, the new genes might be subject to conflicting selective pressures, making the evolution of a collaborative set of gene products more difficult than is possible with the polyprotein.

The Polyproteins Comprise a Small Gene Family

At least three distinct precursor polypeptides are encoded by the ELH gene family. Restriction endonuclease digestion maps of the cloned DNA sequences encoding these proteins are present in Figure 2. The ELH and B peptide genes occur together in a large inverted repeat which may be reminiscent of their common origin while the A peptide genes occur on separate clones. The inverted repeat structure of the clone containing the ELH and B peptide genes indicates homology between these genes. Similarly, because the A and ELH peptide genes are homologous, their complementary strands can hydrogen bond to form a stable inter-

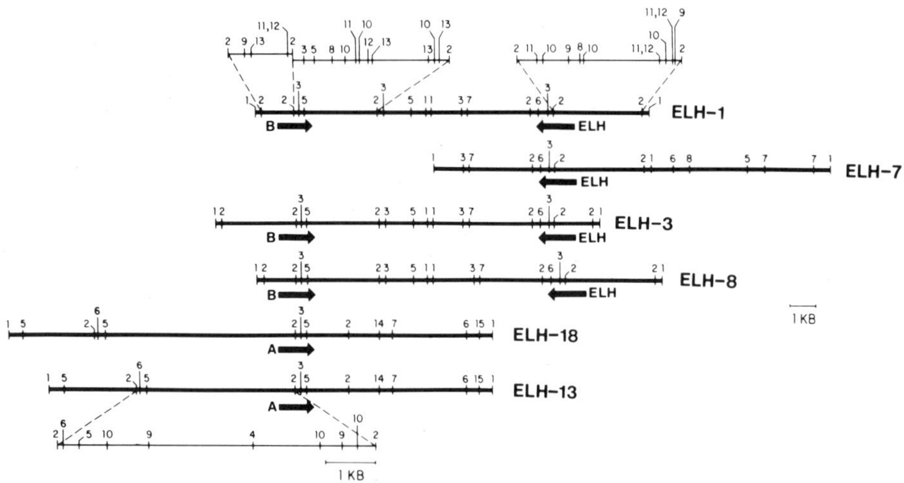

Fig. 2. Restriction enzyme maps of ELH recombinant clones. Sperm DNA isolated from an individual *Aplysia* was used to construct a library of recombinant phage in the lambda cloning vector Charon 4 (Maniatis et al. 1978). High molecular weight DNA was partially digested with the restriction enzyme *Eco* RI and ligated to the cloning vector. Recombinant clones that bear sequences present in the messenger RNA population of the bag cells and the atrial gland but not the hepatopancreas were selected and shown to contain ELH and/or A/B peptide genes.

Restriction maps were determined for the isolated segments of the Aplysia genome by a combination of single, partial, and double restriction enzyme digests of the entire recombinant clone or isolated fragments. The arrows indicate the position of mRNA homologous sequences and point in the direction of transcription as determined from DNA sequencing. (1) *Eco* RI; (2) *Pst* I; (3) *Xho* I; (4) *Stu* I; (5) *Pvu* II; (6) *Hind* III; (7) *Bgl* I; (8) *Xba* I; (9) *Ava* II; (10) *Hinc* II; (11) *Hae* II; (12) *Hha* I; (13) *Hpa* II; (14) *Bam* HI; (15) *Sal* I. (From Scheller et al. 1983.)

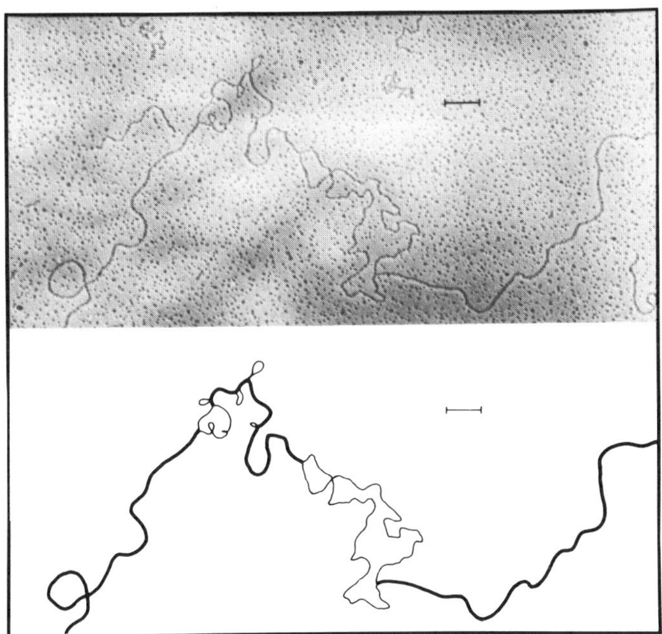

Fig. 3. Heteroduplex formed by λ ELH-7 and λ ELH-18. (DNA was prepared from the two recombinant phage, λ ELH-7 and λ ELH-18.) Equimolar amounts were mixed, denatured in .1M NaOH, neutralized, and allowed to reanneal in 50% formamide at room temperature. The Kleinschmidt technique was used to spread the reaction mixture for visualization in the electron microscope. Thick lines correspond to double-stranded DNA and thin lines to single-stranded DNA. The long stretches of double-stranded DNA at the ends of the hybrid molecule are formed by the homologous arms of the cloning vector. Much of the *Aplysia* DNA remains single stranded yet there is a region of homology which corresponds to the genes. The single-stranded loops in this region are caused by insertions and deletions in the genes. The bar represents 1 kilobase.

molecular hybrid (Fig. 3). DNA sequence analysis confirms that, although these genes are functionally distinct and are expressed in different tissues, they are greater than 90% homologous. The differences among the polyproteins are attributable to a few critical changes (transitions, insertions, and deletions) within the coding sequences which cause a functional change in the proteins produced. For example, stop codons, amidation signals and potential proteolytic cleavage sites have been created or lost by single nucleotide changes (see Fig. 1). Few of the nucleotide changes are silent, suggesting that the non-silent mutations have been selected for and fixed in the population while the silent ones have been lost.

The Specificity of ELH Gene Family Expression

Since many of the structural proteins found in diverse species are similar, it is likely that much of evolution has occurred through the reorganization of genomic regulatory elements. Such processes could account for the tremendous morphological diversity of the biosphere as well as behavioral diversity. The ELH genes are expressed in the bag cells and the A and B peptide genes are expressed in the atrial gland indicating that divergence has occurred in the regulation of expression, as well as the coding sequences, of the individual genes. Two general types of models have been proposed to describe such processes. First it is possible that, following duplication, one of the genes is translocated to a different chromosomal site. Regulatory elements of an unknown nature would then cause the "chromosomal domain" to be in an appropriate configuration for transcription in one tissue but not another. This is unlikely to be the cause of tissue-specific expression in the ELH gene family because the B peptide and ELH genes are found on the same recombinant clone. This linkage makes it improbable that the genes are in separate domains.

The second type of model proposes that differences in the DNA sequence near or within genes are responsible for transcriptional specificity. Some of the various differences noted among the member genes could act in such a fashion. Single base changes which have occurred in the 5′ untranslated region may play a role in these processes. It is possible that the gene-specific insertions (or deletions) direct the tissue-specific gene expression (see Fig. 3). One of the consequences of these insertions and deletions is that each gene is flanked by a different arrangement of interspersed repetitive DNA elements.

While their function remains obscure, repetitive DNA sequences are ubiquitous in the animal kingdom and, although not translated into protein, they are transcribed in a stage- and tissue-specific manner (Scheller et al. 1981). These observations and others have prompted Britten and Davidson to propose a regulatory role for repetitive DNA sequences whereby they provide a communication network within the genome which allows for coordinate expression of unlinked genes in a tissue- and stage-specific manner (Davidson and Britten 1979).

Some repetitive DNA sequence families are shared by several species, yet each species has unique repeat families. All the members of a repeat family are thought to be generated suddenly, in a single (saltatory) replication event. Individual repeats may then be gradually dispersed throughout the genome until they enter a region that is expressed where they may be selected for or against depending on the nature of the integrative relationship they establish. The large size of many repeat families (10^6 members per genome) suggests that such tests of new regulatory relationships might occur relatively frequently and implies that the generations of unexpressed repeats may be selected for because of the evolutionary potential they provide for the organism.

In *Aplysia*, 50% of the total DNA is repetitive. Ten percent of this is interspersed with single copy DNA with an average spacing of 1000 nucleotides (Angerer et al. 1975). This pattern of sequence organization called short period interspersion, is characteristic of most organisms studied. Within the DNA fragment which encodes the ELH and B peptides (ELH-1, Fig. 2), there are three short sequences (x, y, and z; 200-400 nucleotides) which are repeated several times (Fig. 4). In addition, several fragments within this clone are highly repeated in the genome (unpublished data) and are likely, although not proven, to correspond to the internal repeats depicted in Figure 4. Thus, the tissue-specific expression of members of the ELH gene family may have arisen by their incorporation into the appropriate repeat network and consequent linkage with other genes expressed in that cell type.

Once established, the tissue-specific expression of the genes allowed further divergence of the coding sequences. That is to say that mutation in the two sets of genes (atrial gland and bag cell) would be tested out in different microenvironments and therefore subjected to slightly different selective pressures (i.e., different components of the total selective pressure placed on the organism). Thus, the three gene types have diverged, possibly to satisfy the functional requirements of the tissue in which they are expressed.

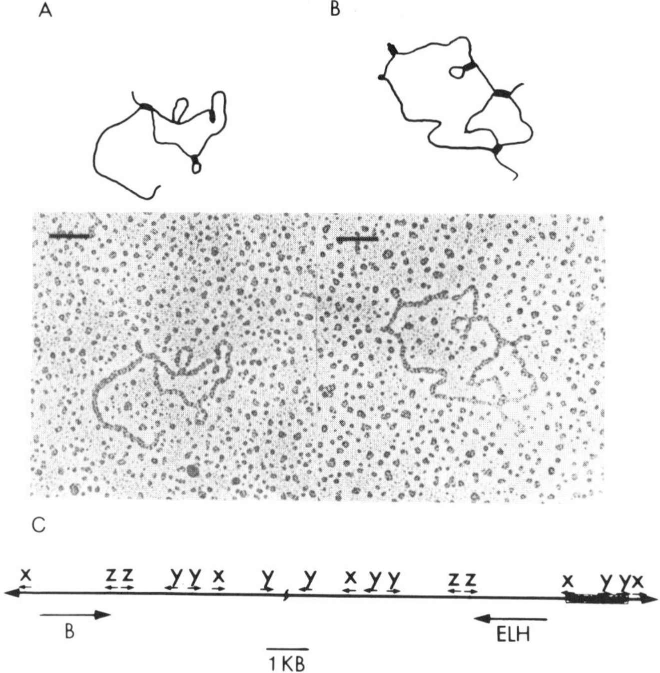

Fig. 4. Repetitive DNA sequences in λ ELH-1. The two large *Eco* RI fragments of λ ELH-1 (see Fig. 2) were isolated, denatured and briefly renatured such that only intramolecular structures were formed. The structure formed by the 6.5 kilobase *Eco* RI fragment of λ ELH-1 (from the left side) is shown in panel A and that seen for the 8.5 kb fragment is shown in panel B. The bar represents 500 nucelotides. Analysis of these structures revealed the presence, position and orientation of three repetitive sequence families (x, y and z) in λ ELH-1 as depicted in panel C. The small arrows indicated the position and orientation of the repetitive sequence elements within this fragment of *Aplysia* DNA. The arrows at the ends of the λ ELH-1 fragment indicate that the entire fragment is an inverted repeat. Beneath this, the position and direction of transcription of the ELH and B peptide genes is indicated. The shaded region is a 1.7 kb element that is present at the 5′ end of the ELH gene but not in the corresponding region of the B peptide gene.

In summary, structural analysis of the gene family that encodes the egg laying peptides suggest that this family descends from a single primordial peptide sequence. The evolution of the ELH polyproteins presumably elicited a corresponding evolution and elaboration of the egg laying behavior. Control of this behavior by a polyprotein family confers numerous selective advantages on the organism. The duplication of a highly selected structural gene to form a polyprotein allows the organism to generate and test new molecules while maintaining the old. The gene family like the polyprotein, allows the organism to test out new gene products and, in addition, the gene family allows for the evolution of new regulatory schemes. Thus the polyprotein gene structure and the gene family organization serve as a readily available source of evolutionary potential which can be utilized to modify behavior in response to natural selection.

Literature Cited

Angerer, R. C., E. H. Davidson, R. J. Britten. 1975. DNA sequence organization in the mollusc, *Aplysia californica*. Cell 6: 29.

Arch, S., J. Smock, R. Gurvis, C. McArthy. 1978. Atrial gland induction of the egg-laying response in *Aplysia californica*. J. Comp. Physiol. 128: 67.

Davidson, E. H., and R. J. Britten. 1979. Regulation of gene expression: Possible role of repetitive sequences. Science 204: 1052.

Herbert, E., and M. Uhler. 1982. Biosynthesis of polyprotein precursors to regulatory peptides. Cell 30: 1.

Kandel, E. R. 1976. Cellular Basis of Behavior. W. H. Freeman and Co., San Francisco.

Kupferman, I. 1972. Studies on the neurosecretory control of egg laying in *Aplysia*. Amer. Zool. 12: 513.

Lorenz, K. Z. 1958. The evolution of behavior. Sci. Amer. 199: 67.

Maniatis, T., R. C. Hardison, E. Lacey, J. Laver, C. O'Connel, D. Quon, G. K. Sim, and A. Efstradiatis. 1978. The isolation of structural genes from libraries of eucaryotic DNA. Cell 15: 687.

McAllister, L. B., R. H. Scheller, E. R. Kandel, R. Axel. 1983. In situ hybridization as a marker for studying the development and location of individual neurons. Cell (in press).

Rothman, B. S., G. Weir, F. E. Dudek. 1983. Direct action of egg laying hormone on ovotestis of Aplysia. Tissue and Cell (in press).

Scheller, R. H., D. M. Anderson, J. W. Posakony, L. B. McAllister, R. J. Britten, E. H. Davidson. 1981. Repetitive sequences of the sea urchin genome II. Subfamily structure and evolutionary conservation. J. Mol. Biol. 149: 15.

Scheller, R. H., J. F. Jackson, L. B. McAllister, J. H. Schwartz, E. R. Kandel, and R. Axel. 1982. A family of genes that codes for ELH, a neuropeptide eliciting a stereotyped pattern of behavior in Aplysia. Cell 28: 707.

Scheller, R. H., J. F. Jackson, L. B. McAllister, B. S. Rothman, E. Mayeri, R. Axel. 1983. A single gene encodes multiple neuropeptides mediating a stereotyped behavior. Cell 32: 7.

Strumwasser, F., L. K. Kacsmarek, A. Y. Chiu, E. Heller, K. R. Jennings, D. P. Viele. 1980. Peptides controlling behavior in *Aplysia. In* F. E. Bloom, ed. Society of General Physiologist Series 35. Peptides: Integrators of Cell and Tissue Function. Raven Press, New York.

The Potential for Genetic Manipulation of the Redbanded Leafroller Moth Sex Pheromone Blend

Wendell L. Roelofs

Jia-Wei Du

Charles Linn

Thomas J. Glover

Louis B. Bjostad

Department of Entomology
New York State Agricultural Experiment Station
Geneva, New York 14466

Introduction

Speciation ultimately must involve "establishment of intrinsic barriers to gene flow between closely related populations by development of reproductive isolating mechanisms" (Bush 1975). In many cases this could involve specific mate recognition systems (Paterson 1978). A discriminant function analysis of the design features of various specific mate recognition signals was used by Lambert and Levey (1979) who found that signals and receivers involved in conspecific interactions have narrow variances (coefficient of variation of 10%). They suggested that one important consequence of narrow variance systems is that selection for efficiency in some new habitat may cause the original tuned receiver and the newly adjusted one to be non-overlapping with only a small change in mean values. Speciation could result from these necessary adjustments to the communication system under the new conditions. However, the genetics of these tightly controlled mating communication systems and their propensity for change in the evolution of new species are poorly understood. These communication systems usually are rigidly canalized via natural selection and the phenotypes often are complex and difficult to quantify.

Consequently, the few studies of the genetics of insect mating communication systems reported to date do not utilize quantitative genetics techniques. Most papers that describe the control of insect communication implicate major genes. Various studies utilizing *Drosophila* indicate both autosomal and sex-linked genetic influences (see for example Tan 1946, Ehrman 1961, Ewing 1969, and Averhoff and Richardson 1976). In sulfur butterflies (*Colias* spp.), Grula and Taylor (1979, 1980) found that some individual pheromone components are under autosomal control and that others are under sex-linked control. Pheromone blend ratio in European corn borer moths (*Ostrinia nubilalis* [Hubner]) was reported to be monofactorially and autosomally controlled (Klun and Maini 1979).

Until very recently, the investigation of lepidopteran communication system genetics has been limited to interspecific crosses or to insects that exhibit a polymorphism in pheromone blend ratio as do the European corn borers. The difficulties in rearing and evaluating individual moths have precluded most extensive selection experiments on Lepidoptera in general and their communication systems in particular. Recently developed technology now allows quantitative genetic methods to be applied to certain moth species. This report details some of the preliminary work done on the redbanded leafroller, *Argyrotaenia velutinana* (Walker)(Lepidoptera: Tortricidae).

Reproductive Isolation in Leafroller Species

The redbanded leafroller moth is one of an unusually high number of sympatric species in the Tortricidae that feed on apple in eastern North America (Chapman and Lienk 1971). While these species may overwinter in many different phenological stages, from eggs to fully formed adults within the pupal case, they are well-synchronized with their plant host in that the penultimate and final instars occur in late May and early June when trees achieve maximum vegetative growth. Since the adult flights of many of these species co-occur, isolating mechanisms such as separation by geography, habitat, or seasonal rhythms, which are important as premating mechanisms in the isolation of some other species, appear not to be involved as mechanisms among the apple-feeding leafrollers in eastern North America.

The information above suggests that reproductive isolation among these apple-feeding leafrollers is effected mainly through specific mating systems. Although interspecific differences in communication can be determined along chemical and temporal axes, which are viewed as niche dimensions (Greenfield and Karandinos 1979), studies of diurnal mating rhythms showed that mating activity of at least twelve sympatric leafroller species occurs within three hours after sunset (Comeau 1971, Comeau et al. 1976). The periodicity of both male and female activity was influenced by temperature (Carde et al. 1975), but slight species variations in these shifts would still leave mating rhythms as a poor mechanism for reproductive isolation among these species. This implies that the sex pheromone signals are the most salient premating isolating mechanism for these leafroller species.

Identification of the sex pheromone blend of several leafroller species (Roelofs and Brown 1982) showed that many of the species utilize the same pheromone components, but in unique combinations of ratios. A common combination is a precise ratio of (*Z*)-11-tetradecenyl acetate (Z11-14:OAc) and (*E*)-11-tetradecenyl acetate (E11-14:OAc), including 97:3, 91:9, 60:40, 50:50, 33:67, 24:76, 15:85, and 12:88. Most of these species also utilize functional group analogs or geometric isomers as additional components. These data suggest that for these species reproductive isolation is maintained through specific mating systems using specific pheromone blends.

Research (Bjostad and Roelofs 1981, Bjostad et al. 1981, Wolf et al. 1981) on the biosynthesis of leafroller pheromones has shown that specific isomer ratios can be controlled in the sequence of steps from $\triangle 11$ 14-carbon acyl intermediates to a precise ratio of Z- and E11-14:OAc's. The usual acid intermediates are biosynthesized in the female pheromone glands via the following sequence: (a) hexadecanoic acid from acetate by way of the fatty acid cycle; (b) chain-shortening of the 16-carbon acid to 14-carbon acid by limited β-oxidation; and (c) $\triangle 11$ desaturation to give a mixture of Z11 and E11 14-carbon acids. In several leafroller species analyzed, this acid Z/E ratio is approximately 1:1 ratio, but is reduced in the last enzymatic sequence to Z and E pheromone acetate ratios that are specific for each species, such as 100:0, 97:3, 91:9, etc. These results indicate that genetic control of the pheromone blends could be keyed on the last reduction sequence, not only for the acetate isomer ratios, but also in yielding reduction products used as additional pheromone components, such as aldehydes and alcohols.

Variability of Redbanded Leafroller Moth Pheromone

The sex pheromone of the redbanded leafroller moth has been identified (Roelofs et al. 1975) as a 91:9 mixture of Z and E11-14:OAc's along with a third component, dodecyl acetate (12:OAc). Analyses (Miller and Roelofs 1980) of individual female pheromone glands from field and laboratory populations showed that each population had E/Z ratios that exhibited little variability. All individuals (ca. 600) had ratios between 4-15% E isomer, with means and standard deviations of 9.1 ±1.8 and 7.0 ±1.4% for the field and laboratory populations, respectively.

A different wild population was brought into the laboratory and several years later became the basis for a project to determine the strength of canalization of this communication system. The objective was to select and mate females producing higher than normal E/Z pheromone ratios. Repeated selection of the progeny will be carried out in an attempt eventually to generate a population utilizing 20% E isomer instead of the normal 9%. Female moths from the new laboratory culture were analyzed to obtain a population profile (Fig. 1) of the E/Z ratios. The range was found to be 4-12% E with a mean of 8.6 ±1.4% E. The analyses were carried out using gas liquid chromatography (GLC) on a 40 m Carbowax 20 M capillary column and were obtained with greater than ±0.4% reproducibility with 5-10 ng of pheromone. Female redbanded leafroller moth pheromone glands were found to contain 107 ±58 ng of these isomers (Miller and Roelofs 1980). Airborne collections of individual female effluvium produced GLC tracings that looked very similar to those from gland extracts.

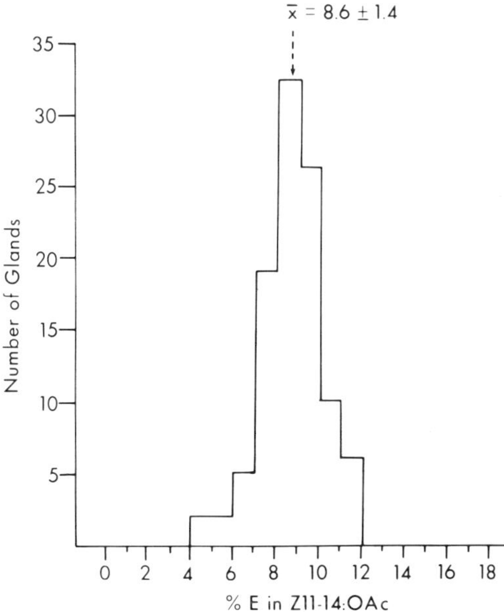

Fig. 1. Gas-liquid chromatographic (GLC) analyses of E and Z11-14:OAc pheromone components.

Progeny from Analyzed Female Moths

Techniques were developed for removing a portion of a female redbanded leafroller moth pheromone gland for GLC analysis. (Unfortunately, there was not enough pheromone for analysis in female moths that had mated and laid eggs, and so the gland had to be analyzed prior to mating.) In brief, a female moth was positioned head first in a piece of rubber tubing so that slight pressure on the tubing would cause ovipositor eversion. A piece of exposed pheromone gland was picked off with fine forceps and placed in 10-20 μl of Skelly B (a hydrocarbon fraction, bp 60°C, consisting mostly of *n*-hexane; 99% *n*-hexane may be substituted) in a small pointed tube and allowed to soak for 30 min. Pheromone component ratios were determined by capillary GLC analysis utilizing the 40 m Carbowax 20 M column. A 6 μl aliquot was injected in the splitless mode with an 80-180°C temperature program @ 10°C/min. The analyzed female moths with some residual pheromone gland were placed singly in small plastic bags with males for mating. Eggs laid on the bag were cut out and held in a shell vial under high humidity. Newly hatched larvae were placed on cups of media and reared under the same conditions as the laboratory colony. Removal of the distal portion of the female pheromone gland has, in some cases, a detrimental effect on mating and egg hatch, but generally the technique proved successful. Single pair matings normally are successful 80% of the time, whereas matings of single analyzed females with single males had a success of ca. 60%. The surgical procedure, therefore, made possible phenotypic evaluation of females prior to mating.

While the E/Z pheromone ratio seems well canalized and exhibits only moderate variability, the population does exhibit variability that is apparently genetic in origin. A summary of E/Z pheromone ratios for 30 sets of progeny produced by single females is presented in Table 1. Analysis of variance was performed on these data. The within family variance component (error) was 1.63 and the between family variance component was 1.14. Variance components, along with the intraclass correlation (t= .41), were calculated according to standard statistical methods (Snedecor and Cochran 1980).

Since these progeny represent full sib families, the heritability of this trait is assumed to be less than or equal to twice the intraclass correlation coefficient (Falconer 1981). Because only females exhibit this phenotype, one might assume that the realized heritability of blend ratio in directional selection experiments will be approximately equal to the intraclass correlation (t= .41). A graphic example of the correlation between female blend ratio and the distribution of blend ratio in her daughters is shown by superimposing typical progenies on an overall population profile (Fig. 2). A few of the progenies exhibited interesting bimodality, possibly indicating segregation of a major gene (Fig. 3).

The above data demonstrate the potential for significant change in the blend ratio utilizing standard directional selection. If males could be evaluated somehow for this phenotype or a characteristic closely linked to it, a higher realized heritability might emerge upon selection.

Selected Males

Techniques for selecting male moths already were available in our laboratory. In previous studies (Baker et al. 1981, Linn and Roelofs 1981) we defined in a wind tunnel the behavioral responses of male Oriental fruit moths, *Grapholitha molesta* (Busck), to an array of pheromone component blends and dosages. In this case the male moths undergo an entire attraction and courtship behavioral sequence to a 3-30 μg dosage of the natural pheromone blend consisting of a 94:6 ratio of Z/E 8-12:OAc's containing 3-30% of the corresponding alcohol, Z8-12:OH. Very few of the males flew upwind and landed with 20% E odor sources. A similar study with redbanded leafroller males from the laboratory culture showed that the

Table 1. Progenies from individual unselected RBLR females mated to unselected laboratory males.

Female Parent % E	Mean % E in Female Progeny	# of Females in Progeny
4.1	6.7 ± 1.0	18
5.9	8.0 ± 1.1	5
5.9	7.2 ± 1.0	14
5.9	8.3 ± 1.0	5
6.2	8.2 ± 1.2	20
6.8	8.0 ± 0.8	26
6.9	8.3 ± 1.0	6
6.9	9.6 ± 0.8	11
7.1	8.8 ± 1.0	11
7.1	7.5 ± 1.0	9
7.5	8.4 ± 0.7	4
7.9	7.6 ± 0.6	14
7.9	8.1 ± 0.9	7
8.1	8.2 ± 1.0	23
8.9	7.2 ± 0.9	11
9.5	9.6 ± 1.2	8
9.5	9.2 ± 1.9	24
10.0	9.0 ± 1.5	4
10.0	10.3 ± 2.0	21
10.4	10.3 ± 1.5	9
10.7	10.9 ± 1.9	5
10.8	9.1 ± 1.8	18
10.9	9.6 ± 2.7	10
11.0	9.4 ± 1.1	4
11.0	10.2 ± 1.6	7
11.0	8.3 ± 1.7	8
11.0	10.2 ± 1.6	7
11.1	11.1 ± 0.8	6
11.2	8.6 ± 0.7	13
11.9	9.5 ± 1.5	18
$\bar{x}$ = 8.8 ± 2.1		Total = 346

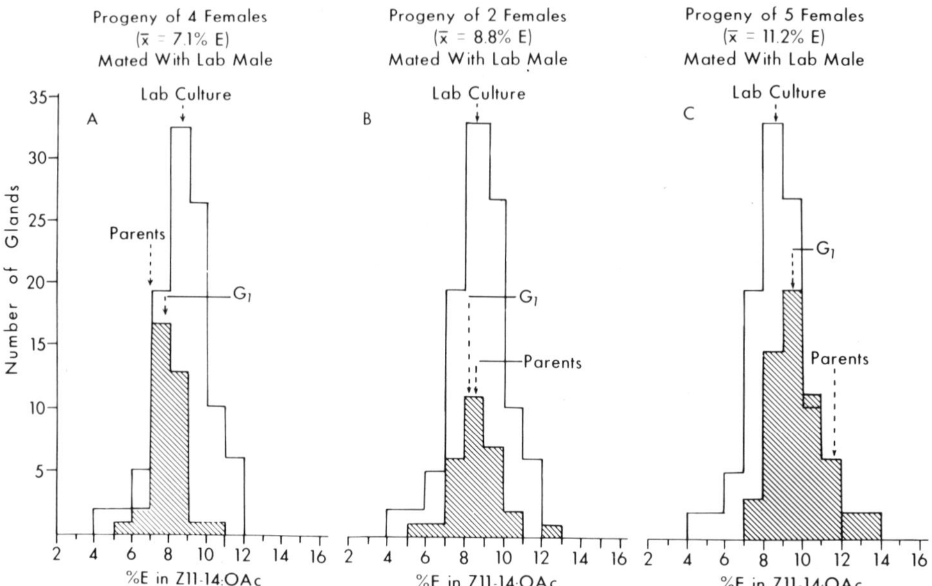

Fig. 2. GLC analyses of E/Z11-14OAc's in progeny of females selected with a low, average, and high E/Z ratio. G_1 = the mean of the resulting progeny.

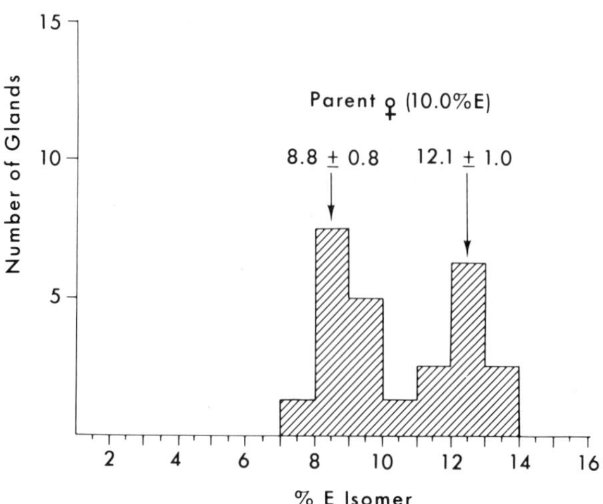

Fig. 3. GLC analyses of the progeny from a female redbanded leafroller moth containing 10% E11-14:OAc in the pheromone blend.

best responses were obtained with the three-component blend of 12:OAc/△11-14:OAc's (3:2) containing the natural 92:8 Z/E ratio of the 14-carbon acetates. Using a 300 μg dosage of this blend, 80% of the males completed the behavioral sequences with an 8% E blend, 40% with a 15% E blend, and 10% with a 20% E blend. The 10% males responding to the 20% E blend was very consistent for over 500 male moths tested throughout a 6-month period. The remaining 90% males would not even activate with this lure containing a high amount of E isomer, although the males responding to the 20% E source also would give full responses when retested with an 8% E blend. This was a phenomenon not seen in the Oriental fruit moth studies. It was hypothesized that because male behavior and female pheromone blend are part of the same communication system, they may be controlled by the same underlying genetic system. If this is the case, males responding to 20% E blends may carry genes enabling their daughters to produce high percent E blends. Several experiments have been designed to elucidate this point.

Directional Selection for Increased Percent E Production

The laboratory population was surveyed again using single female analyses. The distribution of percent E for 366 females had a mean of 8.0% with a range of 4-13% (Table 2). From this baseline population seven females with an average of 10.7% E were selected and mated with males which responded to 20% E in flight tunnel tests. The results of these matings are presented in Table 2. At a later date, four females with an average of 11.6% E were selected from the colony and mated to males labelled "normal" because they responded to 8% E but not to 20% E in flight tunnel tests. The offspring from these matings are also summarized in Table 2. Assuming the mean of the baseline population to be 8.0%, the realized heritability in the offspring of selected females mated with "normal" males was .42. This is remarkably similar to the heritability predicted by the intraclass correlation of the data in Table 1. The selected females mated with high-responding (HR) males produced a greater gain than the previous cross. The realized heritability of .85 here indicates that the high responding males are contributing to the increased percent E production in their daughters. Flight tunnel tests of the 88 males resulting from the selected females X HR male cross showed that 90% responded to the 20% E blend, compared to only 10% responding in the parent laboratory culture.

Initially, the male contribution appeared to be monofactorial, but later crosses indicate that it is considerably more complex. High percent E-producing female progeny from the selected female X HR male crosses were mated with high-responding males from various sources (Table 3). Matings to laboratory colony HR males and to HR males from other selected broods seemed to have little positive effect and, in fact, considerable reversion to the population mean was noted. Only matings of high percent E-producing females with their brothers seemed to hold the selection gain from the previous generation. These results are difficult to interpret and require further experimentation to evaluate the precise nature of the behavioral phenotype exhibited by the high-responding males. The cross of 8% E-producing females and high-responding males, for example, will yield valuable insights into the nature of the male contribution in this population.

Generally, the preliminary results presented in this report indicate that directional selection has considerable potential for producing populations with unique pheromone blends and corresponding male behavioral responses. Such populations will be invaluable as genetic resources for investigation of the canalization, adaptiveness, and evolution of insect sex pheromone communication systems.

Table 2. Distribution of the percent E, Z-11-14:OAc isomer in selected RBLR populations.

E%	Baseline Popula. # Females	High ♀ x "normal" ♂ # Females	High ♀ x HR ♂ # Females
4	1	—	—
5	7	—	—
6	39	—	—
7	99	1	1
8	103	9	9
9	69	8	22
10	34	5	51
11	9	3	34
12	3	2	10
13	2	2	3
14	—	—	4
15	—	—	1
16	—	—	—
	Total = 366	30	135
	Mean = 8.0 ± 1.4	Mean = 9.5 ± 1.6	Mean = 10.3 ± 1.4
		Parent ♀'s Mean = 11.6 N = 4 realized heritability = .42	Parent ♀'s Mean = 10.7 N = 7 realized heritability = .85

Table 3. Results of second generation selection crossing selected females with high-responding (HR) males from various sources.

Female %E	Source of HR male	Resulting female progeny mean + S.D. (N)
12.1	Lab colony	10.8 ± 0.9 (14)
12.1	” ”	9.3 ± 0.9 (61)
10.7	” ”	9.7 ± 0.9 (28)
10.6	” ”	8.6 ± 1.2 (33)
10.2	” ”	8.9 ± 1.2 (28)
11.6	SLBNB[a]	7.4 ± 0.8 (60)
11.2	”	8.5 ± 0.9 (26)
11.0	Brother	11.1 ± 1.0 (37)
11.0	”	11.3 ± 0.9 (23)

[a]SLBNB = selected line but not brothers.

Acknowledgments

We thank K. Poole, F. Wadhams, M. Campbell and L. Child for rearing the thousands of insects.

Literature Cited

Averhoff, W. W. and R. H. Richardson. 1976. Multiple pheromone system controlling mating in *Drosophila melanogaster*, Proc. Natl. Acad. Sci. USA. 73: 591.

Baker, T. C., W. Meyer, and W. Roelofs. 1981. Sex pheromone dosage and blend specificity of response by Oriental fruit moth males. Entomol. exp. Appl. 30: 269.

Bjostad, L. B. and W. L. Roelofs. 1981. Sex pheromone biosynthesis from radiolabeled fatty acids in the redbanded leafroller moth. J. Biol. Chem. 256: 7936.

Bjostad, L. B., W. A. Wolf, and W. L. Roelofs. 1981. Total lipid analysis of the sex pheromone gland of the redbanded leafroller moth, *Argyrotaenia velutinana*, with reference to pheromone biosynthesis. Insect. Biochem. 11: 73.

Bush, G. L. 1975. Modes of animal speciation. Ann. Rev. Ecol. Syst. 6: 339.

Carde, R. T., A. Comeau, T. C. Baker, and W. L. Roelofs. 1975. Moth mating periodicity: temperature regulates the circadian gate. Experientia 31: 46.

Chapman, P. J. and S. E. Lienk. 1971. Tortricid fauna of apple in New York, N. Y. State Agr. Exp. Sta. Geneva, New York.

Comeau, A. 1971. Physiology of Sex Pheromone Attraction in Tortricidae and other Lepidoptera (Heterocera), Ph.D. Thesis. Cornell University. Ithaca, New York.

Comeau, A., R. T. Carde, and W. L. Roelofs. 1976. Relationship of ambient temperatures to diel periodicities of sex attraction in six species of lepidoptera. Can. Entomol. 108: 415.

Ehrman, L. 1961. The genetics of sexual isolation in *Drosophila paulistorum*. Genetics 46: 1025.

Ewing, A. W. 1969. The genetic basis of sound production in *Drosophila pseudoobscura* and *D. persimillis*. Anim. Behav. 17: 555.

Falconer, D. S. 1981. Introduction to Quantitative Genetics. Longman Inc., New York.

Greenfield, M. D. and M. G. Karandinos. 1979. Resource partitioning of the sex communication channel in clearwing moths (Lepidoptera: Sesiidae) of Wisconsin. Ecol. Mono. 49: 403.

Grula, J. W. and O. R. Taylor, Jr. 1979. The inheritance of pheromone production in the sulphur butterflies *Colias eurytheme* and *C. philodice*. Heredity 42: 359.

Grula, J. W. and O. R. Taylor, Jr. 1980. The effect of X-chromosome inheritance on mate selection behavior in the sulfur butterflies *Colias eurytheme* and *C. philodice*. Evolution 34: 688.

Klun, J. A. and S. Maini. 1979. Genetic basis of an insect chemical communication system: the European corn borer. Environ. Entomol. 8: 423.

Lambert, D. M. and B. Levey. 1979. The use of discriminant function analysis to investigate the design features of specific male recognition systems. Proc. Zool. Soc. S. Afr. Symp. on Animal Communications. Capetown.

Linn, C. E. and W. L. Roelofs. 1981. Modification of sex pheromone blend discrimination in male Oriental fruit moths by pre-exposure to (*E*)-8-dodecenyl acetate. Physiol. Entomol. 6: 421.

Miller, J. R. and W. L. Roelofs. 1980. Individual variation in sex pheromone component ratios in two populations of the redbanded leafroller moth, *Argyrotaenia velutinana*. Environ. Entomol. 9: 359.

Paterson, H. E. H. 1978. More evidence against speciation by reinforcement. S. Afr. J. Sci. 74: 369.

Roelofs, W. L. and R. L. Brown. 1982. Pheromones and evolutionary relationships of Tortricidae. Ann. Rev. Ecol. Syst. 13: 395.

Roelofs, W. L., A. Hill, and R. Carde. 1975. Sex pheromone components of the redbanded leafroller, *Argyrotaenia velutinana* (Lepidoptera: Tortricidae) J. Chem. Ecol. 1: 83.

Snedecor, G. W. and W. G. Cochran. 1980. Statistical Methods. The Iowa State University Press, Ames.

Tan, C. C. 1946. Genetics of sexual isolation between *Drosophila pseudoobscura* and *Drosophila persimilis*. Genetics 31: 558.

Wolf, W. A., L. B. Bjostad, and W. L. Roelofs. 1981. Biosynthesis of insect sex pheromones: fatty acid content of pheromone glands of several lepidoptera species. Environ. Entomol. 10: 943.

Generalizing Genetic Dissection of Behavior

Joseph P. Hegmann

Department of Zoology
University of Iowa
Iowa City, Iowa 52242

Introduction

Genetic dissection refers to the use of induced mutations to analyze physiological mechanisms of single behaviors. Defined in this way genetic dissection is a powerful methodological advance for behavioral research; one that promises eventual understanding of mechanisms of behavior from their molecular origins. However, the empirical logic of dissection, using genes that affect behavior as treatment variables to locate their physiological mediation, can be applied to advantage using other genetic methods in addition to mutation. This idea can be developed by referring to an early report that applied dissection to visual system function in *Drosophila* where Hotta and Benzer (1969) suggested altering elements of behavior by specific mutations. Later in the same paragraph they stated the rationale more generally as *"Changes in the genes controlling the development and function of the participating structures affect the resulting behaviour, and can be used, in principle, to dissect the system."* The thesis to be developed here is that generalizing the methods for producing changes in the genes will extend the nature of the system that can be dissected and the questions that can be addressed without altering the principle. Extension will allow analysis of multiple and naturally variable behaviors and it will open up questions of behavior's consequences or function for genetic dissection.

Behavior as a problem area and genetics as an analytical tool are quite beautifully matched. Behavior has invariant features and dependence on physiological systems that raise problems of mechanisms; genetics has methods for enhancing mutation rates to perturb naturally invariant biological systems for analysis. Behavior has variable features and responsiveness to environmental conditions that raise questions of its role in adaptation and evolution; genetics provides methods for amplifying and focusing selection, intensifying drift, or manipulating migration for analysis of biological variability and prediction of long term change. The link between these problems and methods reflects the fact that genes and behavior relate closely as biological phenomena. Behavior is expressed at the organismic level with immediate causes at the system level and consequences at the population level of biological organization. Genes act at the molecular level to influence function at the cellular (cf. Kung 1979, Wu and Ganetzky 1980) and systemic (Hegmann 1975, Batty 1978) levels. They may influence behavior to alter relative reproduction and, thereby, their own frequencies in populations (DeFries and McClearn 1970).

These ties between behavior and genetics link difficult problems to powerful analytical methods. The problems that face behavioral biologists (Tinbergen 1968, Alcock 1975, Wilson 1975) can be expressed generally as how behavior is generated, in the sense both of the immediate function and of the development of its physiological antecedents, and why behavior is expressed, what contribution it makes to organismic fitness now and may have made through evolutionary history. In terms chosen by Jacob (1976) the first of these are *Tomist or Reductionist* and the second, *Integrationist or Evolutionist* questions. Effective synthesis of information on behavior will require merging these reductionist and evolutionist methods and results, so an additional challenge facing behavioral biology is to find means to accomplish that merger. Genetic analytical methods amplify natural forces and they might provide the means if the basic logic of dissection can be extended to put behavioral analyses on a shared biological footing; that of locating the physiological elements genes affect in modifying expression of behavior.

Analytical and Biological Considerations

Generalizing dissection to the use of selection, migration, or genetic drift, in addition to mutation, requires considering the extent to which these other methods accomplish what is needed for the analyses-control of genes as experimental variables. When mutations are induced an alternative gene form is generated which didn't exist in the population prior to treatment with the mutagen, and the form is established with a relative gene frequency of 1 in the mutant stock. The latter point is critical. Single mutant animals don't permit dissection. The experimental power of dissection results from the fact that comparison of a mutant stock to the population from which it was derived is based on comparing animals that have only the mutant gene at the locus involved to those that don't have it. The key is to establish two (or more) groups distinguishable by their relative frequencies of genes with behavioral effects. Groups of animals with radically different gene frequencies can be generated using intensive artificial selection or inbreeding, though only mutation allows fixing one specific allele in one group that is absent in the other. In this sense opening dissection to the use of other genetic methods is formal generalization from the limiting case of an experimental method. Generalizing will weaken the genetic distinction between control and experimental groups. Equal differences in frequencies can be achieved only by using alternatives like inbreeding that alter gene frequencies at all loci. Equal focus on genes that affect behavioral antecedents can be achieved only by using alternative methods like selection that rarely produce total fixation in one group and loss in another.

Extending the methods used for dissection requires evaluating implications of the methods used to control genes as experimental variables. Genes induced by mutation are Model I experimental variables (Sokal and Rohlf 1981) because treatment differences are fixed and are independent of the distribution of similar sorts of effects in the species. That is, mutations represent direct manipulations of gene effects on behavior, not representative samples of the natural array of those effects that characterize Model II experimental variables. In contrast, gene differences imposed by intensive inbreeding or selective breeding are clearly Model II experimental variables. The genes fixed in lines are not repeatable, they sample the alleles available in an initial population. Likewise, differences among separate derivations of selected or inbred lines are not necessarily the same. So, while methods other than induced mutations can be used to treat genes as controlled experimental variables, appropriate adjustments must be made to interpret the results as Model I or Model II variables. Since mutations are Model I variables, results from mutational analyses can be interpreted as changes forced by the mutant gene. Results from analyses using selection, migration, or genetic drift have to be interpreted as differences associated with the array of gene effects sampled. These differences insert genetic components of variance as opposed to causing differences. When more than a single physiological effect is imposed by a mutation it means a simultaneous forced change and is called pleiotropy. The same result from other genetic methods has to be interpreted as simultaneous differences due to the gene differences sampled. These insert components of genetic covariance measurable as genetic correlation.

Extending genetic methods used for dissection of behavior has biological implications that also must be considered. Behavior is similar in form among members of the same species but frequently there is variety in the dynamics of execution of a behavior, in stimuli sufficient to elicit it, and in its context. Both fixed and variable aspects of behavior may depend on fixed and variable genes. Those aspects of behavior which depend on invariant genetic forms for their development can be subjected to genetic dissection only by inducing mutations. In principle, mutations could block the behavior by altering genetic form at any locus involved in development of its antecedents so long as the alteration didn't produce lethality. Eventually all relevant loci can be altered except those at which every mutation is lethal. The constraint that mutations must be nonlethal is the single constraint on dissection by mutation. It makes mutation ideal for analysis of mechanisms of behavior and makes the promise of understanding behavior from the molecular level real. On the other hand, the invulnerability of mutations to natural constraints from adaptation and development makes dissection by mutation sterile for the study of the physiological bases of behavioral variability that is due to genetic variability.

Mutations obscure variance and its causal basis but variation in behavior, its causes and regulation, can be opened to genetic dissection by using methods that allow control of genes as experimental variables subject to constraints imposed by adaptation and development; that is, as Model II experimental variables. So, while migration, selection, and genetic drift are weaker where mutation is of greatest value, they are the option of choice for dissections aimed at questions of the evolutionary significance of behavior. This is one advantage of extending genetic dissection to a broader definition.

"Soft" Genetic Dissection

Methodological generalization of dissection may do violence to the metaphor but it will extend the value of dissections substantially. The variety of organisms that can be subjected to analysis will be extended because some alternative methods for dissection require less prior knowledge of formal genetics. Some also require fewer organisms so each individual can be larger but requirements for short generation intervals will remain. The variety of behaviors open to analysis can increase. That would be a mixed blessing, simply complicating the choice of behaviors for analysis, except for the fact that associations among behaviors can become an integral part of dissections (since adaptive relations among behaviors won't be disrupted) and may provide a guide to choosing which behaviors can be studied profitably in which species.

To protect the metaphor but preserve the meaning of these extensions relative to mutational dissections of behavior it may be useful to refer to the extensions as *soft* genetic dissections. This will emphasize that "development and function of the participating structure" (Hotta and Benzer 1969) is the aim of genetic dissection and still acknowledge the novel power of mutations to override natural genic distributions and natural constraints on development and adaptation. It should also emphasize the need for both sorts of analyses in order to understand any behavior in any biological system: mutational dissections to unravel fundamental gene products necessary for development of antecedents to its invariant features and soft dissections to elucidate genic contributions to the physiological bases of its variation and covariation.

Application to the Cercal Sensory System

The complimentary nature of information from dissection by mutation and from soft genetic dissection can be illustrated using the escape or evasive behavior of orthopteran insects in response to low frequency cercal stimulation from air puffs or sound (Roeder 1948, Bullock and Horridge 1965). This behavior has been subjected to both sorts of analyses, but in two different species of crickets. The jump response to air puffs was cleverly exploited by Bentley (1975) to sort a small number of nonreactive individuals from among 40,000 mutagenized *Teleogryllus oceanicus*. The response is normally expressed following mechanoreception by directional filiform hairs and transmission of impulses through the cercal nerve to synapses on the medial giant interneuron in the 6th abdominal ganglion. Crickets that jumped, in the screening employed, were removed by vacuum. Those that remained included individuals with abnormal filiform hairs. One male was shown by subsequent crossing to be expressing a sex-linked gene, designated *fl*, which, in hemizygous males and homozygous females, blocks the appearance of filiform hairs from earliest development and sequentially strips other classes of cercal hairs later in development. Sections of cercal nerves from a single first instar normal and mutant cricket gave no evidence of cercal nerve loss associated with the mutation and direct stimulation of the cercal nerve in mutants gave rise to action potentials in the medial giant interneurons which are known to be postsynaptic to filiform sensory neurons (Palka and Olberg 1977).

Abnormal growth of interneurons of *fl* mutants in the face of normal connectivity but in the absence of normal excitatory inputs implicates neural activity as a necessary condition for normal neural growth in this system. Absence of normal excitation in *fl* crickets was demonstrated since controlled air puffs which produced graded depolarizations and bursts of following activity in the medial giant interneurons of normal crickets failed to activate the giants of these mutants. Abnormal growth in giant interneurons seems to be the primary cercal physiological deficit of *fl* animals since they showed normal patterns of branching but radically reduced cross-sectional area and overall reduced cell volume relative to normal crickets.

The relationship between sensory filiform hair receptors and interneuron structure and function has been shown using soft genetic dissection of receptor variation to be much more intimate than is obvious from comparing mutant, receptorless crickets to normals. Variation in filiform hair density was examined in detail by Galvin (1976) counting filiform hairs in non-mutagenized stocks of the cricket *Acheta domesticus*. Natural variation in this component of the escape response is illustrated by three aspects of Figure 1. First, different random samples of crickets and different arbitrary groups within random samples have slightly different hair densities. Second, substantial variation within the groups is indicated by the standard errors of the group means reported by Galvin (1976). Finally, all eight possible contrasts between males and females included in random groupings indicate higher densities in males than in females, demonstrating a sexual dimorphism for filiform hair density.

Variation in filiform hair density is accompanied by variation in postsynaptic activity in giant interneurons following stimulation with controlled low frequency sound. An unexpected feature of this relationship was revealed using sound pulses ranging in frequency from 100 to 2000 Hz and in intensity from 65 to 90 db, and monitoring giant interneuron activity extracellularly for 200 msec following each stimulus delivery. Males showed lower levels of giant interneuron activity than females at each stimulus frequency and higher frequency thresholds at each stimulus intensity. Dimorphism for giant interneuron function in response to cercal hair stimulation could emerge directly from differences in interneurons, synaptic function or density, cercal nerve afferent differences, or differences in the receptor hairs. Since filiform

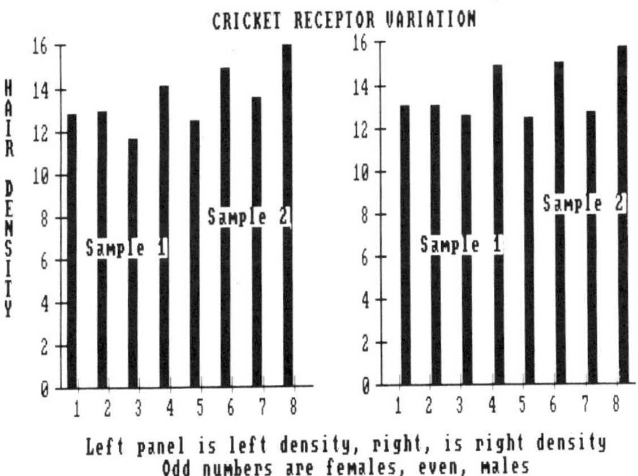

Fig. 1. Hair densities on right and left central appendages of male and female crickets from two random samples each divided into two arbitrary replicates (After Galvin 1976).

hairs are singly innervated (Edwards and Palka 1974), males with higher densities of filiform hairs should have correspondingly higher numbers of cercal nerve fibers and densities of giant interneuron synapses. This suggests a sex difference in cellular function of interneurons, interneurons of males being less responsive than those in females in spite of relatively more afferents per unit area.

Genetic dissection of receptor variation by selection demonstrated that the negative relationship between receptor hair density and interneuron activity is not a sex difference in cellular response. Acting on initial evidence (from comparison of filiform hair densities of parents and offspring) that about 20% of the variation in receptor density was due to gene differences among crickets, Galvin imposed five generations of within-family bidirectional selection based on filiform density of the right cercus only. The response to selection pressure is illustrated in Figure 2 where bars show deviations from matched control line means for hair densities of right (direct response) and left cerci (correlated response) across generations of selection. Those responses were expressed in both males and females of each generation and the density differences within each sex were demonstrated to be associated negatively with interneuron activity in response to standardized low frequency sound stimuli. The negative relationship is apparently regulated within the cellular system and not due to a sex difference in cellular function.

Combining the results from mutational dissection with those from soft dissection using selection suggests an intermediate optimum receptor density for the orthopteran escape response. The mutant *fl* shows no response with no giant interneuron activity when no filiform hairs are present. Take this as one point on a hypothetical plot of interneuron activity as a function of filiform hair density and consider the three additional points that can be deduced from the soft dissection. Activity from giant interneurons of nonselected crickets estimates the likely level of the relationship in crickets under continuous natural selection. The low density lines indicate an increase in interneuron activity with slight decrease in density and the high lines suggest a decrease in interneuron responsiveness with slight increases in density. Since mutants with no hairs show no giant fiber activity on stimulation, the trend to increase giant interneuron responsivity as receptor density drops from mean density must reverse. Soft genetic dissection suggests that this is less trivial than it might seem.

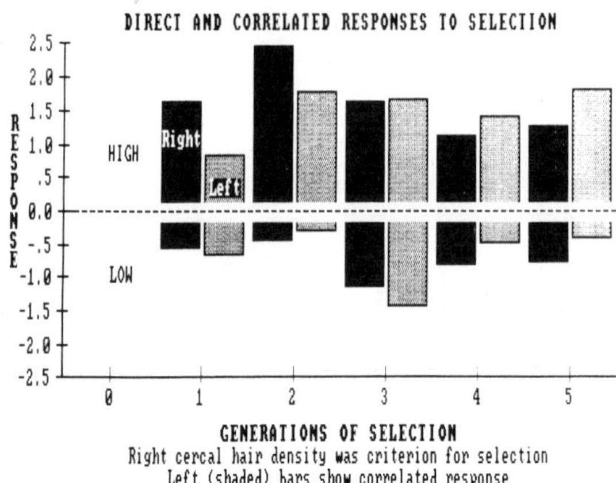

Fig. 2. Response (from control line values) of right and left cercal hair densities to directional selection for high and low right densities (After Galvin 1976).

The dynamics of the response of filiform hair density to dissection by directional selection indicates that crickets deviating from mean receptor density show decreased fitness. Selection differentials for each line and generation are represented in Figure 3. Here the selection differential is the difference between generation average densities and the densities for crickets chosen to breed because they satisfied the criterion of high density in the high line or low density in the low line. These differences weighted by the relative number of offspring each selected animal contributes to the next generation quantify the actual, or realized extent of phenotypic selection pressure. Orderly oscillations in magnitudes of selection differentials, like those obvious in Figure 3 for both lines, indicate that the artificial selection being imposed is counter to natural selection. Initially large differentials are allowed by phenotypic variability which is reduced the following generation if phenotypically extreme animals reproduce poorly. Reduced variability results in choice of breeders less phenotypically extreme. They reconstitute the variability and the oscillation continues. The slight positive selection differentials shown in Figure 3 for the last two generations of the low density lines attest to an especially extreme fitness falloff as density of cercal filiform receptors is forced low.

The two approaches to dissection are complimentary in the most positive sense because findings from each can help with interpretations of outcomes with the other. It is not likely that regulation of receptor density at intermediate levels through associations with fitness would have been demonstrated by mutational dissection and it is certain that the dependence of central neuron growth on genes required for hair maintenance could not have been established by soft dissection. To see the consequences of very sparse receptor densities on physiology or behavior, mutations have to be employed to disrupt fitness relations that constrain natural populations. To reveal regulation of receptor densities with its physiological and behavioral implications, soft genetic dissections are required. Notice that one possible explanation for the mechanism regulating cercal receptor density involves the size of interneurons. Fewer interneuron spikes with more afferent neurons could reflect increased shunting at synaptic terminals and vary in functional relationship to the interneuron cell surface available.

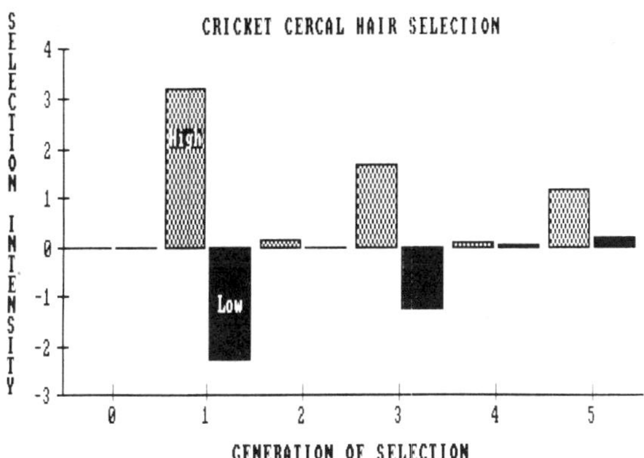

Fig. 3. Deviations of right cercal hair densities of selected breeders in the high and low lines from their population mean densities across five generations (After Galvin 1976).

Generalized Dissection in Behavioral Research

Two broad approaches to using genetics in analysis of behavior will benefit from generalizing dissection and it can provide a common ground between them. For example, genetic dissections of behavioral mechanisms by mutation have a recurrent tendency to divert from analysis of behavior to analysis of physiological systems, especially the nervous system (cf. Breakfield 1979, Hall et al. 1982). The major issues facing systems biologists are relatively clean and the work is exciting and productive; but as important as nervous system structure and function are to behavior, there are skeletal, hormonal, and muscular components which must also be analyzed. Eventually, information from all of these endeavors must be integrated. Mutations, by their nature, provide no help with the matter of integration. An advantage of generalizing is that all genetic methods trace to the same physical causes: mutation manipulates the chemical structure of genes; selection, their frequency. Together, both genetic methods provide a natural arena for synthesis of information from the molecular to the population level because they manipulate characteristics of genes at each level.

Probably the greatest advantage will accrue to the analysts of variation in behavior. Since early compelling evidence accumulated by Thompson (1953) showed that behavior by rodents varied, in part, due to gene differencences, an astonishing volume of literature has accumulated to implicate genes in behavioral variation (McClearn and DeFries 1973, Fuller and Thompson 1978). These studies have provided valuable models of human behavior but there has been a distinct tendency for the work to collect descriptions of gene effects with little or no unified thrust. Each behavior has been chosen for its own intrinsic interest and little synthesis has been attempted. Casting questions of behavioral variance due to gene differences in the context of the search for physiological bases will provide common thrust.

The relationship that can be exploited here is that gene differences that cause variance in antecedents generate covariance among behavioral sequelae (Hegmann 1979 a,b). Changing genes by soft dissection of antecedent systems will simultaneously alter sets of behaviors related by mutual dependence on the antecedent system. This approach to dissection allows using the usual descriptive variance analysis procedures to falsify hypotheses of particular physiological bases to variation in and covariation among behaviors. One advantage is that information from genetic dissections of various behaviors within species will merge to quantify relationships between physiology and behavior. Multiple behaviors will be related by dependence on the same physiological variations and directly in the context of evolutionary potential. These relationships are required to address the "why" questions of behavioral biology. A second advantage is that information from dissections by mutation of the invariant aspects of physiology and behavior of the species can complete the analysis. The physiology will be common ground for two lines of inquiry which would have been virtually unrelatable without generalizing genetic dissection of behavior.

The common ground between genetic dissection of physiological mechanisms of invariant aspects of behavior using induced mutations and genetic dissection of the physiological basis of behavioral variance using selection, migration, or intensified genetic drift will allow a link to the populational level of behavioral analysis. Theoretical extensions (c.f. Lande 1979, 1980) and empirical applications to behaviors of ecological relevance (c.f. Arnold 1981, Hegmann and Dingle 1982) indicate that genetic variance analyses will be pursued with increased intensity for the study of behavior in natural systems. Emphasis will be on identifying and measuring relationships between behaviors and fitness in natural systems. In turn, those relationships are certain to project to the physiological antecedents of behavior.

Literature Cited

Alcock, J. 1975. Animal Behavior, Sinauer Assoc., Inc., Boston.

Arnold, B. J. 1981. Behavioral variation in natural populations. I. Phenotypic, genetic, and environmental correlations between chemoreceptive responses to prey in the garter snake, *Thamnophis elegans*. Evolution 35: 489.

Batty, J. 1978. Plasma levels of testosterone and male sexual behavior in strains of the house mouse *Mus musculus*. Anim. Behav. 26: 339.

Bentley, D. 1975. Single gene cricket mutations: Effects on behavior, sensilla, sensory neurons, and identified interneurons. Science 187: 760.

Breakfield, X. O., ed. 1979. Neurogenetics. Elsevier, New York.

Bullock, T. H., and G. A. Horridge. 1965. Structure and function in the nervous system of invertebrates. Freeman, San Francisco.

DeFries, J. C., and G. E. McClearn. 1970. Social dominance and Darwinian fitness in the laboratory mouse. Amer. Natur. 104: 411.

Edwards, J. S., and J. Palka. 1974. The cerci and abdominal giant fibers of the house cricket *Acheta domesticus* I. Anatomy and physiology of normal adults. Proc. R. Soc. B. 185: 83.

Fuller, J. L., and W. R. Thompson. 1978. The Foundations of Behavior Genetics. Mosley, Saint Louis.

Galvin, N. J. 1976. Genetic variance influencing the cercal hair sensory system in the cricket, *Acheta domesticus*. Ph.D. Thesis. Univ. of Iowa.

Hall, J. C., R. J. Greenspan, and W. A. Harris. 1982. Genetic Neurobiology. MIT Press, Cambridge.

Hegmann, J. P. 1975. The response to selection for altered conduction velocity in mice. Behav. Biol. 13: 413.

Hegmann, J. P. 1979a. A gene-imposed nervous system difference influencing behavioral covariance. Behav. Genet. 9: 165.

Hegmann, J. P. 1979b. Gene influence timing the development of size and nervous system functional differences. Behav. Neur. Biol. 25: 510.

Hegmann, J. P., and H. Dingle. 1982. Phenotypic and genetic covariance structure in milkweed bug life history traits. *In* H. Dingle and J. P. Hegmann, eds. Evolution and Genetics of Life Histories. Springer-Verlag, New York.

Hotta, Y., and S. Benzer. 1969. Abnormal electroretinograms in visual mutants of *Drosophila*. Nature. 222: 354.

Jacob, F. 1976. The Logic of Life. Random House, New York.

Kung, C. 1979. Neurobiology and neurogenetics of paramecium behavior. *In* X. O. Breakfield, ed. Neurogenetics. Elsevier, New York.

Lande, R. 1979. Quantitative genetic analysis of multivariate evolution, applied to brain: body size allometry. Evolution 33: 402.

Lande, R. 1980. The genetic covariance between characters maintained by pleiotropic mutations. Genetics 94: 203.

McClearn, G. E., and J. C. DeFries. 1973. An introduction to behavioral genetics. Freeman, San Francisco.

Palka, J., and R. Olberg. 1977. The cercus to giant interneuron system of crickets. III. Receptive field organization. J. Comp. Physiol. 119: 301.

Roeder, K. D. 1948. Organization of the ascending giant fiber system in the cockroach, *Periplaneta americana*. J. Exp. Zool. 108: 243.

Sokal, R. R., and J. F. Rohlf. 1981. Biometry. Freeman, San Francisco.

Thompson, W. R. 1953. The inheritance of behavior: behavioral differences if fifteen mouse strains. Can. J. Psychol. 7: 145.

Tinbergen, N. 1968. On war and peace in animals and man. Science. 160: 1411.

Wilson, E. O. 1975. Sociobiology. Belknap Press, Cambridge.

Wu, C. F., and B. Ganetzky. 1980. Genetic alteration of nerve membrane excitability in temperature sensitive paralytic mutants of *Drosophila melanogaster*. Nature 286: 814.

Extrapolating Quantitative Genetic Theory to Evolutionary Problems

Montgomery Slatkin*

Mark Kirkpatrick**

Department of Zoology, NJ-15
University of Washington
Seattle, Washington 98195

Introduction

Until recently, a quantitative geneticist at a conference on evolution was like a eunuch in a harem. He was welcome largely because he could not disturb the proceedings. In the past ten years, that situation has changed, at least for conferences on evolution. Evolutionary biologists have come to look to quantitative genetics for predictive models and quantitative geneticists have obliged by directing some of their efforts to evolutionary problems. We will review some of the reasons for the partial integration of quantitative genetics theory and evolutionary biology and discuss some of the consequences for evolutionary biologists.

Background

It is, of course, not a recent discovery that evolutionary biology is concerned primarily with metrical or quantitative characters that are controlled by several genetic loci. Nor is it a recent discovery that quantitative geneticists have been concerned with developing models of such characters. What is new is the extrapolation of the theory of quantitative genetics, which concentrates on the genetic structure of a population and on different modes of artificial selection, to make predictions on an evolutionary time scale. While the first few chapters of the *On the Origin of Species* (Darwin 1859) describe the same extrapolation, evolutionary discussions in population genetics traditionally have been in terms of models of one or two loci.

In the 1910's and 1920's, population geneticists, particularly Wright, Fisher, and Haldane, were concerned both with reconciling Mendelian genetics with the biometrical studies of Galton and Pearson and with putting Darwin's theory of evolution by natural selection on a genetic basis (Provine 1971). These two activities were largely separated by the kinds of models used, even though some papers, such as Fisher's 1922 paper on the dominance ratio, dealt with both topics. Models in quantitative genetics developed by Fisher and Wright assumed a large number of loci with alleles of small effect contributing additively to a trait.

*Present address: Department of Zoology, University of California, Berkeley, CA 94720.

**Present address: Department of Zoology, University of Texas, Austin, Texas 78712.

These models yield formulae for correlations among relatives and for consequences of different schemes of assortative mating and inbreeding. Selection is either ignored or analyzed only for a single generation under the assumption that the genetic structure of the population was already known.

In contrast, theories of natural selection, particularly those of Haldane and Fisher, were based on models involving one genetic locus, usually with two alleles. The goal was to show that such models allow relatively weak natural selection to cause significant evolutionary change in a reasonable time. The relationship between these two approaches to population genetics was largely implicit. The two approaches gave rise to two, often non-overlapping groups: population geneticists who study one- and two-locus models of natural selection and statistical geneticists who study metrical characters and artificial selection.

Maintenance of Quantitative Genetic Variability

In traditional models of quantitative genetics, the question of how the genetic components of the variance of a character are maintained is not addressed. If the components of the variance are known then the standard theory as described by Falconer (1981) and others can be used to predict the evolution of the mean value of the trait under artificial selection until, for some reason, those components change. In applying models of quantitative genetics to evolutionary problems, it is necessary to make some assumption about how genetic variance is maintained in a population because, in such models, predictions about how genetic variance changes under prolonged natural selection are needed.

Lewontin (1974) distinguishes two schools of population geneticists by their views about the roles of selection and mutation in maintaining genetic variation in natural populations. The "classical" school holds the view that most genetic variation is maintained by mutational pressure opposed by natural selection. A simple model that represents this view is of a "wild type" allele favored by selection but opposed by the mutation pressure to deleterious alleles. The "balanced" school holds the view that selection plays the dominant role with opposing selection pressures maintaining variability and with mutation only providing new variants. A simple model that represents this view is of a single locus with selection in favor of heterozygotes. Most evolutionary biologists have views that fall somewhere in between these two extremes.

The extrapolation of quantitative genetics theory to evolutionary problems requires that one of these views (or some other view) be adopted, and the classical view has been favored for several reasons. First, it is very difficult to construct models of continuously variable traits in which genetic variation is maintained by selection alone. Unless rather restrictive assumptions are made, it is difficult to maintain very many alleles at one locus (Lewontin et al. 1978). Additionally, stabilizing selection of the kind thought to be at work on many characters does not readily maintain genetic variation at more than one locus (Wright 1935, Kojima 1959, Lewontin 1964). In contrast, it has been relatively easy to develop quantitative genetic models of a balance between selection and mutation that are both biologically plausible and mathematically tractable. Kimura (1965), Latter (1970), Lande (1975), Bulmer (1980, Chapter 10) and Turelli (1983) have shown under different assumptions that a realistic mutational pressure is sufficient to oppose stabilizing selection and maintain substantial genetic variability in quantitative characters. Unlike models of heterosis, these models have the common feature that there is effectively no limit to the number of alleles that can be maintained at each locus.

A second reason for adopting the classical view is empirical. The few studies of mutational effects on the variance of quantitative characters show that there is a significant increase in the variance each generation (Clayton and Robertson 1964). This mutational input is roughly comparable to the amount of variation lost to stabilizing selection acting on traits in natural populations (Lande 1975, Johnson 1976), suggesting that there is a balance between the forces of mutation and selection.

For an evolutionary biologist, it makes a considerable difference which forces are maintaining the genetic variability of quantitative characters. All of the models of the selection-mutation balance have the common feature that directional selection applied to a population formerly at equilibrium does not greatly disturb genetic or phenotypic variances. Therefore, a population exposed to new environmental conditions could evolve to a new optimal value of a trait without either the variance of the trait or the heritability changing significantly; a pattern of response that has been found in several studies of artificial selection (see Slatkin and Lande 1976). In fact, in the models of the type developed by Kimura (1965) and others, which assume an infinite population size and an infinite range of allelic values, there are no limits to changes in the average value of the trait.

Limits to Selection

When a laboratory population is subject to strong directional selection on a particular trait, the usual course is a rapid and steady response for several generations followed by a cessation of response, with the mean value of the selected trait reaching a plateau (e.g., Mather and Harrison 1949). There can be several, not mutually exclusive, reasons for the lack of further response in such experiments. Which of these reasons is most important affects the ease with which the results from artificial selection experiments can be extrapolated to evolutionary problems. We divide the causes of limits to selection into two broad categories, those that arise only from the trait being selected and those that arise from other traits.

In the models of selection-mutation balance of the kind developed by Kimura (1965) and others, there is no limit to the response to selection because there is, effectively, an infinite range of alleles present in the population at any time. To exceed the limit, the population does not have to wait until the right mutations come along. In a finite population, that is no longer possible, as has been shown for Kimura's model by Latter and Novitski (1969). Once the extreme homozygote is fixed, which can happen quickly, new mutations are necessary for any more response. Hill and Robertson (1966) have shown that, in a finite population, randomly generated linkage disequilibrium can act to decrease the response of a population to selection.

A second reason that the genetic variability of a trait can be exhausted by directional selection is that some of the variability could be due to heterosis of the underlying loci, rather than the selection-mutation balance assumed in most models. If directional selection causes one of the homozygotes to be fixed, the variability due to that locus would be lost. To the extent that the balanced view of the maintenance of genetic variability is correct, we would expect limits on selection to be set by the range of variation of the alleles at the heterotic loci.

Limits on selection can also be set by traits other than the one subject to directional selection. The effects on other traits can be through genetic correlations caused by pleiotropy and linkage disequilibrium or through inbreeding depression. We imagine some other trait being subject to "natural" selection, selection that is caused by the experimental conditions and that does not change when the trait of interest is subject to directional selection.

Pleiotropy can limit the response to directional selection because selection on one trait can be opposed by deleterious effects on correlated traits. One difference between pleiotropy and linkage disequilibrium is in the time for the limit to be exceeded. If linkage disequilibrium is the cause and directional selection is relaxed for a few generations, then recombination will tend to restore linkage equilibrium. Renewed directional selection would then lead to progress beyond the original limit, as has been found in some selection experiments (e.g., Mather and Harrison 1949), which suggests that linkage disequilibrium plays some role. In contrast, if pleiotropy is the cause, then the limit could not be exceeded until new mutations appeared that had different pleiotropic effects. That would be expected to take far longer than a few generations. A second difference between pleiotropy and linkage disequilibrium is in the importance of population size. As discussed later, linkage disequilibrium would be expected to be most important in small populations whereas pleiotropy is likely to be an important constraint even in large populations.

Another way in which linkage disequilibrium and pleiotropy can impose limits to selection is through inbreeding depression caused by the small sizes of most laboratory populations. In real populations (as contrasted with the mathematical kind), selection is performed on excess individuals. If inbreeding depression leads to a sufficiently low absolute fitness of the population, there might be too few excess individuals to continue the directional selection and still maintain the population.

Which of these causes of limit to selection found in studies of artificial selection is most important will determine the extent to which results from artificial selection can be extrapolated to natural selection. If finite population size is important in setting limits, then natural populations, which are generally much larger than laboratory populations, would be less severely limited in their responses to natural selection. On the other hand if pleiotropy is a major cause, then the same limits would be expected in natural populations. Some proponents of the punctuated equilibrium theory of evolution have argued for the importance of constraints on evolution imposed by development (e.g., Gould 1980). What is being attributed to development is in fact a statement about pleiotropy. To say that the developmental system constrains evolution is to say that there are no mutations with the right pleiotropic effects (see Charlesworth et al. 1982). Pleiotropic effects have been found in most characters that have been studied, including behavioral ones (Dobzhansky 1970). But while linkage and pleiotropy have been demonstrated in many populations, we do not yet have experimental evidence that allows us to identify either of these causes as being the most important in natural populations.

Correlated Characters

Most if not all phenotypic traits are correlated with one or more other traits. The correlations may have genetic or non-genetic causes, and, as we have discussed, genetic correlations may result from pleiotropy and linkage disequilibrium. Lande (1980) has shown that, under one set of assumptions, pleiotropy is likely to be much more important than linkage disequilibrium in maintaining genetic correlations in a large population at equilibrium. This result is probably more generally true because a variety of genetic models have shown how difficult it is to maintain permanent linkage disequilibrium in a population in the absence of strong heterotic and epistatic selection (Ewens 1979), which we have argued is probably not at work on most loci controlling quantitative characters. Genetic correlations among traits not only can affect the limits to the response to directional selection but also can affect the course of response to some kinds of natural selection.

The distinction between genetic and non-genetic causes of phenotypic correlation is important because selection on one trait affects the evolution of others only through their genetic correlations, not their phenotypic correlation (Lande 1979, Falconer 1981). As Atchley et al. (1981) have shown, phenotypic correlations may not be a reliable guide to genetic correlations.

To illustrate how correlations among characters affect the course of evolution, we will restate some of the basic elements of the theory. We emphasize that we are not describing any new results but describing existing results in a somewhat different way. We will start by developing the theory for an arbitrary number of traits but then confine our examples to only two traits, for which a graphical description is feasible.

Assume we are modeling the evolution of n quantitative characters which have measurements $z_1,\ldots,z_n$. It is sometimes convenient to express the theory in matrix notation with the phenotypes being represented by column vector **z** with elements $z_1,\ldots,z_n$. Assume that **P** is the matrix of phenotypic covariances among these traits. That is, the ijth element of P is Cov $(z_i z_j)$, where Cov (.) represents the phenotypic covariance in the population being modeled. The matrix **P** is an n x n symmetric matrix which can be measured directly. Assume that **G** is the matrix of genetic covariances of the breeding values of the n characters. The matrix **G** is also an n x n symmetric matrix whose elements can be estimated by using methods described by Falconer (1981).

Given **P** and **G**, the changes in the mean values of the traits between one generation and the next are predicted by the following equation:

$$\triangle \mathbf{z} = \mathbf{G}\mathbf{P}^{-1}\mathbf{s} \tag{1}$$

where **z** is a column vector of mean values of the n traits, and where the elements of the column vector **s** are the selection differentials (that is, the differences in the mean values of the trait before and after selection) of each of the n traits.

Although Equation (1) is familiar to quantitative geneticists (Lande 1979, Falconer 1981), its application to evolutionary problems requires the extrapolation we have already discussed. In this case, the extrapolation takes the form of additional assumptions about processes such as mutation and recombination that can be used to predict how the matrices **P** and **G** change from generation to generation.

Before using Equation (1), we need to elaborate on the selection differentials, the elements of **s**. Our intuition about natural selection is that different traits confer different abilities to survive and reproduce. We could imagine, for example, how a trait like tooth length affects survival in ungulates. Individuals with teeth that are too short might wear their teeth to a degree that they would no longer eat as efficiently as other members of the population and have a higher death rate as a result. Individuals with teeth that are too long might have a higher risk of tooth breakage with subsequent decay leading to more serious infections. Natural selection would be acting like a breeder in choosing those individuals with teeth of intermediate length. We will call this type of selection "direct selection", because it does not depend on genetic or phenotypic correlations.

Direct selection is only one of two components of the selection differentials. Traits that are not exposed to direct selection will change their distributions as a result of selection on correlated characters. We will call this "indirect selection" because it is due to selection on traits other than the one being considered. As Figure 1 illustrates, if two traits are correlated and one is selected in such a way that its mean value changes, as shown, the selection differential of the other trait is non-zero. The ratio of the selection differential of the correlated trait to that of the directly selected trait is just the regression coefficient of the second trait on the first. If only one of n traits is directly selected, say trait 1, with a selection differential of s_1, the selection differentials for the other traits are given by

$$s_i = b_{i1} s_1 \tag{2}$$

where b_{i1} is the phenotypic regression coefficient of trait i on trait 1.

We think this distinction between direct and indirect selection is important to make because discussions by ecologists and evolutionary biologists about selection and adaptation is of the role of direct selection, but genetic evolution depends on both kinds of selection. Furthermore, estimates of selection intensity made by comparing the distributions of a trait among two life history stages (Arnold and Wade 1984) include both kinds of selection. We could observe, for example, a decrease in the variance of scuteller bristle number in *Drosophila* with increasing age of adults. On the basis of that evidence alone, we could conclude that there is stabilizing selection on the distribution of that trait. We could not conclude, however, that scuteller bristle number itself is affected directly by environmental conditions; it could be correlated with physiological traits whose extreme values lead to lower survivorship. To put this another way, the equilibrium number of four that is found in *D. melanogaster* need not be an adaptation to anything.

We will illustrate the potential importance of genetic correlations first with a hypothetical example and second with an example of behavioral traits in a natural population. The hypothetical case concerns two traits. Imagine that one trait is subject to stabilizing selection for a single optimum while the other is subject to bimodal selection for either of two different optima. This is illustrated in Figure 2 where the contours represent combinations of characters that confer equal fitness. This figure represents two "adaptive peaks" separated by an adaptive valley. The three-dimensional surface represented this way is a graph of the relative fitnesses of individuals of different combinations of these traits, which we will call $w(z_1,z_2)$. This is different than Wright's (1932) "adaptive landscape," which graphs the average fitness of a population against the mean values of the traits in he population. Because of the averaging, the shape of the adaptive landscape depends in part on the variances of the traits in the population (Kirkpatrick 1982). The fitness function $w(z_1,z_2)$, however, depends only on the phenotype of the individual.

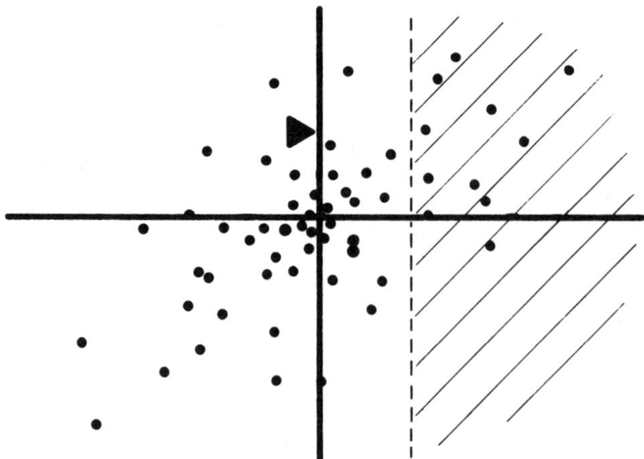

Fig. 1. An illustration of the fact that truncation selection applied to one of two correlated characters will result in a change in the mean value of the other. Before selection, the mean value of both characters is zero. When individuals in the shaded region are selected on the basis of one character, the mean of the other character (indicated by triangle) is also altered.

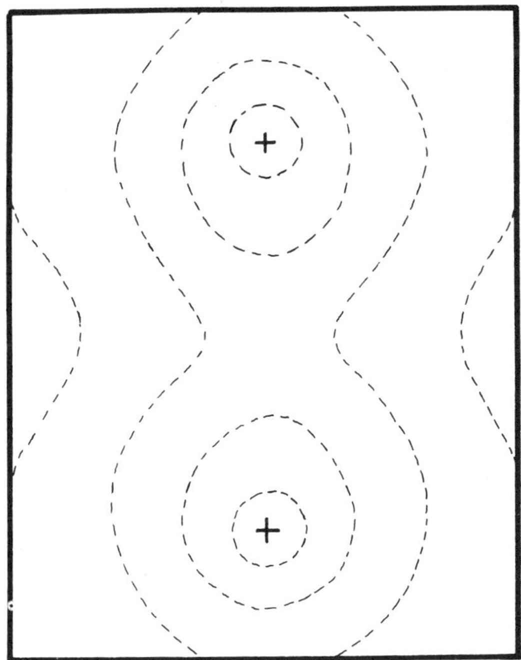

Fig. 2. The surface of relative fitnesses for two characters. The dashed lines indicate values of the characters that have equal fitnesses. The crosses indicate the two values of the characters that represent optimum pairs of values.

We assume the joint distribution of the two traits is given by a function $p(z_1,z_2)$ in a generation. We will assume that this distribution is bivariate normal. The selection differentials, s_1 and s_2, that are needed in Equation (1) are given by

$$s_i = \int \int z\, p(z_1,z_2) w(z_1,z_2) dz_1 dz_2 / W - z_i, \tag{3}$$

where the integrals are taken over all values of the two traits, W is the mean fitness, and z_i the mean value of trait i before selection.

By choosing particular functional forms for $p(z_1,z_2)$ and $w(z_1,z_2)$ and assuming **P** and **G** do not change with time, we can iterate Equation (1) to predict the future course of evolution of these two traits. A simple case will illustrate the possible role of genetic correlations. We assume:

$$G_{11} = G_{22},$$
$$G_{21} = G_{12} = rG_{11},$$
$$P_{11} = P_{22} = G_{11} + V_e,$$
$$P_{12} = P_{21} = G_{12};$$

which is to say that the genetic and phenotypic variances of the two traits are equal and that the only cause of phenotypic correlation is genetic correlation. The extent of genetic correlation is measured by the parameter r, which is, in the symmetric case, the correlation coefficient between the breeding values of the two traits. The parameter V_e is the environmental component of the variance in both traits.

Genetic correlation between the traits affects only slightly the possible equilibria of these equations. If $r<1$, at the equilibria both of the selection differentials must be zero. Assuming that phenotypic variances are less than the widths of the adaptive peaks, i.e., selection is weak, then the two stable equilibria for the population means of z_1 and z_2 are near the two adaptive peaks. The equilibria will not be exactly at the two peaks because of the presence of the other peak.

Genetic correlations can affect which of the two peaks the population will evolve to, as shown in Figure 3. The solid lines in Figure 3 separate the domains of attraction of the two peaks: the initial conditions which will result in the population's evolving to one or the other peak. As we can see, genetic correlations, either positive or negative, can cause a population to evolve "downhill" and across the adaptive valley. This is not to imply that the mean fitness of the population decreases. When the fitness of phenotypes are constant and selection is weak, the mean population fitness cannot decrease (Lande 1979). For many kinds of behavioral characters, however, fitnesses are not constant because of frequency dependence. Dominance heirarchies and sexual selection are two familiar examples involving frequency dependence. The competition or interference between individuals can lead to decreasing population fitness in such cases (Lande 1976).

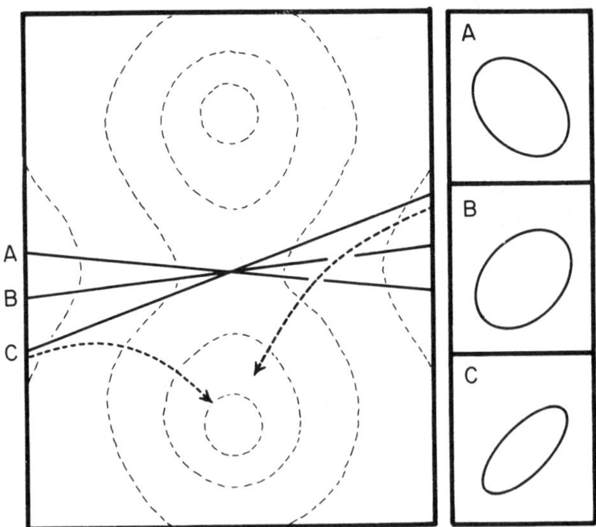

Fig. 3. The domains of attraction of the two stable equilibra for the selection model shown in Fig. 2. For initial values of the two characters above the line, the upper equilibrium is approached and for initial values below the lines, the lower equilibrium is approached. The three lines shown represent three different degrees of genetic correlation between the two characters: A, $r=0.25$; B, $r=0.50$; C, $r=0.75$. The figures on the right show roughly the distribution of the phenotypic values for the three cases. The lines are approximate because they were obtained from a local analysis in the neighborhood of the unstable equilibrium at the saddle point.

An example of genetic correlation between behavioral traits in a natural population comes from the garter snakes studied by Arnold (1981). These snakes show a high genetic correlation between their chemoreceptive responses to two potential types of prey, slugs and leeches. Arnold believes that an inland population has been selected to ignore leeches and consequently has evolved decreased ability to detect slugs as a correlated repsonse. A coastal population, in contrast, preys heavily on slugs and has been selected for chemoreceptive sensitivity to them, and so has acquired sensitivity to leeches as a correlated response.

Artificial Selection as a False Analogy

The role of constraints on genetic evolution has become important recently in discussions of macroevolution. Many paleontologists claim that the fossil record provides strong evidence for the stasis of widespread species (e.g., Eldredge and Gould 1972). Further evidence for stasis is provided by such extant taxa as the salamander genus *Plethodon*, which has been discussed in this context by Wake et al. (1983). This genus contains 23 species that are morphologically very similar to each other, suggesting that none of the species have evolved much since the origin of the group. Apparently only one genus, *Aneides*, has descended from *Plethodon*, and *Aneides* differs primarily in head size. Several different genetic techniques indicate that members of the genus *Plethodon* diverged from one another at least 55 million year ago, making it the oldest extant genus of terrestrial vertebrates. Clearly, the species in this genus have evolved little despite the very long time they have had to do so.

The long term stasis of established species seems incompatible with the results of experiments in which nearly any character can be altered by artificial selection. If natural selection is analogous to artificial selection and if stasis is due to the action of stabilizing selection, then we must conclude that the agents of natural selection, the ecological conditions experienced by species, have not changed. While that possibility is difficult to rule out, it is also difficult to believe that species like those in the genus *Plethodon* have not experienced rather large changes in the environmental conditions they have been exposed to in the past 55 million years. Since the origin of that genus, their vertebrate predators and competitors have certainly changed and probably the same is true for many of the species they feed upon as well.

Artificial selection is not completely analogous to natural selection. When artificial selection is applied, the population exposed to it has no choice but to respond. Natural populations, however, have many alternatives besides responding genetically to new environmental conditions. They can control to some extent the selection they experience. The response can be behavioral or physiological. In addition, the response can be extinction of some populations with a resulting change in the species' geographic distribution. With this view, *Plethodon* salamanders might have been exposed to a wide variety of environmental conditions but persisted only in those for which they were already adapted. The changes in geographic distributions despite morphological stasis has been documented in some beetles by Coope (1979).

We are not implying that genetic evolution does not sometimes result from exposure to new conditions. The evolution of the polar bear and the snowshoe hare quite likely did occur after exposure to arctic conditions in a way exactly analogous to those in an artificial selection experiment. But rapid evolution is not the only result of such exposure. There are no polar tapirs or snowshoe baboons, even though there seem to be no biogeographic reasons why not. And to say that the biology of those groups precludes their acquisition of adaptations to arctic life simply restates the problem.

Conclusion

We will conclude by saying that quantitative genetics and evolution do belong at the same meeting, even if the proceedings are disturbed. But the application of the theory of quantitative genetics to evolutionary problems requires that rather strong assumptions be made about how genetic variation in quantitative characters is maintained. There is a consensus emerging that heritable variation in quantitative characters is maintained by a balance between mutation pressure opposed by stabilizing selection. If that is true for most quantitative characters, then the extrapolation from the short time-scale of quantitative genetics to an evolutionary time-scale is justified. If not, and it is found that additional mechanisms, not now a part of quantitative genetics theory, are necessary to describe long-term changes in quantitative characters, then the current theory of quantitative genetics will be of limited applicability to evolution. While there is no evidence in favor of the latter possibility, there is also insufficient evidence to exclude it completely.

Acknowledgments

We thank P. Fuerst, S. Levin and M. Wade for numerous useful comments on an earlier version of this manuscript.

Literature Cited

Arnold, S. J. 1981. Behavioral variation in natural populations. I. Phenotype, genetic and environmental correlations between chemoreceptive responses to prey in the garter snake, *Thamnophis elegans*. Evolution 35: 489.

Arnold, S. J. and M. J. Wade. 1984. On the measurement of natural and sexual selection : applications. Evolution 38:720.

Atchley, W. R., J. J. Rutledge, and D. E. Cowley. 1981. Genetic components of size and shape. II. Multivariate covariance patterns in the rat and mouse skull. Evolution 35: 1037.

Bulmer, M. G. 1980. The Mathematical Theory of Quantitative Genetics. Oxford University Press, Oxford.

Charlesworth, B., R. Lande, and M. Slatkin. 1982. A neo-Darwinian commentary on macroevolution. Evolution 36: 474.

Clayton, G. A., and A. Robertson. 1964. The effects of X-rays on quantitative characters. Genet. Res. 5: 410.

Coope, G. R. 1979. Late Cenozoic Coleoptera: Evolution, biogeography and ecology. Ann. Rev. Ecol. Syst. 10: 247.

Darwin, C. 1859. On the Origin of Species. John Murray, London.

Dobzhansky, T. 1970. Genetics of the Evolutionary Process. Columbia University Press, New York.

Eldredge, N., and S. J. Gould. 1972. Punctuated equilibria: an alternative to phyletic gradualism. *In* T.J.M. Schopf, ed. Models in Paleobiology. Freeman, Cooper and Co., San Francisco.

Ewens, W. J. 1979. Mathematical Population Genetics. Springer-Verlag, New York.

Falconer, D. S. 1960. Introduction to Quantitative Genetics, 2nd ed. Longman, New York.

Fisher, R. A. 1918. The correlation between relatives on the supposition of Mendelian inheritance. Trans. Roy. Soc. Edin. 52: 399.

Fisher, R. A. 1922. On the dominance ratio. Proc. Roy. Soc. Edin. 42: 321.

Gould, S. J. 1980. Is a new and general theory of evolution emerging? Paleobiology 6: 119.

Hill, W. G., and A. Robertson. 1966. The effect of linkage on limits to artificial selection. Genet. Res. 60: 615.

Johnson, C. 1976. Introduction to Natural Selection. Univ. Park Press, Baltimore.

Kimura, M. 1965. A stochastic model concerning the maintenance of genetic variability in quantitative characters. Proc. Natl. Acad. Sci. USA 54: 731.

Kirkpatrick, M. 1982. Quantum evolution and punctuated equilibrium in continuous genetic characters. Amer. Natur. 119: 833.

Kojima, K. 1959. Stable equilibria for the optimum model. Proc. Natl. Acad. Sci. USA 45: 989.

Lande, R. 1975. The maintenance of genetic variability by mutation in a polygenic character with linked loci. Genet. Res. 26: 221.

Lande, R. 1976. Natural selection and random genetic drift in phenotypic evolution. Evolution 30: 314.

Lande, R. 1980. The genetic covariance between characters maintained by pleiotropic mutations. Genetics 94: 203.

Lande, R. 1981. The minimum number of genes contributing to quantitative variation between and within populations. Genetics 99: 541.

Latter, B.D.H. 1970. Selection in finite populations with multiple alleles. II. Centripetal selection, mutation and isoallelic variation. Genetics 66: 165.

Latter, B.D.H., and C. E. Novitski. 1969. Selection in finite populations with multiple alleles. I. Limits to directional selection. Genetics 62: 859.

Lewontin, R. C. 1964. The interaction of linkage and selection. II. Optimum models. Genetics 50: 757.

Lewontin, R. C. 1974. The Genetic Basis of Evolutionary Change. Columbia University Press, New York.

Lewontin, R. C., L. R. Ginsburg, and S. Taljaprukar. 1978. Heterosis as an explanation for large amounts of genic variability. Genetics 88: 149.

Mather, K., and B. J. Harrison. 1949. The manifold effects of selection. Heredity 3: 1.

Provine, W. B. 1971. The Origins of Theoretical Population Genetics. University of Chicago Press, Chicago.

Slatkin, M., and R. Lande. 1976. Niche width in a fluctuating environment—density independent selection. Amer. Natur. 110: 31.

Turelli, M. 1984. Heritable genetic variation via mutation-selection balance: Lerch's zeta meets the abdominal bristle. Theor. Pop. Biol. 25:138.

Wake, D., G. Roth, and M. Wake. 1983. On the problem of stasis in organismal evolution. J. Theor. Biol. 101: 211.

Wright, S. 1932. The roles of mutation, inbreeding, crossbreeding, and selection in evolution. *In* Proc. 6th International Cong. Genetics. Vol. 1.

Wright, S. 1935. The analysis of variance and the correlation between relatives with respect to deviations from an optimum. J. Genetics 30: 243.

The Role of Behavior in Host-Associated Divergence in Herbivorous Insects

Douglas J. Futuyma

Department of Ecology and Evolution
State University of New York at Stony Brook
Stony Brook, New York 11794

In this essay, I will bring some genetic considerations to bear on the question of why host specificity evolves in herbivorous insects. This specificity is evident in the behavior of the feeding stages, that accept certain plants and not others, and often in the behavior of ovipositing females, which lay eggs on the plants on which their offspring will develop. Because individual insects will sometimes fail to find an appropriate host, we might expect selection to favor, to at least some extent, the capacity to accept a wider spectrum of plants than they often do. Why, then, are so many species so extraordinarily host-specific?

The frequent answer is that behavior (host preference) is molded by selection to fit the physiological capacities of the animal. (In this paper, I use "physiology" in a narrow sense, referring primarily to biochemical processes, and exclude behavior, which of course has a physiological basis.) Thus, for example, larvae of the black swallowtail (*Papilio polyxenes* F.) are poisoned by the secondary compounds of crucifers and of certain umbellifers (Erickson and Feeny 1974, Berenbaum and Feeny 1981), and indeed do not feed on these plants. Following the popular ecological doctrine that "a jack of all trades is master of none," we may suppose that it is difficult for an insect to adapt to a wide variety of toxic plant compounds, and that a physiologically specialized genotype may have highest fitness on a particular host because it is more effective at detoxifying a particular class of compounds. Because of this physiological trade-off, selection will favor specialized host preference. A variant of this theme is that different genotypes may be specially adapted to the compounds of different plants, so that disruptive selection favors polymorphism in attraction to different hosts, on which mating and feeding occur. This could result in the generation of host-specific species by sympatric speciation. Finally, optimal foraging theory suggests that species should specialize on those hosts that provide the greatest reward in fitness.

Implicit in these ideas, which are surely all relevant in some cases, is the notion that selection molds a species to the challenges posed by the external environment, and that the evolution of behavior is guided by an organism's physiological and morphological capacities. But the critical aspect of animal behavior is that it consists, very largely, of mechanisms by which the species constructs its own environment. Behavior can dictate the context in which physiological capacities are exercised, and can obviate the need for certain physiological traits altogether. Monarch butterfly larvae do not need to cope with the glucosinolates of crucifers, because the behavior of monarchs makes crucifers irrelevant to their existence. Consequently, whether behavior is molded to physiology or vice versa can depend on the nature of genetic variation in the two kinds of traits.

Consider, for example, the butterfly *Pieris napi macdunnoughii* Remington, which oviposits on several native crucifers in the Rocky Mountains, as well as on two introduced crucifers on which the larvae do not survive (Rodman and Chew 1980). Rodman and Chew believe that the females confuse the alien plants with a native host (*Descurainia*) on which the larvae grow well. *Descurainia* and the aliens share certain glucosinolates that may act as oviposition stimulants.

Under such circumstances, selection might favor alleles (say a recessive allele a) that enable the larva to survive on a toxic host **T**. It also could favor an allele (say a recessive mutant b) such that bb females avoid host **T**. Such females might avoid **T** only, or they also might avoid a suitable host (like *Descurainia*) that they cannot distinguish from **T**, and restrict their oviposition to other normal, suitable hosts (**N**). Suppose, then, that in a population with prevalent genotype $AABB$, recessive mutations a and b enter, with frequencies q_A and q_B respectively (the frequencies of A and B are $1-q_A$ and $1-q_B$ respectively). Assign the constant fitness values 1 to $A\_\_\ B\_\_$ (accepts host **T**, has high mortality on it), $1+s$ to Abb (rejects **T**, and would have high mortality if it did accept it), $1+r$ to $aaB\_\_$ (accepts T, survives on it), and 1+t to $aabb$ (rejects T, but would survive on it if it did accept it). We assume that $s = t > 0$; that is, $A\_\_\ B\_\_$ has lower fitness because it lays some eggs on a plant to which it is not adapted. If the viability of aa on host **T** is as great as that of $A\_\_$ on host N, the $aaB\_\_$ genotype, because it has access to a resource (**T**) that is unavailable to bb genotypes, may have highest fitness ($r > s,t$).

Assuming weak selection and linkage equilibrium, Wright's equations (1969, p. 67) may be used to find $\triangle q_A$ and $\triangle q_B$:

$$\triangle q_A = q_A^2(1-q_A)\ [r-(r+s-t)q_B^2]/\bar{w}$$

$$\triangle q_B = q_B^2(1-q_B)\ s-(r+s-t)q_A^2]/\bar{w}.$$

If $s = t$,

$$\triangle q_A = q_A^2(1-q_A)\ (r-rq_B^2)/\bar{w}$$

$$\triangle q_A = q_B^2(1-q_B)\ (s-rq_A^2)/\bar{w}.$$

Then $\triangle q_A > 0$ if $r(1-q_B^2) > 0$ and $\triangle q_B > 0$ if $s/r > q_A^2$. If both mutants are present at the same low frequency, $\triangle q_A > \triangle q_B$ if $r > s$: the "physiological mutant" a increases faster than the "avoidance mutant" b. Mutant a will increase as long as any individuals in the population oviposit on host **T**. However, the "rejection mutant" b will spread as long as physiological adaptation to **T** is not widespread in the population ($q_A^2 < s/r$).

If we assume that physiological adaptation to **T** and behavioral avoidance of **T** are caused by dominant (rather than recessive) mutations A and B respectively, we can assign the fitnesses $1+t$, $1+r$, $1+s$, and 1 to genotypes $A\_\_\ B\_\_$, $A\_\_\ bb$, $aaB\_\_$ and $aabb$ (the latter being the prevalent "wild type"). Then if p_A, q_A, p_B and q_B are the frequencies of A, a, B, and b respectively, and if $s=t$, it is easy to show that $\triangle p_A = p_A q_A^2(rq_B^2)/\bar{w}$ and $\triangle p_B = p_B q_B^2 [t-r(1-q_A^2)]/\bar{w}$. Allele A will increase as long as $rq_B^2 > 0$, i.e. as long as any individuals in the population accept the toxic host. Allele B increases ($\triangle p_B > 0$) if $q_A^2 < (r-\tau)/r$. For example, if $r = .6$ and $t = .5$, $\triangle p_B > 0$ if $q_A^2 > .167$ ($q_A > .408$). The "rejection mutant" B

will increase faster than the "physiological mutant" A if A is quite rare (e.g. if $p_A = .01$, $\triangle p_B > \triangle p_A$ if $p_B > .013$). (This formulation again assumes linkage equilibrium; I have not worked out the conditions for spread when there is gametic disequilibrium.)

If, then, the mutation for avoiding the toxic host (e.g. B) enters the population before the mutation for physiological adaptation (A), it may reach fixation while A is still increasing in frequency. Since the population then no longer uses the toxic host, A experiences no further selection (except very weak selection on the order of the rate of back mutation from B to b). If, as is perhaps unlikely, A carries no pleiotropic disadvantage (i.e., if detoxification ability has no cost) there will be a neutral polymorphism at the A locus, and A can even drift to fixation, so that populations could be capable of detoxifying the compounds of a host (**T**) that they never use. Thus populations such as *Pieris napi macdunnoughii* could avoid inferior hosts, yet in some instances have the physiological capacity to survive on them. Moreover, the host range of such a species may become narrowed if the $B\_$ genotype avoids not only the toxic host, but suitable hosts (such as *Descurainia*) with which it is confused (Levins and MacArthur 1969). Discrimination between the hosts may be accomplished, moreover, by raising the threshold for acceptance of a suitable host on the basis of some quite subtle cue (e.g., leaf shape, pubescence) that allows unambiguous recognition. Many seemingly trivial features of plants, then, will act as "defensive" characters even if they did not evolve because of their defensive function.

In this example, the course of evolution depends on whether a population harbors more genetic variation for physiological or behavioral traits. There is little information on this question. In selection experiments for adaptation to inferior hosts, Wasserman and Futuyma (1981) observed a more pronounced response in host preference in a bruchid beetle. Gould (1979) found that survival of spider mites on an unsuitable host increased rapidly under selection. Although it is possible that mortality was caused by unwillingness to feed, the control mites did not avoid this host when given a choice, suggesting that the response to selection was based on genetic variation in physiological rather than behavioral characters. Physiological adaptation to alfalfa, but not oviposition preference, appears to have evolved in pest populations of a species of *Colias* (Tabashnik 1983 and this volume). In contrast, asexual genotypes of the fall cankerworm *Alsophila pometaria* (Harris), (Lepidoptera: Geometridae) that are associated with stands of oak *vs.* maple differ in host preference and in phenology, but differ little, if at all, in the nutritional indices widely used to measure physiological adaptation (Futuyma et al. 1984).

These considerations lead to the conclusion that the correlation between what plants an insect population actually feeds on, and those that it is physiologically capable of using successfully, might often be quite low. The range of host plants that potentially could support growth is likely to be broader than the range actually used for several reasons. Alleles that provide physiological adaptation to a plant may drift as neutral polymorphisms in a population that has evolved behavioral avoidance of the plant. Alternatively, the insect may avoid a palatable plant that presents the same repellant cue as an unsuitable plant. In addition, the existence of a "search image"-like phenomenon in some insects (Rausher 1980) suggests that selection could favor genetically encoded behavioral specialization on locally abundant hosts. Biochemically suitable but rare hosts could be excluded from the menu simply because searching for them may lower search efficiency (Futuyma 1983).

Not only the amount of genetic variation in behavioral and physiological traits, but also the mode of inheritance, is important in some models of host utilization. A commonly cited model that would help to account for the vast number of host-specific species of insects is sympatric speciation in association with divergence in host utilization (Thorpe 1930, Maynard Smith 1966, Bush 1975). This model is of particular interest because it implies that host specialization is prevalent because of species selection rather than individual selection alone. If high speciation rates are generated by the property of host specialization, so that host-specialized species speciate faster than generalized species, it is not necessary to suppose that host-specialization is prevalent because of ubiquitous selection for specialization within species. It should be noted, however, that host specialization could promote high speciation rates whether one favors a sympatric or an allopatric model. A shift of an allopatric population onto a different host could provide a reduction in gene flow between populations after they regain sympatry, and so initiate the speciation process.

In practice it will be difficult, if not impossible, to tell if a speciation event associated with divergence in host use occurred in sympatry or entailed some geographic separation. Moreover, since speciation could be initiated by brief isolation of a population in a very localized area where a novel food plant is prevalent, the difference between the allopatric (or parapatric) and the sympatric models may be trivial in terms of rates and patterns of speciation. The distinction between the models is interesting insofar as it impinges on our understanding of the precise nature of the genetic changes that usually result in speciation. No one denies that single mutations (e.g., in genes that regulate key developmental processes) *could* engender some of the morphological differences that distinguish higher taxa; but there is considerable argument about how likely such mutations are to be viable and how frequently they have actually been incorporated into populations. Similarly, we would like to know if the genetic conditions that indubitably could cause truly sympatric speciation are common or not, because the genetic differences between sympatrically generated species may well differ from those between allopatrically generated species.

In the most commonly cited scenario for sympatric speciation by host shift (Bush 1975), genotypes *A*__ and *aa* have higher fitness on hosts **I** and **II** respectively, and genotypes *B*__ and *bb* at a second locus choose hosts **I** and **II** respectively as their exclusive sites of mating and oviposition. Futuyma and Mayer (1980) have argued (as did Mayr 1947) that this model yields sympatric speciation only if there is essentially full dominance and penetrance of the *B* locus; otherwise there will be persistent gene flow between the subpopulations. In a somewhat analogous model, Felsenstein (1981) has shown that sympatric speciation requires strong linkage disequilibrium between a locus governing fitness in two niches (hosts) and a locus governing assortative mating, which in the case of host-specific insects is presumed to be a consequence of host preference. Felsenstein shows that pronounced linkage disequilibrium (i.e., progress toward speciation) is favored by tight linkage between these loci and by strong selection (about a two-fold difference in fitness) at the locus (*A*) governing fitness. These are fairly exacting conditions. Indeed, a major point of Felsenstein's paper is to show that recombination can be important in retarding or preventing sympatric speciation.

It is important to consider what the consequences will be if host preference is controlled by two or more loci, rather than a single locus (*B*). Incomplete dominance or penetrance at the *B* locus could be modified by a second locus (*D*) to give complete dominance or penetrance, and thus further reduce gene flow between the host-associated moieties. Similarly, it is possible that the switch from exclusive preference for host **I** to exclusive preference for host **II** requires mutations in two (or more) receptors (or pathways in the central nervous system). Thus genotype *B*__ *D*__ might be attracted to host **I**, *bbdd* to host **II**, and the recombinant genotypes to both hosts. In this instance, two host-restricted populations (species) will form only if selection at the *A* locus (affecting fitness) is strong enough to engender linkage disequilibrium among the three loci *A*, *B*, and *D*. Felsenstein (1981) has considered this case, and has concluded that the gene frequency changes at the *D* locus that are necessary to give speciation will be very slow, because the strength of selection at this locus depends on the magnitude of linkage disequilibrium between loci *B* and *D*, which in turn depends on the magnitude of linkage disequilibrium between loci *A* and *B*. Felsenstein's model may not be applicable to host-specific insects in all its details (Bush and Diehl 1982). However, if both traits, host preference and physiological adaptation to the hosts, were controlled by several or many variable loci, sympatric speciation would be difficult to achieve, since it is unlikely that linkage disequilibrium could be established among a multitude of loosely linked loci.

For these reasons, it is important to know if a virtually complete switch in host preference is likely to be caused by a single mutation (as in Bush's [1975] scenario), or requires multiple substitutions. A complete shift in preference is certainly conceivable if, for example, a mutation alters a receptor so as to change the pattern of signals sent to the central nervous system in response to different ratios of compounds presented by different plants. (See Cardé, this volume, for analogous possible cases in the control of responses to sex pheromones.) But in at least some cases in which mutations of taste receptors have been identified, the effect is to reduce the specificity of feeding preference, rather than to switch from one exclusive preference to another. For example, silkworms carrying a mutation of the bitter receptor are not deterred from feeding on plants that are normally rejected (Dethier, this volume).

It is possible that host specificity is based partially on the pattern of response of several or many sensory receptors to the complex of compounds that distinguish one plant from another. Suppose, for example, that one locus (*B*) programs attraction to a chemical of host **I**, and another locus (*D*) controls the response to a chemical of host **II**. Then a switch to exclusive use of host **II** may require two substitutions, so that, for example, *B*__ *D*__ is attracted to **I** and ignores **II**, and vice versa for *bbdd*. This is not implausible. For example, the small ermine moth *Yponomeuta cagnagellus* (Hübner) feeds on *Euonymus* (believed to be the ancestral host), and *Y. malinellus* Zeller feeds on *Malus*. *Yponomeuta cagnagellus* is stimulated to feed by dulcitol (a constituent of *Euonymus*) and is deterred by phloridzin (a constituent of *Malus*). *Yponomeuta malinellus* is not stimulated by dulcitol, and is not deterred by phloridzin (Gerrits-Heybroek et al. 1978, van Drongelen 1980, van Drongelen and van Loon 1980). Presumably another constituent of *Malus* acts as a stimulant. If these responses are under separate genetic control, the switch to *Malus* would have required at least three gene substitutions (affecting responses to dulcitol, phloridzin, and an unidentified compound). Sympatric speciation would then require linkage disequilibrium among (at least) these three loci and a locus affecting survival on the hosts. If Felsenstein's (1981) analysis applies, such linkage disequilibrium would be difficult to achieve. Host preference is presumably not a unitary trait, but a neurophysiologically complex trait with a correspondingly complex genetic basis. The genetic basis of host preference has, to my knowledge, never been determined in detail, but can be critical to the course of evolution.

As I indicated earlier in this paper, a common assumption in explanations of host specificity (including, but not limited to, sympatric speciation) is that there is a trade-off in fitness. On each host species a different physiologically specialized genotype enjoys a selective advantage. This could explain either the derivation of two host-specific species from one, or the derivation of a host-specific population from a more polyphagous ancestor. However, insects can sometimes feed without apparent ill effect on abnormal hosts (e.g., Waldbauer and Fraenkel 1961). Moreover, host specialists do not always grow more efficiently on their typical host than closely related generalists do (Smiley 1978, Futuyma and Wasserman 1981). On Long Island, New York, certain genotypes of the fall cankerworm *Alsophila pometaria* are more prevalent in stands of maple than of oak. One such genotype, although it shows greater behavioral preference for maple than do oak-associated genotypes, appears not to grow on maple any more efficiently than the most common oak-associated genotype does (Futuyma et al. 1984). I would not wish to extrapolate from these few instances to all Lepidoptera, much less from Lepidoptera to all phytophagous insects, but although different mechanisms could possibly engender host specificity in Lepidoptera than in, say, Homoptera or Diptera, the phenomenon of host specificity is ubiquitous in all these groups. Thus it is reasonable to seek a general explanation for it until it becomes evident that different explanations are required in different instances. Special adaptations to the host are common in host-specific insects, but if we imagine that the origin of such a specially adapted genotype within a population drives the population to become restricted to the host to which that genotype is adapted, we should be able to show that such genotypes exist within populations and have a selective advantage. We have not found host-associated polymorphism in physiological adaptation in *Alsophila*, but Rausher (1984) and Via (1984) have described possible instances of such genotype x environment interactions in a cassidine beetle and an agromyzid fly, respectively.

An alternative explanation of physiological and morphological adaptations to the host in host specific species is that specialized adaptations evolve *after* the population already has become specialized at the behavioral level. For example, the geometrid moth larva *Patalene olyzonaria puber* Grote and Robinson, that feeds almost exclusively on red cedar (*Juniperus virginiana*), has a specialized adaptation in the form of a color pattern that exquisitely matches the foliage of *Juniperus* (D.J.F., personal observation). It would have low fitness on virtually any other plant, on which it would not be cryptic. This adaptation is presumably not a cause, but a consequence of specialization; a population that is already restricted to a particular host is freed from conflicting selection pressures that other plants may impose, and can develop special adaptations at leisure. The same could account for specialized detoxification abilities.

In this view, host specialization may evolve first at the behavioral level, without concomitant physiological specialization, which may follow. I have argued elsewhere (Futuyma 1983) that specialized host preference may be most likely to evolve in a population isolated in a locality in which only a subset of its traditional hosts is abundant. For example, M. D. Bowers (pers. comm.) has found that a population of *Euphydryas chalcedona* Dbldy. & Hew., situated in a high altitude locality where only one host species is present, refused to feed on plants used by low altitude populations. The loss of recognition of a traditional host could be the consequence either of mutation and drift (as may occur for vestigial traits in general), or of selection for a genetically programmed specific searching image that improves foraging efficiency (Futuyma 1983). Thus local abundance of a host may be one of the ecological, rather than physiological, factors accounting for restricted diets (Gilbert 1978). The acquisition of ethological reproductive isolation by a host-restricted local population will then constitute the evolution of a specialized species. Evolution of behavioral traits (host preference and assortative mating) will then have overriding importance in the evolution of host specificity.

In summary, if genetic variation in host preference is more abundant than variation in detoxifying mechanisms or other physiological properties, there are circumstances in which insects will evolve to avoid unsuitable hosts rather than achieving physiological adaptation. Genetic variation in detoxifying ability may then persist as neutral polymorphism. If host preference has a complex genetic and neurobiological basis, the conditions for evolution of host-specialized species by sympatric speciation may be stringent. Physiological specialization for a particular host may be the consequence rather than the cause of specialized preference behavior. Insects need not display fine coadaptation between behavioral and physiological adaptations to plants.

Since the completion of this manuscript (July 1983), F. Gould (*Environmental Entomology* 15:1 (1986)) has published results of a two-locus simulation model similar to the preliminary analysis described in this paper.

Acknowledgments

I am grateful to the members, present and past, of the plant/herbivore discussion group at Stony Brook for discussion and criticism of these ideas. I gratefully acknowledge the National Science Foundation (DEB 76-20232) for support of research during the preparation of this paper. This is contribution No. 455 in Ecology and Evolution from the State University of New York at Stony Brook.

Literature Cited

Berenbaum, M. R., and P. Feeny. 1981. Toxicity of angular furanocoumarins to swallowtails: escalation in the coevolutionary arms race? Science 212: 927.

Bush, G. L. 1975. Sympatric speciation in phytophagous parasitic insects. P. W. Price, ed. *In* Evolutionary Strategies of Parasitic Insects and Mites. Plenum Press, New York.

Bush, G. L., and S. R. Diehl. 1982. Host shifts, genetic models of sympatric speciation and the origin of parasitic insect species. Proc. 5th. Int. Symp. Insect-Plant Relationships, Wageningen. Pudoc, Wageningen.

van Drongelen, W. 1980. Behavioural responses of two small ermine moth species (Lepidoptera: Yponomeutidae) to plant constituents. Ent. Exp. Appl. 28: 54.

van Drongelen, W., and J. A. van Loon. 1980. Inheritance of gustatory sensitivity in F_1 progeny of crosses between *Yponomeuta cagnagellus* and *Y. malinellus* (Lepidoptera). Ent. Exp. Appl. 28: 199.

Erickson, J. M., and P. Feeny. 1974. Sinigrin: a chemical barrier to the black swallowtail butterfly, *Papilio polyxenes*. Ecology 55: 103.

Felsenstein, J. 1981. Skepticism towards Santa Rosalia, or why are there so few species of animals? Evolution 35: 124.

Futuyma, D. J. 1983. Selective factors in the evolution of host choice by phytophagous insects. S. Ahmad, ed. *In* Herbivorous Insects: Host-Seeking Behavior and Mechanisms. Academic Press, New York.

Futuyma, D. J., and G. C. Mayer. 1980. Non-allopatric speciation in animals. Syst. Zool. 29: 254.

Futuyma, D. J., and S. S. Wasserman. 1981. Food plant specialization and feeding efficiency in the tent caterpillars *Malacosoma disstria* Hübner and *M. americanum* (Fabricius). Ent. Exp. Appl. 30: 106.

Futuyma, D. J., R. P. Cort, and I. van Noordwijk. 1984. Adaptation to host plant in the fall cankerworm (*Alsophila pometaria*) and its bearing on the evolution of host affiliation in phytophagous insects. Amer. Natur. 123: 287.

Gerrits-Heybroek, E. M., W. M. Herrebout, S. A. Ulenberg, and J. T. Wiebes. 1978. Host plants preferences of five species of small ermine moths (Lepidoptera, Yponomeutidae). Ent. Exp. Appl. 24: 360.

Gilbert, L. E. 1978. Development of theory in the analysis of insect-plant interactions. *In* D. Horn, R. Mitchell, and G. Stairs, eds. Analysis of Ecological Systems. Ohio State University Press, Columbus.

Gould, F. 1979. Rapid host range evolution in a population of the phytophagous mite *Tetranychus urticae* Koch. Evolution 33: 791.

Levins, R., and R. MacArthur. 1969. An hypothesis to explain the incidence of monophagy. Ecology 50: 910.

Maynard Smith, J. 1966. Sympatric speciation. Amer. Natur. 100: 637.

Mayr, E. 1947. Ecological factors in speciation. Evolution 1: 263.

Rausher, M. D. 1980. Host abundance, juvenile survival, and oviposition preference in *Battus philenor*. Evolution 34: 342.

Rausher, M. D. 1984. Trade-offs in performance on different hosts: evidence from within - and between-site variation in the beetle *Deloyala guttata*. Evolution 38: 582.

Rodman, J. E., and F. S. Chew. 1980. Phytochemical correlates of herbivory in a community of native and naturalized Cruciferae. Biochem. Syst. Ecol. 8: 43.

Smiley, J. 1978. Plant chemistry and the evolution of host specificity: new evidence from *Heliconius* and *Passiflora*. Science 201: 745.

Tabashnik, B. E. 1983. Host range evolution: the shift from native legume hosts to alfalfa by the butterfly, *Colias philodice eriphyle*. Evolution 37: 150.

Thorpe, W. H. 1930. Biological races in insects and allied groups. Biol. Rev. 5: 177.

Via, S. 1984. The quantitative genetics of polyphagy in an insect herbivore. I. Genotype - environment interaction in larval performance on different host plant species. Evolution 38: 881.

Waldbauer, G. P., and G. Fraenkel. 1961. Feeding on normally rejected plants by maxillectomized larvae of the tobacco hornworm, *Protoparce sexta* (Lepidoptera, Sphingidae). Ann. Entomol. Soc. Amer. 54: 477.

Wasserman, S. S., and D. J. Futuyma. 1981. Evolution of host plant utilization in laboratory populations of the southern cowpea weevil, *Callosobruchus maculatus* Fabricius (Coleoptera: Bruchidae). Evolution 35: 605.

Wright, S. 1969. Evolution and the Genetics of Populations. Vol. 2. The Theory of Gene Frequencies. University of Chicago Press, Chicago.

The Role of Pheromones in Reproductive Isolation and Speciation of Insects

Ring T. Cardé

Department of Entomology
University of Massachusetts
Amherst, Massachusetts 01003

Introduction

Reproductive isolation is a benchmark of the biological species concept and, as the term was coined by Dobzhansky (1937), it was designed to embrace the mechanisms that hinder or prevent the interbreeding of species. The concept of reproductive isolation is, of course, relevant only to sexually reproducing species and typically excludes geographical isolation from the behavioral, ecological, morphological or genetic devices that can serve as barriers to interspecific mating (Littlejohn 1981). Among insects, the diverse behavioral mechanisms that prevent hybridization typically are the same mechanisms that function in pair formation, that is the communication systems that result in the recognition and spatial convergence of potential mates. The principal communication channels emphasized vary among insect groups; for example, acoustical cues predominate in many Orthoptera, visual cues are emphasized in firefly beetles (Lampyridae) and butterflies, and chemical cues (pheromones) are a nearly generic solution in moths. But it is perhaps misleading to characterize these systems so simply. The communication system employed by moths, as an example, involves more than the chemical communication modality; to locate the conspecific chemical emitter, the responders use visual and wind cues to navigate a course upwind (Cardé 1984) and, often, visual and tactile (and in a few species, auditory inputs) for recognition of the "calling" (pheromone-emitting) conspecific.

Aside from the difficulties of considering reproductive isolating mechanisms by single communication modalities, when most species typically integrate information from a variety of sensory inputs, is the importance of characterizing communication systems as they relate to the success of *individuals* in finding and selecting a mate. The latter feature of chemical communication systems, namely selection for efficient, "narrowly tuned" emitters and receivers, the process of sexual selection, and the use of environmental conditions that maximize an individual's energetic efficiency, have been given relatively scant consideration in describing the factors that mold the chemical communication channel (Cardé and Baker 1984). Instead, the supposed species partitioning effects attributed to differences in the chemical channel have remained the dominant theme in interpreting why such differences exist and how they are maintained (e.g., Roelofs and Cardé 1974, Cardé et al. 1977, Greenfield and Karindinos 1979).

The mechanisms that could promote distinctive chemical communication channels may be listed as follows:

1. Lowering or eliminating response to and production of pheromone similar to that of a closely related, coexisting species, thereby reducing the chance of hybridization.

2. Lowering or eliminating response to and production of pheromone similar to that of a coexisting species, thereby reducing the cost of responding to another species.
3. Narrow tuning of a chemical blend so that sensitivity of the responder to the signal is enhanced (and the probability of response to non-optimal ratios is lowered), and variance in ratio of production of components is reduced, parsimonious emission of the pheromone by the sender to select sensitive responders (Cardé and Baker 1984).
4. Sexual selection of mates involving close-range courtship pheromones that enhance mating success (Baker and Cardé 1979).
5. Environmental effects which under certain daily (or seasonal) periods may enhance transmission of the chemical message or directly affect the energetic cost of mate finding (Cardé and Baker 1984).
6. "Secondary" effects by parasites, predators or other exploiters of the communication channel (Otte 1974).
7. Stochastic effects promoting drift.

Selection, of course, operates to maximize inclusive fitness, but given the number of factors listed above, and the unobservable ones that may have been important in the past history of a species, how reliably can we divine which factors have dictated the current form of the channel? A related problem with reproductive isolating mechanisms, noted by White (1978) among others, is whether such isolating mechanisms are involved in speciation itself. Again, it may be impossible to distinguish if differences evolved during reinforcement or with genetic divergence. Templeton (1981) and Paterson (1978) argue that there is currently little theoretical reason to suppose that reinforcement of differences in the premating pheromone communication channel is involved in the initial stages of speciation.

Two purposes of this review are to describe the nature of the differences in the sex pheromone systems among closely related insect species and the intraspecific variation within this communication channel. Such an analysis, even at a cursory level, cannot be accomplished without a thorough understanding of the chemical components involved in communication. The examples discussed below generally meet this criterion and they are taken mostly from the Lepidoptera, the insect group for which we have the most extensive knowledge. Some speculations on the possible role of sexual communication pheromones in speciation process also are considered.

Reproductive Isolation by Pheromones Among Closely Related Species

Species within a genus that co-occur synchronously in the same habitat provide useful indicators of the relative importance of pheromones as reproductive isolating mechanisms. Although a number of species' pairs could be used as examples, the pheromone blends of many tortricid moths are particularly well defined. As in most moths, females emit a pheromone that induces male attraction. In northeastern North American *Platynota idaeusalis* (Walker) and *P. flavedana* (Clemens) overlap broadly in host plants and seasonal cycles (Chapman and Lienk 1971). Indeed, these two species, and the *Archips* species discussed next, often emerge on the same host trees and females can be expected to call in close proximity. The *Platynota idaeusalis* pheromone (Fig. 1) is approximately a 1:1 blend of (*E*)-11-tetradecenyl acetate and (*E*)-11-tetradecenyl-1-ol (Hill et al. 1974), whereas *P. flavendana* employs a 85:15 mix of the (*E*)- and (*Z*)-11-tetradecen-1-ol (Hill et al. 1977). If the natural blend of one of these species is altered by addition of the unique component from the other species' pheromone blend, the attractiveness of the blend is abolished.

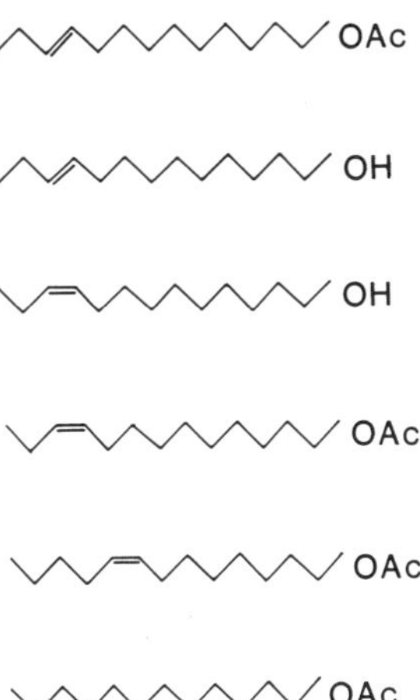

Fig. 1. Attractant pheromone structures in *Playnota* and *Archips* moths. Top to bottom the structures are (*E*)-11-tetradecenyl acetate, (*E*)-11-tetradecen-1-ol, (*Z*)-11-tetradecen-1-ol, (*Z*)-11-tetradecenyl acetate, (*Z*)-9-tetradecenyl acetate and dodecyl acetate.

Similar pheromone differences exist among several *Archips* species that coexist with *P. idaeusalis* and *P. flavedana* (Fig. 1). *Archips semiferanus* (Walker) utilizes a 30:70 blend of (*Z*)-and (*E*)-11-tetradecenyl acetates (Miller et al. 1976) and *A. cervasivoranus* (Fitch) employs a 20:80 blend of the same components (Roelofs et al. 1980). *Archips argyrospilus* (Walker) possesses a 90:10 mix of these two components, whereas the sibling species *A. mortuanus* (Kearfott) is optimally attracted to a 60:40 blend. In the latter two species attraction is enhanced by a small proportion of (*E*)-9-tetradecenyl acetate and dodecyl acetate added to the 11-tetradecenyl acetate mix (Cardé et al. 1977). Further investigation of the pheromones of these leafrollers could well reveal additional pheromone components, but it is clear from their highly specific attraction to their own pheromone that differences in the pheromone channel provide highly effective premating isolation devices. Similar examples are summarized by Cardé and Baker (1984) and Roelofs and Brown (1982) and in the following sections.

Character Displacement

Dobzhansky (1940) viewed what is now termed in part reproductive character displacement as a process whereby genetically controlled factors that would decrease (or prevent) hybridization of two or more adaptive complexes would acquire positive selective value. In the case of pheromone communication, we can include development of tuning mechanisms that prevent attraction to a non-conspecific emitter as a selective advantage. Character displacement (Brown and Wilson 1956) "is the situation in which, when two species overlap geographically, the differences between them are accentuated in the zone of sympatry and weakened or lost entirely in the parts of their ranges outside this zone." It has been delineated as reproductive when it reduces reproductive interference and ecological when it encompasses competition (Grant 1972). In the pheromone channel, this distinction in practice may be difficult to make, because it rests on judgments of the likelihood of hybridization after attraction and the cost of responders following pheromone trails of non-specifics, or even of non-specific responders disrupting the calling behavior of an emitter. Ecological character displacement in the chemical channel often has been confused with its reproductive counterpart. The former situation should apply to interspecific competition among species which are not closely allied and thus would not hybridize.

Reproductive character displacement of isolating mechanisms is not widely documented. Waage (1981 has shown reproductive character displacement in the wing patterns—used in mate recognition—in two damselfly species *(Calopteryx). In contrast, Walker (1974) examined its potential occurence in 254 species of gryliids and tettigoniids which use acoustical communication for pair formation; only five species pairs showed some* evidence of displacement. A particularly vexing problem of validating character displacement is that examples may be selected because they fit the hypothesis, whereas numerous cases that show no such shifts are ignored (Walker 1974). In contrast, Waage (1981) has shown reproductive character displacement in the wing patterns—used in mate recognition—in two damselfly species (*Calopteryx*).

In the case of pheromones used for sexual recruitment, this procedural difficulty has not yet proven troublesome because so few potential cases of character displacement have been documented. First, for most species there is relatively little information on geographical variation in pheromone production and response. Most of the information that is available documents the response of individuals to variations on the optimal lure blend; such data are inherently difficult to interpret because nearly all field trapping schemes rely upon traps which lose some efficacy as they capture responders. This process can obscure differences among treatments and make reliable comparisons of trap catch between different geographical areas difficult unless the optimal blend (rather than the degree of broad tuning) is shifted.

One of the most intriguing and well documented cases of geographical divergence occurs in the bark beetle *Ips pini* (Say) (Lanier et al. 1980). Beetles from California and Idaho produce and respond to (−)-ipsdienol. The (+)-enantiomer of the pheromone is a strong behavioral antagonist, lowering the response to (−)-ipsdienol when only 5% of the (+)-enantiomer is added (Birch et al. 1980). *Ips pini* from New York, in contrast, release a 65:35 mix of the (+) and (−) enantiomers and are lured to this blend much more than to either enantiomer alone (Fig. 2).

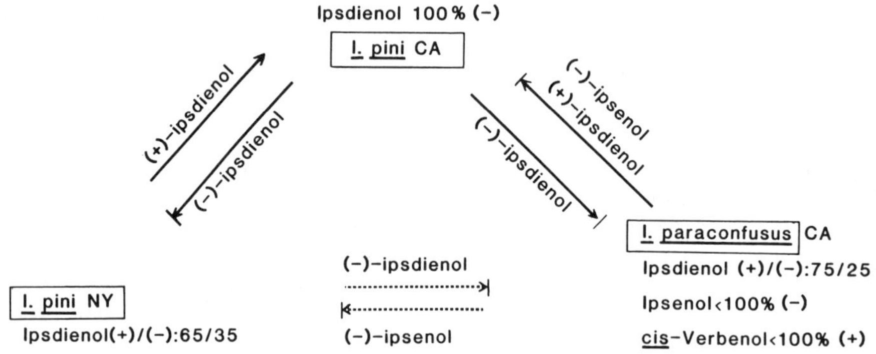

Fig. 2. Relationships in attractive and antagonistic chemicals among *Ips pini* from New York and California and *Ips paraconfusus* from California (figure after R. M. Silverstein).

In California, *I. pini* co-exists with *I. paraconfusus* Lanier, which utilizes a blend of (+)-and (−)-ipsdienol (3:1) plus (−)-ipsenol and *cis*-verbenol. The presence of boring beetles of the latter species interferes with the attraction of boring *I. pini*. Some of the antagonistic effect is caused by the release of (−)-ipsenol by *I. paraconfusus* (Birch et al. 1980). The case involves more than sexual recruitment, because for a bark beetle invasion of a tree to be successful (overwhelming the tree's defense tactic of sap secretion) mass attack by thousands of beetles must occur. The shift of the channel in California *I. pini* would seem due to character displacement, but both reproductive and ecological explanations are possible.

Hybrids of *I. pini* from Idaho (which appear to be pheromonally similar to the California populations) and *I. pini* from New York appear similar to New York beetles, suggesting that (+)-ipsdienol is produced by the hybrids. Reciprocal crosses did not differ in their patterns of attraction and response, indicating that the genes governing the communication system are not sex-linked (Lanier 1980).

In *A. argyrospilus*, New York males may be more narrowly tuned to the female's pheromone than those from British Columbia (Roelofs et al. 1974, Cardé et al. 1977). Although there are many co-occurring tortricine moths in both localities, in New York *A. argyrospilus* occurs with its sibling species *A. mortuanus*. In British Columbia, males appeared to be attracted to mixtures lower in (*E*)-11-tetradecenyl acetate or dodecyl acetate or higher in (*Z*)-9-tetradecenyl acetate than in New York. The evident narrowing of the communication channel in *A. argyrospilus* from New York thus could be due to reproductive character displacement. Because these apparent differences are based on the breadth of response and not a shift in the optimal blend, this interpretation assumes that trapping efficiency (how well traps ensnared arriving males) was unaffected by population levels. To be sure of these population differences, verification of these trends by additional field trials is desirable.

A more complex situation exists in *Spodoptera littoralis* (Boisduval), the Egyptian cotton leafworm moth (Campion et al. 1980, Dunkelblum et al. 1982), and one which is difficult to evaluate in part because of differences in the purities of the synthetic chemicals used in the field tests performed by different research groups. The female pheromone consists of three components with known contributions to attraction: (*Z,E*)-9,11-tetradecedienyl acetate (**I**), tetradecyl acetate (**II**) and (*Z,E*)-9,12-tetradecadienyl acetate (**III**) (Nesbitt et al. 1973, Campion et al. 1980, Dunkelblum et al. 1982). **I** elicits attraction by itself and the addition of **II** to **I** augments attraction in Crete (Campion et al. 1980), whereas the addition of 0.2 to 1% of **III** to **I** is important to attraction in Israel (Dunkelblum et al. 1982). Quantification of the volatiles emitted by females of corresponding geographical origins (Table 1) follows this trend.

The pheromone of *S. litura* (F.) consists of **I** and **III** (Tamaki et al. 1973) and these two *Spodoptera* species overlap in distribution in Oman (Wilshire 1977). The evident differences in the pheromone channel between these two populations of *S. littoralis* from Crete and Israel may be related to reproductive (communication) isolation of *S. littoralis* and *S. litura* in areas of sympatry. Obviously, more comparative behavioral data using synthetic chemicals of identical purity are needed to validate the geographical differences in pheromone evident in *S. littoralis*.

Interpopulational divergence in sexual recruitment and recognition with pheromones is theoretically optimum for reinforcement because ecological divergence is implied and "hybrid zones" of relatively long duration (without introgression or extinction of one of the populations) are possible (Templeton 1981). Doubtless the number of species in which the sexual

Table 1. Composition of behaviorally active components emitted by *Spodoptera littoralis* females. The figures are percentages (and ranges) of components relative to **I**, (Z,E)-9,11-tetradecadienyl acetate. **II** is tetradecyl acetate and **III** is (Z,E)-9,12-tetradecadienyl acetate (Campion et al. 1980).

Female origin	II	III
Crete	241 (115-430)	0
Egypt	43 (11- 74)	9
Israel	7 (1- 17)	6

recruitment pheromones will be found to differ over their geographical ranges will increase once such comparative studies are undertaken widely.

Pheromone "Strains": the European Corn Borer

The European corn borer, *Ostrinia nubilalis* (Hübner), was introduced in North America in separate propagules in the early 1900's. It was discovered in eastern Massachusetts near Waltham and a few years later in several localities near Lake Erie and in the vicinity of Schenectady, New York. Several lines of evidence suggest that these introductions originated from at least two distinct geographical regions in Europe. The Massachusetts population was bivoltine and widely polyphagous whereas the western populations were univoltine originally and had a narrow host range (see Caffrey and Worthley 1927). It is now evident that there are two pheromone strains (hereafter termed the *Z*- and *E*-strains) which occur allopatrically in some regions of North America and Europe, while there are other areas in eastern North America and Europe where both strains co-occur and where they hybridize.

The pheromone component eliciting attraction was proposed by Klun and Brindley (1970) on the basis of gas-liquid chromatography (GLC) retention times to be (*Z*)-11-tetradecenyl acetate. This component was documented (Klun and Robinson 1971) to be attractive to males in Iowa. In 1973, Klun et al. reported that an isomeric mixture (96% Z:4% E) was optimal for luring males in Iowa. An analysis of the isomeric constituents of female abdominal tips for the population from London, Ontario, where the moth is also attracted to the *Z* isomer blend, found that females produced a 97:3 blend of the *Z* and *E* isomers (Kochansky et al. 1975).

In central New York, on the other hand, field screening of known pheromone compounds and pheromone analogues revealed that male *O. nubilalis* were lured to (*E*)-11-tetradecenyl acetate (Roelofs and Comeau 1971, Roelofs et al. 1972). Chemical studies of the isomeric blends extracted from female abdominal tips of a Geneva, New York strain showed that the female abdominal tip produced a 96:4 blend of the *E* and *Z* isomers. Field tests in New York in 1973 demonstrated that "pure" *E* isomer (<0.5% Z) caught from ½ to ¼ as many males as the *E* isomer plus 1.8% *Z* isomer. Addition of 3.5% *Z* isomer produced male catch equivalent to the pure *E* isomer whereas isomeric blends containing 30% or more of *Z* isomer produced catches statistically indistinguishable from unbaited traps (Kochansky et al. 1975).

Field tests of male attraction to *Z:E* blends of 11-tetradecenyl acetates in a number of North American and European localities (Klun and Cooperators 1975) showed that the majority of survey sites had *Z*-responding males. In Europe, localities in Austria, France, Germany, Romania, Spain, Poland and Switzerland were documented to have *Z*-responding males, whereas in northern Italy, the *E*-type appeared predominant. In the United States, the populations in Ontario, New Brunswick, Quebec, Iowa, Missouri and Minnesota were *Z*-responding. In Geneva, New York, males were lured to the *E* isomer.

In central Pennsylvania, there appeared to be a mix of *Z*-responding and *E*-responding types along with males responding to 1:1 mix of these compounds (Klun and Cooperators 1975, Cardé et al. 1978). Cardé et al. (1975b) had previously trapped the same Pennsylvania site and found males lured to predominantly *Z* or predominantly *E* blends, but these 1973 field tests did not detect males in significant numbers (above unbaited traps) at mixtures ranging from 90% *Z* to 90% *E*. The two pheromone strains have been detected in New Jersey by Klun and Maini (1979), in North Carolina by Kennedy and Anderson (1980), and in Massachusetts by Fletcher-Howell et al. (1983).

The identification (Kochansky et al. 1975) of two different pheromone blends in females from Ontario and New York and the documentation (Klun and Cooperators 1975) of the distribution of male response types in North America and Europe support the existence of distinct pheromone strains which originated in North America as separate introductions from Europe. Although the precise sites from which these populations originated will not be verified easily, Italy and another population from elsewhere in Europe remain logical deductions.

Thus the following questions naturally arise:

1. To what degree is there assortative (i.e., intrastrain) mating where the strains cooccur?
2. How frequently do *Z*-type females lure *E*-males and *E*-type females lure *Z*-males?
3. What are the genetic mechanisms involved in production of the *Z* and *E* isomers and the male responsiveness to isomer blends of this compound?
4. Are the hybrids at any genetic advantage or disadvantage?
5. Are there differences (besides in the chemical communication channel itself) between the two strains (e.g., time of sexual activity, food plant range, diapause characteristics)?
6. Might the three communication "types" (i.e., the *Z*-strain, the *E*-strain and the hybrid) remain in a given locality over a long period, or is the hybridization seen now in some localities a "short-term" phenomenon?

These questions have been answered in part by recent studies. Liebherr and Roelofs (1975) studied the degree of reproductive isolation that might exist between the two strains confined for mating in the laboratory. Mating success (transfer of spermatophore) was markedly lower in interstrain *vs.* intrastrain crosses, but hybrid progeny exhibited heterosis (faster development and higher survivorship than in the pure strains). Thus, from these tests and similar findings of Buechi et al. (1982) using Swiss populations of the two pheromone strains (probably the interstrain cross situation reported by Arbuthnot [1944] represents the same phenomenon) hybrids would be expected to be relatively infrequent in nature given the extent of reproductive isolation that seems to occur at "close range" and the additional isolation that would be afforded by differences in the "long distance" attraction communication channel. (Klun and Maini [1979], in contrast, noted little isolation between the strains in laboratory crosses under a somewhat different protocol.) That hybrids of two strains exhibit heterosis suggests that their basic genetic systems are quite similar (Liebherr and Roelofs 1975). Other biotypes of the corn borer exist, particularly in terms of voltinism, but these have not been associated with either of the two pheromone strains. Most literature on the biology of *O. nubilalis* presumably refers to the Z-strain.

The Klun and Maini (1979) study of the genetics of pheromone production and laboratory response in hybrid crosses of these strains indicates that the major features of these traits are controlled by codominant alleles at a single locus. Hybrid females that arise from laboratory crosses of the two strains produce a pheromone blend approximating a 35:65 Z:E ratio. Males of this cross appear to respond preferentially to this ratio of components.

The extent to which "genetic coupling" between pheromone production and response occurs is critical to modeling the evolution of mating isolating (Sved 1981). In an accoustical system, Hoy et al. (1977) have shown that F_1 hybrid females of two *Teleogryllus* cricket species were most attracted to the songs of F_1 hybrid "sibling" males, that is males from crosses with the same maternal and paternal species as parents, rather than males of "reciprocal" crosses in which the species of the maternal and paternal parents were reversed. These elegant experiments suggest that the genetic control of male song and female acceptance are not independent, perhaps because of a common neural basis for control of the male's calling and the "feature" detectors of the female's song receptor (Hoy et al. 1977). The possible association of responder and emitter traits in pheromone systems should be a focus of future genetic studies.

Evidence that the *O. nubilalis* pheromone strains in sympatry are not panmictic comes from studies of allozymes of males attracted to various blends at the central Pennsylvania site. Harrison and Vawter (1977) studied ten enzyme-synthesizing loci and documented statistically significant differences in allele (*D*) frequencies between the two strains, but the genetic distance between the strains using Nei's measure (1972) was only 0.002. However, there are comparable situations where "good" sibling species of Lepidoptera possess similar degrees of electrophoretic differentiation (Harrison and Vawter 1977).

Cianchi et al. (1980) studied genetic divergence on the basis of 30 gene-enzyme systems in *laboratory-bred* populations of the two strains (from Iowa and New York) and found *D* to be 0.024, rather than the 0.002 measured in the Harrison and Vawter study. The differences in the two studies might be a consequence of (1) studying field-attracted strains in sympatry *vs.* laboratory-bred, allopatric populations; (2) the larger number of gene-enzyme systems investigated by Cianchi et al.; or (3) introgression in the Pennsylvania site. Clearly, the magnitude of the differentiation between the two pheromone strains is slight, and to date no fixed allelic differences have been uncovered that would allow unambiguous identification of a field or laboratory individual as *Z* or *E* strain on the basis of its allozyme profile.

Evidence that some field hybridization between the strains is occurring comes from pheromone gland analysis of feral females from New Jersey (4 F_1 hybrids of 40) and Italy (4 of 10) (Klun and Maini 1979). In Southern Switzerland, Buechi et al. (1982) documented 14-15% evident F_1 hybrid females from two localities.

Klun and Maini (1979) have characterized *O. nubilalis* as a "monotypic species that exhibits pheromonal polymorphism" but the strong degree of intrastrain pheromone response specificity and interstrain reproductive isolation found by Liebherr and Roelofs (1975) and Buechi et al. (1982) suggest that their status remains unresolved. It is plausible that their current areas of sympatry in Europe and certainly the United States are of very recent origin. Following the fate of these strains in sympatry and determining their current degree of genetic divergence will reveal the potential of reproductive character displacement in reinforcing these barriers. One prerequisite for reinforcement is that "the two populations must maintain their populations independently, implying considerable ecological divergence" (Templeton 1981), a condition that is unlikely in the *O. nubilalis* strains.

Within Population Variation in Pheromone Communication

Little is known of within population variation in pheromone production and responsiveness to blends. Variation in the ratio of two of the components has been studied in the redbanded leafroller, *Argyrotaenia velutinana* (Walker), by Miller and Roelofs (1980). The mean ratios of the (*Z*)-11- and (*E*)-11-tetradecenyl acetate components from gland extracts were 9.1 ± 1.8% SD for a field-collected population and 7.0 ± 1.4% SD for a 30-year-old laboratory strain (which had undergone about 250 generations of culture). The range of ratios of individual females was 4 to 15% and 4 to 14%, respectively, for the two populations and their coefficients of variance were identical (9.7%).

Characterization of the variance in male response to these ratios is difficult. Males attracted to a given ratio (and captured in a trap baited with that blend) may have an optimal response to the natural blend ratio. This renders catch across a spectrum of ratios only an imperfect reflection of the individual variation in response to blends, a phenomenon demonstrated with an attraction mark-capture technique. Males of *Grapholita molesta* (Busck), the Oriental fruit moth, lured to either 3, 8 or 11% mixtures of (*E*) in (*Z*)-8-dodecenyl acetate, marked themselves with a blend-coded color dust. Males were recaptured at traps baited with the same blends on the following day. If disparate phenotypes responsive to each blend existed, then males should tend to be reattracted to the same blends they visited previously. Instead males were attracted to the three blends in the same proportions as in their original visits to the dishes containing the color dyes (Cardé et al. 1975a). Thus, insects responding to nonoptimal blend cannot be assumed to be optimally attuned to that specific ratio. Stabilizing selection in which the responding sex would be most sensitive to the pheromone blend emitted by the majority of emitters and the emitters would release a blend to which responders are most attuned is discussed more fully by Cardé and Baker (1984). We may speculate that one of the factors promoting stabilizing selection could be pleiotropic effects associated with pheromone response and production.

Environmental Effects

Timing of sexual behavior is another mechanism of partitioning the pheromone channel. Such differences are responsible, in part, for communication specificity among *Callosamia* silkworm moth species that are cross-attractive (Brown 1972) and among many sesiid (clearwing) moths that employ structurally similar attractants (Greenfield and Karandinos 1979). Energetics, pheromone dispersion, and predation, however, also play important roles in the timing of sexual events. Many insects possess a daily rhythm of sexual activity that is modified by current temperature conditions so that sexual activity tends to occur under the most favorable daily conditions (Cardé and Baker 1984). These attributes, as well as the obvious partitioning function, could well lead to divergence in the timing of sexual behavior.

Sexual Selection

Sexual selection involves, as Darwin (1871) proposed, the differential success of individuals in securing mates. Typically, the success of the resource-limited sex (the sex with the least parental investment) is characterized by the number of matings that can be procured. Pheromone responders that are most able to locate and successfully court emitters, usually the females, will be most successful.

Such factors have been proposed as responsible for the evolution of attractant pheromones in scorpionflies (Mecoptera) by Thornhill (1979) and courtship pheromones by Baker and Cardé (1979). Once initiated, highly directional sexual selection can proceed very rapidly (O'Donald 1972, Lande 1981), leading to divergence in the pheromone communication channel.

Exploitation of the Communication Channel

Parasites and predators may orient to the very chemicals used for sexual communication by their host insects. This strategy is prevalent among predaceous exploiters of bark beetles (e.g., Borden 1977, Bakke and Kramme 1981, Tilden et al. 1983) but seemingly absent among exploiters of the Lepidoptera (the reports that egg parasites of *Heliothis* moths are influenced by the female moth's pheromone [Lewis et al. 1982] cannot yet be verified to occur under natural concentrations of host pheromone). It is doubtful that changes in the communication channel of the exploited species would be rapid enough to allow it to "escape" the exploiter in evolutionary time, but the exploiter must be viewed as a selective agent in the communication channel of the host, nonetheless.

Genetics of Pheromone Production and Response

In addition to the pioneering study of Klun and Maini (1979) on *O. nubilalis*, Grula and Taylor (1979) have dissected the genetics of the male-produced "recognition" pheromone in two sulfur butterfly species by hybridization. In *Colias eurytheme* Boisduval, the X chromosome carries most of the genes controlling production of the most important component, 13-methyl heptacosane (in Lepidoptera the female is the heterogametic sex). Another major male courtship signal is the ultraviolet reflectance pattern, also carried in the X chromosome. In *C. philodice* Godart, production of the principal wing compounds, three n-hexyl esters, are evidently controlled by one or more codominant autosomal genes.

In studies of hybrids of budworm moths (*Choristoneura fumiferana* (Clemens) and *C. pinus* Freeman, the attractiveness of F_1 hybrid females in the field was used to characterize their pheromone phenotypes (Sanders et al. 1977). In some autosomally identical crosses, the paternally derived X chromosome appeared to influence the relative attractiveness of the female to males of the two species. In hybrids of tussock moths (female *Hemerocampa pseudotsugata* McDunnough X male *H. leucostigma* [J. E. Smith]) the F_1 female's pheromone was most like the pheromone of the female parent and the male's pheromone receptors were most like those of the male parent, as judged from behavioral assays in the laboratory and whole antenna electrophysiology (Grant et al. 1975). In neither of these studies was it possible to characterize the pheromone chemistry of the hybrids, and consequently precise genetic analyses are not feasible. Teal et al. (this volume) summarize a detailed study of the pheromone chemistry of *Heliothis* hybrids and backcrosses, concluding that there is a maternal influence on the total quantity of pheromone produced and that four of the pheromone components are controlled autosomally.

In *Drosophila melanogaster* (Meigen), Averhoff and Richardson (1976) found that genes controlling a multicomponent courtship pheromone system were dispersed throughout the genome. An independent assortment of loci would enhance the diversity of genotypes and promote the outbreeding of this species by negative assortative mating (Averhoff and Richardson 1976). Possibly, outbreeding might also be mediated by sensory habituation in the responding sex to its own volatiles and hence its own genotype. The systems in *C. eurytheme* and *D. melanogaster* raise "the interesting possibility that insect communication systems have both variable and invariable components and that different sets of genes with different modes of inheritance give rise to this dichotomy" (Grula and Taylor 1979).

Speciation

Although differences in the pheromone communication channel are highly effective premating isolating mechanisms for many insect species, the role of such recognition systems in speciation remains speculative. Models proposing rapid evolution of new pheromone communication systems (and hence reproductive isolation) rely upon genetic traits for production of and response to novel components or blends. A novel emitter might have a good probability of mating with a similarly novel responder, but these traits presumably would be recessive. Resulting offspring thus would possess a normal phenotype. If some of these siblings mate, however, then recombination of the novel traits in emitters and responders could carry the new communication system to fixation within a relatively few generations (Roelofs and Comeau 1969). Carrying either or both of these traits on the sex chromosome would simplify the problems associated with the initial recombination of emitter and responder traits. There is, however, no empirical evidence that mutational differences in the pheromone channel initiate speciation events and in such cases the incipient species presumably would lack ecological divergence necessary to coexist with the ancestral species.

Among the mechanisms that promote rapid divergence of the pheromone location-recognition system in a given population are reduction in communication interference with other organisms (not closely related) sharing the communication system, and sexual selection. Reproductive character displacement could be inferred from cases where sympatric populations show greater premating isolation than allopatric ones, although it could be argued that enhanced premating isolation in sympatry could be the consequence of pleiotropic effects for characters reducing competition (Templeton 1981). The current evidence for *I. pini* is suggestive of reproductive character displacement. Some insight into this process will be gained by following the fate of the pheromone strains of *O. nubilalis* which, as discussed earlier in this chapter, are hybridizing to some degree in several areas in the northeastern United States and in Switzerland.

Summary

Differences in the pheromone channel serve as barriers to interspecific mating among many insect species. In the case of congenerics, the specificity of the message is typically due to the ratio of the components, their stereoconfiguration and their functional group. The effectiveness of these barriers is due to the "narrowly-tuned" production and response spectrum and the antagonistic effects of slight modifications of the pheromone structures, ratio of components, or in some cases its emission rate. In some species the daily timing of pheromone communication, rather than differences in pheromone chemistry, serves as the isolating mechanism.

Whether such differences in the pheromone communication channels arise in isolation, through reproductive character displacement, or through other mechanisms obviously will be difficult to discern. Study of species such as *I. pini* and *O. nubilalis*, populations of which possess different pheromone systems, will provide some insight into these processes.

The importance of these communication systems to reproductive success would suggest that they would be under strong selective pressure and consequently their heritabilities should be low (Falconer 1981). This is the paradox that "evolution by natural selection destroys the genetic variance upon which it feeds" (Lewontin 1979). The nature of the natural variability and inheritance of pheromone production and response is, as yet, known but for a few insects. Description of the genetic architecture of the pheromone channel would do much to reveal the structure upon which the many selective forces can act.

Acknowledgment

Drs. T. C. Baker, S. R. Diehl, M. D. Greenfield, R. G. Harrison and M. D. Huettel offered invaluable critiques of this paper.

Literature Cited

Arbuthnot, K. D. 1944. Strains of the European corn borer in the United States. U. S. of Agric. Tech. Bull. 869.

Averhoff, W. W. and R. H. Richardson. 1976. Multiple pheromone system controlling mating in *Drosophila melanogaster*. Proc. Nat. Acad. Sci. USA 73: 591.

Baker, T. C. and R. T. Cardé. 1979. Courtship behavior of the oriental fruit moth (*Grapholita molesta*): experimental analysis and consideration of the role of sexual selection in the evolution of courtship pheromones in the Lepidoptera. Ann. Entomol. Soc. Am. 72: 173.

Bakke, A. and T. Kvamme. 1981. Kairomone response in *Thanasimus* predators to pheromone components of *Ips tyographus*. J. Chem. Ecol. 7: 305.

Birch, M. C. D. M. Light, D. L. Wood, L. E. Browne, R. M. Silverstein, B. J. Bergot, G. Ohloff, J. R. West, and J. C. Young. 1980. Pheromonal attraction and allomonal interruption of *Ips pini* in California by the two enantiomers of ipsdienol. J. Chem. Ecol. 6: 703.

Borden, J. H. 1977. Behavioral response of Coleoptera to pheromones, allomones, and kairomones. *In* H. H. Shorey and J. J. McKelvey, Jr., eds. Chemical Control of Insect Behavior. John Wiley and Sons, New York.

Brown, L. N. 1972. Mating behavior and life habits of the sweet-bay silk moth (*Callosomia carolina*). Science 176: 73.

Brown, W. L. and E. O. Wilson. 1956. Character displacement. Syst. Zool. 5: 49.

Buechi, R., E. Priesner, and R. Brunetti. 1982. Das sympatrische Vorkommen von zwei Pheromstämmen des Maiszünslers, *Ostrinia nubilalis* Hbn., in der Südschweiz. Mitt. Schweiz. Entomol. Ges. 55: 33.

Caffrey, D. J. and L. H. Worthley. 1927. A progress report on the investigations of the European corn borer. USDA Bull. 1476.

Campion, D. G., P. Hunter-Jones, L. J. McVeigh, D. R. Hall, R. Lester, and B. F. Nesbitt. 1980. Modification of the attractiveness of the primary pheromone component of the Egyptian cotton leafworm, *Spodoptera littoralis* (Boisduval) (Lepidoptera: Noctuidae), by secondary pheromone components and related chemicals. Bull. Entomol. Res. 70: 417.

Cardé, R. T. 1984. Chemorientation: Flying insects. *In* W. J. Bell and R. T. Cardé, eds. Chemical Ecology of Insects. Chapman and Hall, London.

Cardé, R. T., and T. C. Baker. 1984. Sexual communication with pheromones. *In* W. J. Bell and R. T. Cardé, eds. Chemical Ecology of Insects. Chapman and Hall, London.

Cardé, R. T., T. C. Baker, and W. L. Roelofs. 1975a. Sex attractant responses of male oriental fruit moths to a range of component ratios: Pheromone polymorphism? Experientia 32: 1406.

Cardé, R. T., J. Kochansky, J. F. Stimmel, A. G. Wheeler, Jr., and W. L. Roelofs. 1975b. Sex pheromone of the European corn borer (*Ostrinia nubilalis*): *cis*- and *trans*-responding males in Pennsylvania. Environ. Entomol. 4: 413.

Cardé, R. T., A. M. Cardé, A. S. Hill, and W. L. Roelofs. 1977. Sex pheromone specificity as a reproductive isolating mechanism among the sibling species *Archips argyrospilus* and *A. mortuanus* and other sympatric tortricine moths (Lepidoptera: Tortricidae). J. Chem. Ecol. 3: 71.

Cardé, R. T., W. L. Roelofs, R. G. Harrison, A. T. Vawter, P. F. Brussard, A. Mutuura, and E. Monroe. 1978. European corn borer: Pheromone polymorphism or sibling species? Science 199: 555.

Chapman, P. J. and S. E. Lienk. 1971. Tortricid fauna of apple in New York (Lepidoptera: Tortricidae); including an account of apples' occurrence in the state, especially as a naturalized plant. Spec. Publ. N.Y.S. Agr. Exp. Sta. Geneva, New York.

Cianchi, R., S. Maini, and L. Bullini. 1980. Genetic distance between pheromone strains of the European corn borer *Ostrinia nubilalis*: Different contribution of variable substrate, regulatory and non-regulatory enzymes. Heredity 45: 383.

Darwin, C. 1871. The descent of man, and selection in relation to sex. John Murry, London.

Dobzhansky, T. 1937. Genetic nature of species differences. Amer. Natur. 71: 404.

Dobzhansky, T. 1940. Speciation as a stage in the evolutionary process. Amer. Natur. 74: 312.

Dunkelblum, E., M. Kehat, S. Gothilf, S. Greenburg, and B. Sklarsz. 1982. Optimized mixture of sex pheromoneal components for trapping of male *Spodoptera littoralis* in Israel. Phytoparasitica 10: 21.

Falconer, D. S. 1981. Introduction to quantitative genetics. 2nd ed. Longman. Harlow. U. K.

Fletcher-Howell, G. F., D. N. Ferro, and S. Butkewich. 1983. Pheromone and blacklight trap monitoring of adult European corn borer (Lepidoptera: Pyralidae) in western Massachusetts. Environ. Entomol. 12: 531.

Grant, G. G., D. Frech, and D. Grisdale. 1975. Tussock moths: pheromone cross stimulation, calling behavior and effect of hybridization. Ann. Entomol. Soc. Am. 68: 519.

Grant, P. R. 1972. Convergent and divergent character displacement. Biol. J. Linn. Soc. 4: 39.

Greenfield, M. D. and M. G. Karandinos. 1979. Resource partitioning of the sex communication channel in clearwing moths (Lepidoptera: Sesiidae) of Wisconsin. Ecol. Mon. 49: 403.

Grula, J. W. and O. R. Taylor. 1979. The inheritance of pheromone production in the sulfur butterflies *Colias eurytheme* and *C. philodice*. Heredity 42: 359.

Harrison, R. G. and A. T. Vawter. 1977. Allozyme differences between pheromone strains of the European corn borer, *Ostrinia nubilalis*. Ann. Entomol. Soc. Am. 70: 717.

Hill, A., R. Cardé, A. Comeau, W. Bode, and W. Roelofs. 1974. Sex pheromones of the tufted apple bud moth (*Platynota ideausalis*). Environ. Entomol. 3: 249.

Hill, A., R. Cardé, W. Bode, and W. Roelofs. 1977. Sex pheromone components of the varigated leafroller moth, *Platynota flavedana*. J. Chem. Ecol. 3: 369.

Hoy, R. R. J. Hahn, and R. C. Paul. 1977. Hybrid cricket auditory behavior: evidence for genetic coupling in animal communication. Science 195: 82.

Kennedy, G. G. and T. E. Anderson. 1980. European corn borer trapping in North Carolina with various sex pheromone component blends. J. Econ. Entomol. 73: 642.

Klun, J. A. and T. A. Brindley. 1970. *cis*-11-Tetradecenyl acetate, a sex stimulant of the European corn borer. J. Econ. Entomol. 63: 779.

Klun, J. A. and J. F. Robinson. 1971. European corn borer moth: sex attractant and sex inhibitors. Ann. Entomol. Soc. Am. 64: 1083.

Klun, J. A., D. L. Chapman, K. C. Mattes, P. W. Wojtkowski, M. Beroza, and P. E. Sonnet. 1973. Insect sex pheromones: minor amount of opposite geometrical isomer critical to attraction. Science 181: 661.

Klun, J. A. and Cooperators. 1975. Insect sex pheromones intraspecific pheromonal variability of *Ostrinia nubilalis* in North America and Europe. Environ. Entomol. 4: 891.

Klun, J. P. and S. Maini. 1979. Genetic basis of an insect chemical communication system. Environ. Entomol. 8: 423.

Kochansky, J., R. T. Cardé, J. Liebherr, and W. L. Roelofs. 1975. Sex pheromone of the European corn borer, *Ostrinia nubilalis* (Lepidoptera: Pyralidae) in New York. J. Chem. Ecol. 1: 225.

Lande, R. 1981. Models of speciation by sexual selection on polygenic traits. Proc. Nat. Acad. Sci. USA 78: 3721.

Lanier, G. N., A. Classon, T. Stewart, J. J. Piston, and R. M. Silverstein. 1980. *Ips pini*: the basis for interpopulational differences in pheromone biology. J. Chem. Ecol. 6: 677.

Lewis, W. J., D. A. Nordlund, R. C. Gueldner, P. E. A. Teal, and J. H. Tumlinson. 1982. Kairomones and their use for management of entomophagous insects. XIII. Kairomonal activity for *Trichogramma* spp. of abdominal tips, excretion and a synthetic sex pheromone blend of *Heliothis zea* (Boddie) moths. J. Chem. Ecol. 8: 1323.

Lewontin, R. C. 1979. Sociobiology as an adaptationist program. Behav. Sco. 24: 5.

Liebherr, J. and W. Roelofs. 1975. Laboratory hybridization and mating studies using two pheromone strains of *Ostrinia nubilalis*. Ann. Entomol. Soc. Am. 68: 305.

Littlejohn, M. J. 1981. Reproductive isolation: A critical review. *In* W. R. Atchley and D. S. Woodruff, eds. Evolution and Speciation. Cambridge Univ. Press, Cambridge.

Miller, J. R., T. C. Baker, R. T. Cardé, and W. L. Roelofs. 1976. Reinvestigation of oak leaf roller sex pheromone components and the hypothesis that they vary with diet. Science 192: 140.

Miller, J. R. and W. L. Roelofs. 1980. Individual variation in sex pheromone component ratios in two populations of the redbanded leafroller moth, *Argyrotaenia velutinana*. Environ. Entomol. 9: 359.

Nei, M. 1972. Genetic distance between populations. Amer. Natur. 106: 283.

Nesbitt, B. F., P. S. Beevor, R. A. Cole, R. Lester, and R. G. Poppi. 1973. Sex pheromones of two noctuid moths. Nature 244: 208.

O'Donald, P. 1962. The theory of sexual selection. Heredity 17: 541.

Otte, D. 1974. Effects and functions in the evolution of signaling systems. Ann. Rev. Ecol. Syst. 5: 385.

Paterson, M. F. M. 1978. More evidence against speciation by reinforcement. S. Afr. J. Sci. 74: 369.

Roelofs, W. L. and A. Comeau. 1969. Sex pheromone specificity: Taxonomic and evolutionary aspects in Lepidoptera. Science 165: 398.

Roelofs, W. L. and A. Comeau. 1971. Sex attractants in Lepidoptera. *In* A. S. Tahori, ed. Chemical Releasers in Insects. III. Gordon and Breach, New York.

Roelofs, W. L., R. T. Cardé, R. J. Bartell, and P. J. Tierney. 1972. Sex attractant trapping of the European corn borer in New York. Environ. Entomol. 1: 606.

Roelofs, W. L. and R. T. Cardé. 1974. Sex pheromones in the reproductive isolation of lepidopterous species. *In* M. C. Birch, ed. Phermones. North Holland, Amsterdam.

Roelofs, W., A. Hill, R. T. Cardé, J. Tette, H. Madsen, and J. Vakenti. 1974. Sex pheromone of the fruittree leafroller moth, *Archips argyrospilus*. Environ. Entomol. 3: 747.

Roelofs, W. L., A. J. Tamhanker, A. Comeau, A. S. Hill, and E. F. Taschenberg. 1980. Moth activity periods and identification of the sex pheromone of the uglynest caterpillar, *Archips cerasivoranus*. Ann. Entomol. Soc. Am. 73: 631.

Roelofs, W. L. and R. L. Brown. 1982. Pheromones and the evolutionary relationships of the Tortricidae. Ann. Rev. Ecol. Syst. 13: 395.

Sanders, C. J., G. E. Daterman, and T. J. Ennis. 1977. Sex pheromone responses of *Christoneura* spp. and their hybrids (Lepidoptera: Tortricidae). Can. Entomol. 109: 1203.

Sved, J. A. 1981. A two-sex polygenic model for the evolution of premating isolation. I. Deterministic theory for natural populations. Genetics 97: 197.

Tamaki, Y., H. Noguchi, and T. Yushima. 1973. Sex pheromone of *Spodoptera litura* (F.) (Lepidoptera: Noctuidae) Isolation, identification and synthesis. Appl. Entomol. Zool. 8: 200.

Teal, P. E. A., J. H. Tumlinson, and M. D. Huettel. 1983. Genetics of sex pheromone production and pheromone-mediated behavior in noctuid moths. *In* M. D. Huettel, ed. Evolutionary Genetics of Invertebrate Behavior. Plenum Press, New York.

Templeton, A. R. 1981. Mechanisms of speciation — a population genetic approach. Ann. Rev. Ecol. Syst. 12: 23.

Thornhill, R. 1979. Male and female sexual selection and the evolution of mating strategies in insects. *In* M. S. Blum, and N. A. Blum, eds. Sexual Selection and Reproductive Competition in Insects. Academic Press, New York.

Tilden, P. E., W. D. Bedard, K. Q. Lindahl, Jr., and D. L. Wood. 1983. Trapping *Dendroctonus brevicomis*: changes in attractant release rate, dispersion of attractant and silhouette. J. Chem. Ecol. 9: 311.

Waage, J. 1979. Reproductive character displacement in *Calopteryx* (Odonata: Calopterygidae). Evolution. 33: 104.

Walker, T. J. 1974. Character displacement and acoustic insects. Amer. Zool. 14: 1137.

White, M. J. 1978. Modes of Speciation. W. H. Freeman, San Francisco.

Wilshire, E. P. 1977. Middle East Lepidoptera, XXXVII. Notes on the *Spodoptera litura* (F.) group (Noctuidae: Trifinae). Proc. Brit. Entomol. Natur. Hist. Soc. 10-92.

Analyzing Proximate Causes of Behavior

V. G. Dethier

Department of Zoology
University of Massachusetts
Amherst, MA 01003

Introduction

Of the several ways of describing behavior the one most frequently employed by students of evolution and ecology is that based on consequences or goals. Within this framework behavior can be classified in terms of immediate causation or by particular functional (adaptive) consequences (e.g., courtship, parental behavior, egg-laying, territorial behavior, etc.). Many of the papers presented in this colloquium fall into this category. They include discussions of: desiccation survival, territoriality, cannibalism, courtship, mating success, foraging, oviposition preference, life history differences, host utilization, maternal defense, pheromone production, and feeding. Experiments are described in which sophisticated genetic analyses and techniques were employed to gain insight to the relative contributions of genetic, environmental, and experiential factors to the expressed behaviors. In many cases the experiments were undertaken with a view toward understanding the forces directing fitness, selection, speciation, and evolution.

Another way of describing and classifying behavior is in terms of proximate causes. Functional classification, as outlined above, is not always a true guide to causation (Beer 1963, Hinde 1970). In seeking to understand the evolutionary genetics of invertebrate behavior there may be some advantages in attacking problems directly from the point of view of proximate causes. To do this, however, requires a more intimate knowledge of the machinery of behavior than we now possess. A particular handicap derives from the fact that causal factors are not restricted in our thinking to events in the real world (e.g., stimuli) but often include hypothetical constructs such as learning, aggression, emotion, etc. (Hinde 1970, Tully et al. 1982).

Behavior is a concatenation of hierarchically organized patterns of action. These patterns are basically motor events arising from different kinds of sequential and simultaneous sensory input. This input arises in response to occurrences in the external environment and within the body itself. It is orchestrated against a background of multiple physiological variables and experiential influences, but always within the constraints of a set genetic endowment. In short, behavior emerges as an outcome of the integrated actions of sensory input, central neural integration, fluctuations in hormonal milieu, the setting and pacing of circadian clocks, the status of changing reproductive and age periods, ambient environmental conditions, and the impact of experience. In the final analysis, however, central neural circuitry and integration are the proximate determinants of behavior.

Genetic change can be introduced at any point in this biological system and its milieu and so becomes a limiting factor as regards functional and, hence, evolutionary outcomes. Knowledge of the point at which genetic control is exerted will undoubtedly enrich our understanding of evolutionary consequences and our ability to predict outcomes. Genetic analyses at the unit level can also be invaluable for unraveling the intricacies of behavioral complexity. Unfortunately the number of genetic studies directed to causal elements of invertebrate behavior are fewer than might be hoped. The brief review that follows is designed to complement the presentations at this symposium by calling attention to several series of genetic analyses not previously discussed and more directly concerned with mechanisms of behavior. They concern feeding behavior and learning.

Feeding Behavior

Feeding behavior is currently best understood in phytophagous insects and in the flies *Phormia* and *Drosophila*. In order to demonstrate the full significance of genetic analyses of feeding behavior, especially as it relates to evolutionary matters, it is desirable to present a montage combining elements of information about both groups of insects. It is crucial at the outset to make a clear distinction between factors initiating and maintaining an individual insect/food relationship on the one hand and the outcome in terms of survival and fecundity on the other. In the latter connection, especially with regard to herbivores, it should be borne in mind that the plant is not merely food; it is habitat as well.

Since the initial behavioral interactions are those of an individual, most causal analyses begin there rather than with populations. The behavioral interface between insect and plant consists of the chemical and physical characteristics of the plant and the sensory system (predominantly chemoreception) of the insect. Circumstances may vary, but the sequence of behavior patterns leading to ingestion are basically similar among species.

Some species of herbivorous insects are exposed to the plant during all stages of their lives. Plant feeding Coleoptera, such as the Colorado potato beetle, feed on the plant in larval and adult stages. Thus, the adult usually oviposits on the species of plant which nourished it in earlier developmental stages. Lepidoptera characteristically have different dietary regimens as adults and larvae; consequently, adults oviposit on plants that may have nourished them as larvae but are not food for adults. Some species do not even oviposit on the plants that are best for larvae and may not oviposit on plants at all (Dethier 1959).

The complex patterns of feeding behavior of biting herbivores, as exemplified by caterpillars, are linked in the following sequence. First, the volatile profile of the plant is assessed by olfactory receptors. If the leaf passes this test, the caterpillar bites and assesses the leaf sap with maxillary gustatory receptors. If the pattern of receptor response is accepted by the central nervous system, the leaf material is taken into the pre-oral cavity where a final monitoring is made by the epipharyngeal receptors. Ingestion continues or terminates at this point depending on the characteristics of the sensory information, internal states such as degree of inanition, and previous experience.

At each checkpoint, receptors encode two characteristics of the stimulus: its quantity (intensity or concentration) and its quality. Limits on quantitative and qualitative estimation are built into the receptors. The former have not been thoroughly studied in phytophagous species, but extrapolation from studies of blowflies suggests possible mechanisms. The limits of intensity discrimination (differential threshold) seem to be set in the receptor at the molecular level by the nature of the putative receptor protein (Dethier and Bowdan, 1984). A strict linear relation between enzyme kinetics at the receptor membrane and behavioral differential thresholds has been demonstrated. Theoretically, a genetic change at the level of the receptor protein could have a significant effect on behavioral discrimination of quantitative differences.

Qualitative encoding is also a prime function of receptors. The following characteristics of receptors influence the nature of an insect's perceived world: (1) the number and kinds of primary receptors (sugar receptor, salt receptor, etc.); (2) the width of the tuning curve (how widely or narrowly specific it is, i.e., how many different chemicals it responds to); and (3) the point of maximal response in the action spectrum (Dethier 1978). Change in one or more of these receptor characteristics could alter radically thc kind of information given to the central nervous system concerning the identity and characteristics of a plant.

A change in any one type of receptor could result in a behavioral change irrespective of whether that behavior depended on "labeled line" information or "across-fiber patterning." By "labeled line" is meant a receptor the activity of which is alone sufficient to trigger a specific kind of behavior (e.g., the sinigrin receptor in crucifer-feeders). Any stimulus that could activate this receptor would be interpreted as sinigrin. Thus, the crucial feature is the manner in which this receptor is "wired" into the central nervous system. "Across-fiber patterning" refers to the total pattern of response of multiples of receptors that differ in one or more respects. The current view of the basis of host-plant selection is that it depends more on central nervous system integration of complex sensory patterns than on cues from single narrowly specific receptors (cf. Dethier and Crnjar 1982). Thus, mutational changes at the receptor level or at the integrative level in the central nervous system could drastically alter food-plant preference.

To date, significant genetic analyses of feeding behavior in insects have been carried out only with flies, but they already give substance to some of the possibilities just mentioned. Behavioral mutants with altered gustatory responses have been found in *Drosophila* by Isono and Kikuchi (1974), Falk and Atidia (1975), Rodrigues and Siddiqi (1978), Tompkins et al. (1979), and Tanimura et al. (1982). With the exception of the last case, the foci of the mutations were in the central nervous system. Falk and Atidia (1975), by means of blastoderm fate mapping (Hotta and Benzer 1972), located three mutations conferring reduced sensitivity to salt (Lot—94, Lot—114, Lot—235) and having no effect on sensitivity to sugars. Tanimura et al. (1982) have found trehalose-low-sensitive laboratory strains of *Drosophila* and have produced trehalose-low-sensitive mutants by treatment with ethylmethane sulfonate. Evidence in these cases suggests that a single gene is responsible for the change and that the change occurs in the receptor (neuron) membrane. The receptor is sensitive to several sugars and mediates a behavioral feeding response (Falk 1979). The different sugars probably act at different molecular sites on the receptor membrane as in *Phormia*. Since the trehalose mutant retains its sensitivity to glucose, fructose, and sucrose, it may be presumed that only one specific molecular site is affected.

Extrapolating these findings to phytophagous insects, one can envision how one food preference might be changed to another and even monophagy to polyphagy, or vice versa. The possibility hinges on the presence of multiple sites of action even in specialized receptors. For example, a receptor might have special membrane sites *a*, *b*, and *c*, for compounds *A*, *B*, and *C*, with site *a* having the lowest threshold in terms of molarity. If this receptor was the one mediating acceptance, the animal would ingest any food containing threshold concentrations of *A*, *B*, or *C*; however, since *A* would be the most stimulating of the three compounds mole for mole, food containing *A* would be preferred. If a mutation occurred at site *a*, the receptor would no longer respond to *A* but would respond maximally to *B* or *C*, whichever was the more stimulating. This scheme would operate whether the receptors involved were labeled lines or were components of an across-fiber patterning system. In the latter instance, a change in the characteristics of any of the cooperating receptors would alter the total pattern relayed to the central nervous system and, hence, change the behavioral response.

In a similar fashion, a single mutation could change monophagy to polyphagy or vice versa. It has been demonstrated many times in lepidopterous larvae that surgical inactivation of a receptor or group of receptors can transform a specialist feeder to a generalist feeder (Torii and Morii 1948, Dethier 1953, Ito et al. 1959, Waldbauer and Fraenkel 1961, Waldbauer 1962). The same effect might be expected from genetic extirpation of the receptors or the central interneurons that received the input. Ishikawa et al. (1963) described a "non-preference" mutant of the silkworm *Bombyx mori* (L.) which presumably was insensitive to deterrents and as a consequence did not restrict its feeding to mulberry. Electrophysiological studies of the maxillae suggested that the taste deficit occurred at a central nervous system locus rather than in the receptors themselves.

Learning

Although genetically determined receptor characteristics and hard-wiring in the central nervous system set the constraints within which the actual behavioral mechanisms controlling feeding (and oviposition) behavior operate, the expression, that is, the behavioral output, is subject to modification by environmental and experiential influences. In insects, the second of these two modulators is currently the subject of many genetic studies. It is becoming increasingly clear that insects are not as quintessentially instinctive as they have been described. Learning plays a significant role in their encounters with the world.

Learning, however, is not homogeneous. It is defined principally by exclusion (Hinde 1970). Insofar as feeding behavior is concerned, eight different categories of experience have been described and, although extended genetic analysis of only three have been undertaken, the following summary provides helpful background information. The categories are: central excitatory state (CES), central inhibitory state (CIS), induction, habituation, aversion learning, pre-imaginal conditioning, classical conditioning, and operant conditioning. The term CES was coined to designate a transitory state of increased excitation (lower threshold) established in the central nervous system by momentary sensory stimulation which of itself caused a brief positive behavioral response (Dethier et al. 1965). The consequence of establishing CES is that subsequent sensory stimuli that are below behavioral threshold now elicit responses. CIS refers to a corresponding perseverating inhibitory state distinct in its characteristics from sensory adaptation and habituation (Dethier et al. 1968). These phenomena are of short duration (minutes) and are related in their intensity and perseveration to the strength of the initiating stimulus and the state of satiety.

Induction refers to a change in preference for particular food plants (although it could also apply to oviposition preference) caused by ingestion of one to the exclusion of others. It is, in a manner of speaking, an acquired preference. Brief references to changes in feeding preference have appeared in the literature since 1951 (Johansson 1951, Kuznetzov 1952, Getzova and Lozina-Lozinskii 1955, Kozhanicikov 1958, Yamamoto and Fraenkel 1960, Iwao and Machida 1961, and Stride and Straatman 1962). The term induction was coined by Jermy et al. (1968) because, as presently understood, the phenomenon does not conform to other well delineated forms of learning. It means nothing more or less than that exposure to one food (or host) predisposes an insect to select that one over others in a choice test. It most closely resembles imprinting but differs from that in a number of characteristics, one of which is the absence of a critical period. Its neural basis is not fully understood. Schoonhoven (1969) presented evidence indicating that a change occurs in the receptors themselves. Stadler and Hanson (1976) concluded that changes occur both peripherally and centrally. Considering the enormous variability found in responses of the gustatory receptors of caterpillars (Dethier and Crnjar 1982), the evidence for peripheral change appears equivocal. There is a possibility that induction might actually arise from habituation to normally unacceptable compounds present in less preferred food.

Habituation by phytophagous insects to deterrents has recently been demonstrated and studied extensively by Jermy et al. (1982) and A. Szentesi and E. A. Bernays (pers. comm.). Larvae of *Pieris brassicae* (L.) and *Mamestra brassicae* L. and nymphs of *Schistocerca gregaria* become behaviorally habituated to deterrents placed on food-plants. Central neural mechanisms are involved.

Aversion learning of the Garcia type is in some respects the opposite of induction and habituation. First studied extensively in vertebrates (Garcia and Koelling 1966), it was subsequently demonstrated in a terrestrial slug (*Limax maximus* L.) by Gelperin (1975) and in polyphagous lepidopterous larvae (Arctiidae) by Dethier (1980). It is a type of learning in which an association is made between food-related sensory cues that are not themselves deterrent and internal consequences (malaise) of ingestion. It is characterized by rapid onset, high resistance to extinction, specific association with olfactory and gustatory cues, and long delay (>1 hour) between sensory input and developing malaise. One trial may be sufficient to establish the association.

Classical, operant, and pre-imaginal conditioning are well known phenomena and need not be described in detail at this point. As recent studies show, the first two occur in insects more commonly than once believed. Pre-imaginal conditioning (Thorpe 1938, 1939, Thorpe and Jones 1937) has not been proven insofar as feeding is concerned but may possibly be a component of oviposition behavior (cf. Cushing 1941, Hershberger and Smith 1967, Manning 1967). On the other hand, associative learning on the part of the adult is clearly an important component of oviposition (Monteith 1963, Arthur 1966, 1971, 1981, Taylor 1974, van Lenteren and Bakker 1975, Vinson et al. 1977, Jaenike 1982, Prokopy et al. 1982).

Although associative learning is quite sophisticated in social and non-social Hymenoptera, most genetic analyses of learning by insects have been undertaken with the flies *Drosophila* and *Phormia*. This choice was motivated by awareness of the accumulated wealth of genetic information about *Drosophila* and of behavioral physiological information about *Phormia*. Compared to Hymenoptera, however, flies appear to be notoriously poor learners. Although numerous attempts to condition them had been made, no conclusive evidence of conditioning occurred until Nelson (1971) proved that *Phormia* could associate salt stimulation of the tarsi with reward by a meal of sugar. This opened the way for genetic analysis. Two laboratories concentrated on this problem — Hirsch and his associates with *Phormia* and Benzer and Quinn and their associates with *Drosophila*. Each group approached the problem differently. Hirsch's group worked with individuals from genetically heterogeneous pools; Benzer's group worked with populations from genetically uniform strains.

Hirsch began by progressively selecting bidirectional lines for a particular behavior. Employing Nelson's technique of conditioning, he developed "bright" and "dull" learners by selecting over 27 generations (Hirsch and McCauley 1977, McGuire and Hirsch 1977). Also by selection, McGuire (1978, 1981) produced high and low CES lines (CES has recently been demonstrated in *Drosophila* by Vargo and Hirsch, 1981). From the same stock, Tully and Hirsch (1982a, 1982b) produced, by single-pair matings, pure breeding high and low CES lines and presented evidence that there was a single major-gene correlate of CES.

Conclusion

The purpose in selecting two categories of behavior, one associated with real events (feeding) and the other with a concept (learning), for examination in terms of proximate causes was two-fold. It aimed first at emphasizing the linked and hierarchical nature of behavior and second at showing that genetic analyses of component parts can be most helpful in providing an understanding of behavior described in functional, that is, adaptive terms. Consider, for example, oviposition preference. Successful accomplishment depends upon ability to discriminate and, in many species, on a modicum of associative learning. It is extremely unlikely that one gene controls both abilities. It is even unlikely that a single gene controls the two basic components of learning: information storage and retrieval. Or, consider feeding behavior. It is essential that the behavior of the insect be correlated with the suitability of the plant in terms of nutrients, toxins, availability, microclimate, growth pattern, morphology, and parasite load. Response to each of these characteristics may be under quite different genetic control. Evidence has been presented at this symposium to the effect that preference and suitability are controlled differently. There is also evidence that genes for detoxication mechanisms are not the same ones that control sensory processes. To understand fully how genetics influences behavioral outcomes, a major desideratum is an understanding of how genes control individual components of the behavioral machinery.

Literature Cited

Aceves-Pina, E. O., and W. G. Quinn. 1979. Learning in normal and mutant *Drosophila* larvae. Science 206: 93.

Arthur, A. P. 1966. Associative learning in *Itoplectis conquisitor* (Say) (Hymenoptera:Ichneumonidae). Can. Entomol. 98: 213.

Arthur, A. P. 1971. Associative learning by *Nemeritis canescens* (Hymenoptera: Ichneumonidae). Can. Entomol. 103: 1137.

Arthur, A. P. 1981. Host acceptance by parasitoids. *In* D. A. Norlund, R. L. Jones, and W. J. Lewis, eds. Semiochemicals: Their Role in Pest Control. Wiley, New York.

Beer, C. G. 1963. Incubation and nestbuilding behavior of blackheaded gulls: III. The prelaying period. Behaviour 21. 13.

Benzer, S. 1967. Behavioral mutants of *Drosophila* isolated by countercurrent distribution. Proc. Natl. Acad. Sci. USA 58: 1112.

Benzer, S. 1973. Genetic dissection of behavior. Sci. Amer. 229: 24.

Booker, R., and W. G. Quinn. 1981. Conditioning of leg position in normal and mutant *Drosophila*. Proc. Natl. Acad. Sci. USA 78: 3940.

Byers, A. D., R. Davis, and J. Kiger. 1981. Defect in cyclic AMP phosphodiesterase due to the *dunce* mutation of learning in *Drosophila melanogaster*. Nature 289: 79.

Cushing, J. E. 1941. An experiment on olfactory conditioning in *Drosophila guttifera*. Proc. Natl. Acad. Sci. USA 27: 496.

Dethier, V. G. 1953. Host plant perception in phytophagous insects. Trans. IXth Int. Congr. Entomol. 2: 81.

Dethier, V. G. 1959. Egg-laying habits of Lepidoptera in relation to available food. Can. Entomol. 41: 554.

Dethier, V. G. 1978. Other tastes, other worlds. Science 201: 224.

Dethier, V. G. 1980. Food-aversion learning in two polyphagous caterpillars, *Diacrisia virginica* and *Estigmene congrua*. Physiol. Entomol. 5: 321.

Dethier, V. G., and E. Bowdan. 1984. Relation between differential thresholds and sugar receptor mechanisms in the blowfly. Behavioral Neuroscience. 98:791-803.

Dethier, V. G., and R. M. Crnjar. 1982. Candidate codes in the gustatory system of caterpillars. J. Gen. Physiol. 79: 549.

Dethier, V. G., R. L. Solomon, and L. H. Turner. 1965. Sensory input and central excitation and inhibition in the blowfly J. Comp. Physiol. Pyschol. 60: 303.

Dethier, V. G., R. L. Solomon, and L. H. Turner. 1968. Central inhibition in the blowfly. J. Comp. Physiol. Psychol. 60: 144.

Dudai, Y., Y-H. Jan, D. Byers, W. G. Quinn, and S. Benzer. 1976. *Dunce*, a mutant of *Drosophila* deficient in learning. Proc. Natl. Acad. Sci. USA 73: 1684.

Falk, R. 1979. Taste responses of *Drosophila melanogaster*. J. Insect. Physiol. 25: 87.

Falk, R., and J. Atidia. 1975. Mutation affecting taste perception in *Drosophila melanogaster*. Nature 254: 325.

Garcia, J., and R. A. Koelling. 1966. Relation of cue to consequences in avoidance learning. Psychonom. Sci. 4: 123.

Gelperin, A. 1975. Rapid food-aversion learning by a terrestrial mollusk. Science 189: 567.

Getzova, A. B., and L. K. Lozina-Lozinskii. 1955. Rol' povedeniia nasekomykh v protzesse prisposoblenia ikh k rastitel' nor pischce (Role of behavior in the process of insects' adaptation to plant diet). Zool. Zhurn. 34: 1066.

Hershberger, W. A., and M. P. Smith. 1967. Conditioning in *Drosophila melanogaster*. Animal Behav. 15: 259.

Hinde, R. A. 1970. Animal Behaviour. 2nd ed. McGraw-Hill Book Co., New York.

Hirsch, J. 1962. Individual differences in behavior and their genetic basis. *In* E. L. Bliss, ed. Roots of Behavior. Paul B. Hoeber, New York.

Hirsch, J., and L. A. McCauley. 1977. Successful replication of, and selective breeding for, classical conditioning in the blowfly, *Phormia regina*. Anim. Behav. 25: 784.

Horridge, G. A. 1962a. Learning of leg position by headless insects. Nature 193: 696.

Horridge, G. A. 1962b. Learning of leg position by the ventral nerve cord in headless insects. Proc. Roy. Soc. London Ser. B 157: 33.

Hotta, Y., and S. Benzer. 1972. Mapping of behavior in *Drosophila* mosaics. Nature 240: 527.

Ishikawa, S., Y. Tazima, and T. Hirao. 1963. Responses of the chemoreceptors of maxillary sensory hairs in a ''non-preference'' mutant of the silkworm. J. Seric. Sci. Tokyo 32: 125.

Isono, K., and T. Kikuchi. 1974. Autosomal recessive mutation in sugar response of *Drosophila*. Nature (London) 248: 243.

Ito, T., Y. Horie, and G. Fraenkel. 1959. Feeding on cabbage and cherry leaves by maxillectomized silkworm larvae. J. Seric. Sci. Tokyo 28: 107.

Iwao, S., and A. Machida. 1961. Further experiments on the host-plant preferences in a phytophagous lady-beetle, *Epilachna pustulosa* Kono. The Insect Ecology (Tokyo) 9: 9.

Jaenike, J. 1982. Environmental modification of oviposition behavior in *Drosophila*. Amer. Natur. 119: 784.

Jermy, T., E. A. Bernays, and A. Szentesi. 1982. The effect of repeated exposure to feeding deterrents on their acceptability to phytophagous insects. *In* H. Visser and A. Minks, ed. Proc. 5th Int. Symp. Insect-Plant Relationships. Pudoc, Wageningen.

Jermy, T., F. E. Hanson, and V. G. Dethier. 1968. Induction of specific food preferences in lepidopterous larvae. Entomol. Exp. Appl. 11: 211.

Johansson, A. S. 1951. The food plant preference of the larva of *Pieris brassicae* L. Norsk. Ent. Tid. B8: 187.

Kozhanicikov, I. V. 1958. Biilogiceskie osobennosti evropeiskikh vidov roda *Galerucella* i usloviia obrazovaniia biologiceskikh form u *Galerucella lineola* F. (Biological pecularities of the European species of the genus *Galerucella* and the differentiation of biological forms in *Galerucella lineola* L.) Trudy Zool. Inst. 24: 271.

Kuznetzov, V. I. 1952. Voprosy prisposobleniia ceshuekrylykh k novym pishcevym usloviiam. (The question of adaptation in lepidopterous species to new feeding conditions.) Trudy Zool. Inst. 11: 166.

van Lenteren, J. C., and K. Bakker. 1975. Discrimination between parasitized and unparasitized hosts in the parasitic wasp *Pseudocoila bochei*: A matter of learning. Nature 254: 417.

McGuire, T. R. 1978. Behavior-genetic analysis of *Phormia regina*: conditioning central excitatory state, and selection. Ph.D. Thesis. University of Illinois, Urbana-Champaign.

McGuire, T. R. 1981. Selection for central excitatory state (CES) in the blowfly *Phormia regina*. Behav. Genet. 11: 331.

McGuire, T. R., and J. Hirsch. 1977. Behavior-genetic analysis of *Phormia regina*: Conditioning, reliable individual differences, and selection. Proc. Natl. Acad. Sci. USA 74: 5193.

Manning, A. 1967. Preimaginal conditioning in *Drosophila* using geraniol. Nature 216: 338.

Medioni, J., and G. Vaysse. 1977. Suppression of the tarsal reflex by associative conditioning in *Drosophila melanogaster*. C. R. Seances Soc. Biol. Paris 169: 1386.

Monteith, L. G. 1963. Habituation and associative learning in *Drino bohemica* Mesn. (Diptera: Tachinidae) Can. Entomol. 95: 418.

Nelson, M. C. 1971. Classical conditioning in the blowfly. J. Comp. Physiol. Psychol. 77: 353.

Prokopy, R. J., A. L. Averill, S. S. Cooley, and C. A. Roitberg. 1982. Associative learning in egglaying site selection by apple maggot flies. Science 218: 76.

Quinn, W. G., and Y. Dudai. 1976. Memory phases in *Drosophila*. Nature 262: 576.

Quinn, W. G., W. A. Harris, and S. Benzer. 1974. Conditioned behavior in *Drosophila melanogaster*. Proc. Natl. Acad. Sci. USA 71: 708.

Quinn, W. G., P. P. Sziber, and R. Booker. 1978. The *Drosophila* memory mutant *amnesiac*. Nature 277: 212.

Rodrigues, V., and O. Siddiqi. 1978. Genetic analysis of chemosensory pathway. Proc. Indian Acad. Sci. 87(B): 147.

Schoonhoven, L. M. 1969. Sensitivity changes in some chemoreceptors and their effect on food selection behaviour. Proc. K. ned Akad. Wet. (C) 72: 491.

Stadler, E., and F. E. Hanson. 1976. Influence of induction of host preference on chemoreception of *Manduca sexta*: behavioral and electrophysiological studies. Symp. Biol. Hung. 16: 267.

Stride, G. E., and R. Straatman. 1962. The host plant relationship of an Australian swallowtail, *Papilio aegus*, and its significance in the evolution of host plant selection. Proc. Linn. Soc. N.S.W. 87: 69.

Szentesi, A., and E. A. Bernays. A study of behavioural habituation to a feeding deterrent in nymphs of *Schistocerca gregaria*.

Tanimura, T., K. Isono, T. Takamura, and I. Shimada. 1982. Genetic dimorphism in the taste sensitivity to trehalose in *Drosophila melanogaster*. J. Comp. Physiol. A. 147: 433.

Taylor, R. J. 1974. Role of learning in insect parasitism. Ecol. Monogr. 44: 89.

Tempel, B. L., N. Bonini, D. R. Dawson, and W. G. Quinn. 1983. Reward learning in normal and mutant *Drosophila*. Proc. Natl. Acad. Sci. USA 80: 1482.

Thorpe, W. H. 1938. Further experiments on olfactory conditioning in a parasitic insect. The nature of the conditioning process. Proc. Roy. Soc. London, B 126: 370.

Thorpe, W. H. 1939. Further experiments on pre-imaginal conditioning in insects. Proc. Roy. Soc. London, B 127: 424.

Thorpe, W. H., and F. G. W. Jones. 1937. Olfactory conditioning in a parasitic insect and its relation to the problem of host selection. Proc. Roy. Soc. London, B 124: 56.

Tompkins, L., M. J. Cadosa, F. V. White, and T. G. Sanders. 1979. Isolation and analysis of chemosensory behavior mutants in *Drosophila melanogaster*. Proc. Natl. Acad. Sci. USA 76: 884.

Torii, K., and K. Morii. 1948. Studies on the feeding habit of silkworms. Bul. Res. Inst. Seric. Sci. 2: 3.

Tully, T., and J. Hirsch. 1982a. Behavior-genetics analysis of *Phormia regina*. I. Isolation of pure breeding lines for high and low levels of central excitatory state (CES) from an unselected population. Behav. Genetics 12: 395.

Tully, T., and V. Hirsch. 1982b. Behavior-genetic analysis of *Phormia regina* II. Detection of a single, major-gene effect
from behavioural variation for central excitory state (CES) using hybrid crosses. Anim. Behav. 30: 1193.

Tully, T., S. Zawistowski, and J. Hirsch. 1982. Behavior-genetic analysis of *Phormia regina*: III. A phenotypic correlation between the central excitatory state (CES) and conditioning remains in replicated F_2 generations of hybrid crosses. Behav. Genetics 12: 181.

Vargo, M., and J. Hirsch. 1982. Central excitation in the fruit fly (*Drosophila melanogaster*) J. Comp. Physiol. Psychol. 96: 452.

Vinson, S. B., C. S. Barfield, and R. D. Hensen. 1977. Oviposition behavior of *Bracon mellitor*, a parasite of the boll weevil (*Anthonomus grandis*). II. Associative learning. Physiol. Entomol. 2: 157.

Waldebauer, G. P. 1962. The growth and reproduction of maxillectomized tobacco hornworms feeding on normally rejected non-solanaceous plants. Ent. Exp. Appl. 5: 147.

Waldbauer, G. P., and G. Fraenkel. 1961. Feeding on normally rejected plants by maxillectomized larvae of the tobacco hornworm, *Protoparce sexta* (Lepidoptera, Sphingidae). Ann. Entomol. Soc. Amer. 54: 477.

Yamamoto, R. T., and G. Fraenkel. 1960. The physiological basis for the selection of plants for egg-laying in the tobacco hornworm, *Protoparce sexta* (Johan.) Proc. 11th Int. Congr. Entomol. Vienna 3: 127.

ADDENDUM

Since this manuscript was submitted the following two reviews on learning in Diptera have appeared:

McGuire, T. R. 1984. Learning in three species of Diptera: the blowfly *Phormia regina,* the fruitfly *Drosophila melanogaster,* and the housefly *Musca domestica.* Behav. Genetics, 14: 479-526.

Tully, T. 1984. *Drosophila* learning: behavior and biochemistry. Behav. Genetics, 14: 527-557.

INDEX OF ORGANISMS

SUBJECT INDEX